Nuclear Reactor Physics

Nuclear Reactor Physics

Trenton Hensley

NY RESEARCH PRESS
New York

Published by NY Research Press
118-35 Queens Blvd., Suite 400,
Forest Hills, NY 11375, USA
www.nyresearchpress.com

Nuclear Reactor Physics
Trenton Hensley

International Standard Book Number: 978-1-63238-707-3 (Hardback)

Cataloging-in-Publication Data

Nuclear reactor physics / Trenton Hensley.
p. cm.
Includes bibliographical references and index.
ISBN 978-1-63238-707-3
1. Nuclear reactors. 2. Nuclear physics. 3. Physics. I. Hensley, Trenton.
QC786 .N83 2019
539.73--dc23

Printed in China.

Contents

Preface

Nuclear reactor physics is concerned with the study of chain reactions in nuclear reactors and their application for producing energy in a controlled fission reaction. The initiation and control of a self-sustained nuclear chain reaction occurs in a nuclear reactor. The key components in a nuclear power plant can be grouped under the classes of reactor assembly, steam generation, power generation, fuel handling, safety systems and controls. The reactor assembly consists of nuclear fuel, moderator, neutron source, reactor core, coolant and control rods, besides others. Boiler feedwater pump and steam generators are the components in steam generation. Nuclear reactors are classified according to the type of nuclear reactions they undergo, the moderator, coolants used, etc. Nuclear reactors are used for electricity generation and for the propulsion of ships. Heat generated due to nuclear fission is transferred to water or gas, which then drives steam turbines for varied applications. This book outlines the processes and applications of nuclear reactors in detail. While understanding the long-term perspectives of the topics, the book makes an effort in highlighting their impact as a modern tool for the growth of nuclear reactor physics. This book is meant for students who are looking for an elaborate reference text on nuclear reactor physics.

Given below is the chapter wise description of the book:

Chapter 1- A nuclear reactor is a device that is designed to start, control and sustain a nuclear chain reaction. The aim of this chapter is to give an introduction to the working principles and functioning of the different types of nuclear reactors, such as thermal-neutron reactor, breeder reactor, research reactor, subcritical- critical- supercritical reactor, gaseous fission reactor, etc.

Chapter 2- Nuclear reactors are used for the generation of electric power and for the propulsion of ships. These are classified according to the type of nuclear reaction sustained or the type of moderator material or the coolant used. They can divided into light-water reactors, heavy water reactors, graphite-moderated reactors and molten salt reactors, etc. This chapter closely examines the different kinds of nuclear power reactors for providing a comprehensive understanding.

Chapter 3- Nuclear power reactor has a number of components, which are essential for its functioning, such as nuclear reactor core, neutron moderator, nuclear fuel, neutron reflector, etc. This chapter discusses in detail about such components for an understanding of the working of a nuclear power reactor.

Chapter 4- The field of nuclear reactor physics is concerned with the study of controlled chain reactions for generating a desired rate of fission for the production of energy in a nuclear reactor. The aim of this chapter is to give an overview of the principles and fundamental concepts of nuclear reactor physics, such as nuclear reactions and nuclear chain reaction.

Chapter 5- In nuclear physics and chemistry, a nuclear reaction is the process by which two nuclei undergo a collision to produce one or more nuclides. Nuclear reactions are of two types, nuclear fission and nuclear fusion. The topics elaborated in this chapter explore the concepts central to the understanding of nuclear fission and fusion such as thermonuclear fusion, inertial confinement fusion, nuclear fission product, delayed neutron, fissile material, etc.

Chapter 6- An unstable nucleus loses energy in the form of radiation in a process called nuclear decay. The emission of radiation can be in the form of an alpha or beta particle or as gamma radiation. Therefore, an understanding of the different forms of nuclear decay, such as alpha, beta and gamma decay as well as spontaneous fission, proton emission, neutron emission, etc. is vital to the understanding of nuclear reactions in a nuclear reactor. These topics have been extensively discussed in the following chapter.

Indeed, my job was extremely crucial and challenging as I had to ensure that every chapter is informative and structured in a student-friendly manner. I am thankful for the support provided by my family and colleagues during the completion of this book.

Trenton Hensley

1

An Introduction to Nuclear Reactors

A nuclear reactor is a device that is designed to start, control and sustain a nuclear chain reaction. The aim of this chapter is to give an introduction to the working principles and functioning of the different types of nuclear reactors, such as thermal-neutron reactor, breeder reactor, research reactor, subcritical- critical- supercritical reactor, gaseous fission reactor, etc.

A nuclear reactor is a system used to initiate and contain a nuclear chain reaction, and they have many useful applications. These nuclear reactions produce thermal energy through either nuclear fission (in practice) or nuclear fusion (in development). Nuclear reactors are primarily used for the generation of electricity, however they can be used for propulsion in vehicles such as submarines or naval vessels, for production of useful isotopes or neutrons, and for research and training.

The core of a nuclear research reactor.

Nuclear reactors can be found all over the world. Used in nuclear power plants, fission reactors contribute about 11% of the world's total electricity. Although there are many different designs for fission reactors, most consist of the same components for their operation. The difference between each type of fission reactor comes from the different approaches used to satisfy these requirements.

Functioning of Nuclear Reactor

The basic operating principles of a nuclear reactor for the production of power are as follows: Nuclear chain reactions within the reactor produce heat, which is transferred to a coolant (most commonly light water), the coolant either boils to steam directly or heats another loop of water into steam, it then passes through a turbine which spins a generator, and produces electricity. Although the basic principles appear simple, the process is fairly complex.

Fuel

Nuclear reactors require the use of nuclear fuels, elements that can be readily altered and will release thermal energy. Uranium is the most common element used as a nuclear fuel, although thorium is also possible. The naturally occurring isotopes are found in countries such as Kazakhstan, Canada and Australia.

The uranium fuel is manufactured into small fuel pellets and are packed into fuel rods and surrounded by cladding to avoid leaking into the coolant. These fuel rods are assembled into a fuel bundle, as seen below. There can be hundreds of fuel bundles in a nuclear reactor, meaning there can be tens of thousands of fuel rods.

Nuclear Fuel

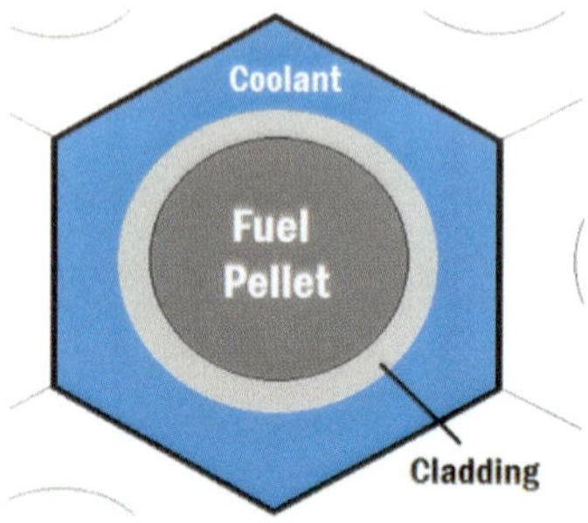

Fuel rods are surrounded by coolant to avoid overheating.

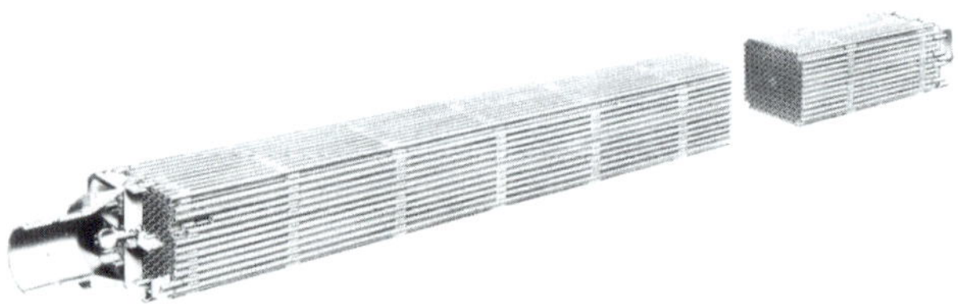

These assemblies are used in boiling and pressurized water reactors, and are packed into the core to produce heat in a nuclear reaction.

Fuel Enrichment

Not all nuclei of a given element are built exactly the same. An element is defined by the number of protons within the nucleus, and varying numbers of neutrons in the nucleus can cause it to behave differently. Natural uranium is mostly made up of uranium-238 (99.3%), with uranium-235 (0.7%) and very small amount of uranium-234 (0.0055%). Most reactors require a higher percentage of uranium-235 in order to sustain nuclear fission reactions, which can be done through uranium enrichment processes.

Moderator

Moderators are used to slow down the neutrons produced from fission. This is necessary because many nuclear fuels (uranium-235, for instance) require neutrons to be slow-moving in order to absorb them. Nuclei with low mass numbers are most effective at doing so, therefore materials like water or graphite are often used.

Most reactors use light water as a moderator, such as pressurized water reactors and boiling water reactors. Carbon works similarly and is used in reactors such as the RBMK. A third type of moderator used in CANDU reactors is heavy water, which is water composed of heavy hydrogen, called deuterium, rather than normal hydrogen.

The simulation below should help in visualizing how a moderator does its job: neutrons that are going too fast are absorbed by uranium-238 and do not yield fission (green) and the moderated neutrons are absorbed by uranium-235 which splits into smaller atoms and produces excess neutrons to continue the reaction (red).

Coolant

The coolant, as its name implies, is used to remove heat from the core and move it to somewhere that it is useful. This keeps the fuel from overheating and melting down, at the same time as transferring the heat to water to make steam. Light water, heavy water, and various gases are the most common coolants for nuclear reactors. Coolants may also serve as the moderator, as is the case in many water-moderated reactors.

Control Rods

Control rods can be inserted into the reactor core to reduce the amount of fuel which undergoes fission reactions. The rods contain neutron absorbing atoms such as gadolinium or cadmium. By absorbing neutrons within the core, it prevents those neutrons from reacting with the fuel. Control rod movement can be used to adjust the number of reactions occurring at the core, or fully inserted to shut down the reactor completely.

Safety Systems

Safety systems are those in place to shut down the reactor and prevent radioactive material from being released. Some systems are passive, such as the dropping the control rods into the reactor core in CANDU reactors. The control rods are suspended above the core and held there by an electromagnet (a magnet which requires a constant supply of electricity to operate). In the event that a loss of power occurs, the control rods work to stop the reactions in the core. Strong containment buildings must also surround the reactor to prevent any radioactive leaks or external damage to the reactor.

Other safety systems require activation. An example of such a system is the release of large quantities of water to surround the reactor core. This provides cooling for the core to disperse the thermal energy and avoid a melt down.

Economics

The building of nuclear reactors is economically intensive. The initial capital costs are high compared to fossil fuel plants with similar output. Nuclear power requires a high amount of additional safety, and is completely responsible for all possible nuclear waste. What makes nuclear power economically feasible is the large amount of energy that comes from a small volume of fuel. This relationship is known as energy density, and provides a cost advantage to using nuclear fuels. The cost of fuel is relatively lower for a nuclear power plant compared to fossil fuels. This is what makes nuclear reactors competitive despite high initial capital costs.

Electrical Power Generation

The energy released in the fission process generates heat, some of which can be converted into usable energy. A common method of harnessing this thermal energy is to use it to boil water to produce pressurized steam which will then drive a steam turbine that generates electricity.

The People in a Nuclear Power Plant

Nuclear power plants typically employ just under a thousand people per reactor (including security guards and engineers associated with the plant but working elsewhere).

- Nuclear engineers
- Reactor operators
- Health physicists

In the United States and Canada, all non-management and non-security workers are members of the International Brotherhood of Electrical Workers.

Reactor Types

Classifications

Nuclear Reactors are classified by several methods; a brief outline of these classification schemes is provided.

NC State's PULSTAR Reactor is a 1 MW pool-type research reactor with 4 percent enriched, pin-type fuel consisting of UO_2 pellets in zircaloy cladding.

Classification by Type of Nuclear Reaction

- Nuclear fission. Most reactors, and all commercial ones, are based on nuclear fission. They generally use uranium as fuel, but research on using thorium is ongoing (an example is the liquid fluoride reactor). Fission reactors can be divided roughly into two classes, depending on the energy of the neutrons that are used to sustain the fission chain reaction:
    - Thermal reactors use slow or thermal neutrons. Most power reactors are of this type. These are characterized by neutron moderator materials that slow neutrons until they approach the average kinetic energy of the surrounding particles, that is, until they are *thermalized.* Thermal neutrons have a far higher probability of fissioning uranium-235, and a lower probability of capture by uranium-238 than the faster neutrons that result from fission. As well as the moderator, thermal reactors have fuel (fissionable material), containments, pressure vessels, shielding, and instrumentation to monitor and control the reactor's systems.
    - Fast neutron reactors use fast neutrons to sustain the fission chain reaction. They are characterized by an absence of moderating material. Initiating the chain reaction requires enriched uranium (and/or enrichment with plutonium 239), due to the lower probability of fissioning U-235, and a higher probability of capture by U-238 (as compared to a moderated, thermal neutron). Fast reactors have the potential to produce less transuranic waste because all actinides are fissionable with fast neutrons, but they are more difficult to build and more expensive to operate. Overall, fast reactors are less common than thermal reactors in most applications. Some early power stations were fast reactors, as are some Russian naval propulsion units. Construction of prototypes is continuing.
- Nuclear fusion. Fusion power is an experimental technology, generally with hydrogen as fuel. While not currently suitable for power production, Farnsworth-Hirsch fusors are used to produce neutron radiation.
- Radioactive decay. Examples include radioisotope thermoelectric generators and atomic batteries, which generate heat and power by exploiting passive radioactive decay.

Classification by Moderator Material

Used by thermal reactors:

- Graphite moderated reactors
- Water moderated reactors
    - Heavy water reactors
    - Light water moderated reactors (LWRs). Light water reactors use ordinary water to moderate and cool the reactors. When at operating temperatures if the temperature of the water increases, its density drops, and fewer neutrons passing through it are slowed enough to trigger further reactions. That negative feedback stabilizes the reaction rate. Graphite and heavy water reactors tend to be more thoroughly thermalized than light

water reactors. Due to the extra thermalization, these types can use natural uranium/ unenriched fuel.

- Light element moderated reactors. These reactors are moderated by lithium or beryllium.
  - Molten salt reactors (MSRs) are moderated by a light elements such as lithium or beryllium, which are constituents of the coolant/fuel matrix salts LiF and BeF_2.
  - Liquid metal cooled reactors, such as one whose coolant in a mixture of Lead and Bismuth, may use BeO as a moderator.
- Organically moderated reactors (OMR) use biphenyl and terphenyl as moderator and coolant.

Classification by Coolant

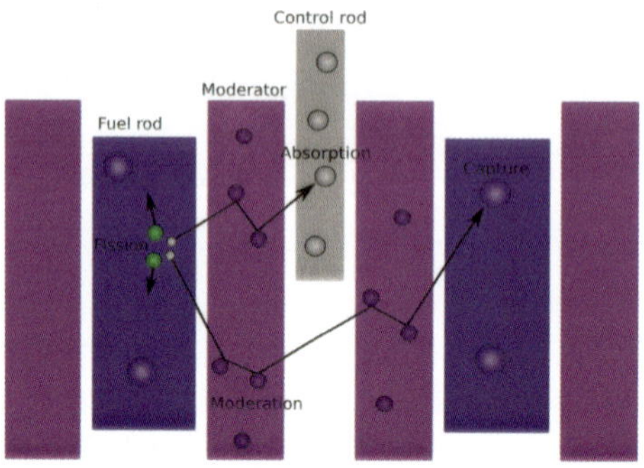

In thermal nuclear reactors (LWRs in specific), the coolant acts as a moderator that must slow down the neutrons before they can be efficiently absorbed by the fuel.

- Water cooled reactor
  - Pressurized water reactor (PWR)
    - A primary characteristic of PWRs is a pressurizer, a specialized pressure vessel. Most commercial PWRs and naval reactors use pressurizers. During normal operation, a pressurizer is partially filled with water, and a steam bubble is maintained above it by heating the water with submerged heaters. During normal operation, the pressurizer is connected to the primary reactor pressure vessel (RPV) and the pressurizer "bubble" provides an expansion space for changes in water volume in the reactor. This arrangement also provides a means of pressure control for the reactor by increasing or decreasing the steam pressure in the pressurizer using the pressurizer heaters.
    - Pressurized channels. Channel-type reactors can be refueled under load.
  - Boiling water reactor (BWR)
    - BWRs are characterized by boiling water around the fuel rods in the lower portion of primary reactor pressure vessel. During normal operation, pressure control is accomplished by controlling the amount of steam flowing from the reactor pressure vessel to the turbine.

- Pool-type reactor

- Liquid metal cooled reactor. Since water is a moderator, it cannot be used as a coolant in a fast reactor. Liquid metal coolants have included sodium, NaK, lead, lead-bismuth eutectic, and in early reactors, mercury.
    - Sodium-cooled fast reactor
    - Lead-cooled fast reactor
- Gas cooled reactors are cooled by a circulating inert gas, usually helium. Nitrogen and carbon dioxide have also been used. Utilization of the heat varies, depending on the reactor. Some reactors run hot enough that the gas can directly power a gas turbine. Older designs usually run the gas through a heat exchanger to make steam for a steam turbine.
- Molten Salt Reactors (MSRs) are cooled by circulating a molten salt, typically a eutectic mixture of fluoride salts, such as LiF and BeF2. In a typical MSR, the coolant is also used a matrix in which the fissile material is dissolved.

Classification by Generation

- Generation I reactor
- Generation II reactor
- Generation III reactor
- Generation IV reactor

The "Gen IV"-term was dubbed by the DOE for developing new plant types in 2000. In 2003, the French CEA was the first to refer to Gen II types in Nucleonics Week; "Etienne Pochon, CEA director of nuclear industry support, outlined EPR's improved performance and enhanced safety features compared to the advanced Generation II designs on which it was based." First mentioning of Gen III was also in 2000 in conjunction with the launch of the GIF plans.

Classification by Phase of Fuel

- Solid fueled
- Fluid fueled
- Gas fueled

Classification by use

- Electricity
    - Nuclear power plants
- Propulsion

- Nuclear marine propulsion
- Various proposed forms of rocket propulsion

- Other uses of heat
    - Desalination
    - Heat for domestic and industrial heating
    - Hydrogen production for use in a hydrogen economy
- Production reactors for transmutation of elements
    - Breeder reactors. Fast breeder reactors are capable of enriching Uranium during the fission chain reaction (by converting fertile U-238 to Pu-239) which allows an operational fast reactor to generate more fissile material than it consumes. Thus, a breeder reactor, once running, can be re-fueled with natural or even depleted uranium.
    - Creating various radioactive isotopes, such as americium for use in smoke detectors, and cobalt-60, molybdenum-99 and others, used for imaging and medical treatment.
    - Production of materials for nuclear weapons such as weapons-grade plutonium.
- Providing a source of neutron radiation (for example with the pulsed Godiva device) and positron radiation) (for example, neutron activation analysis and potassium-argon dating.
- Research reactor: Typically reactors used for research and training, materials testing, or the production of radioisotopes for medicine and industry. These are much smaller than power reactors or those propelling ships, and many are on university campuses. There are about 280 such reactors operating, in 56 countries. Some operate with high-enriched uranium fuel, and international efforts are underway to substitute low-enriched fuel.

Current Technologies

There are two types of nuclear power in current use:

- The Radioisotope thermoelectric generator
    - These systems produces heat through passive radioactive decay. Some radioisotope thermoelectric generators have been created to power space probes (for example, the Cassini probe), some lighthouses in the former Soviet Union, and some pacemakers. The heat output of these generators diminishes with time; the heat is converted to electricity utilizing the thermoelectric effect.
- Nuclear fission reactors
    - The nuclear fission reactor produces heat through a controlled nuclear chain reaction in a critical mass of fissile material. All current nuclear power plants are critical fission reactors. The output of fission reactors is controllable. There are several subtypes of critical fission reactors, which can be classified as Generation I, Generation II and

Generation III. All reactors will be compared to the Pressurized Water Reactor (PWR), as that is the standard modern reactor design.

Diablo Canyon—a PWR

- Pressurized Water Reactors (PWR)

- These reactors use a pressure vessel to contain the nuclear fuel, control rods, moderator, and coolant. They are cooled and moderated by high pressure liquid water. The hot radioactive water that leaves the pressure vessel is looped through a steam generator, which in turn heats a secondary (non-radioactive) loop of water to steam that can run turbines. They are the majority of current reactors, and are generally considered the safest and most reliable technology currently in large scale deployment. This is a thermal neutron reactor design, the newest of which are the Advanced Pressurized Water Reactor and the European Pressurized Reactor. United States Naval reactors are of this type.

Laguna Verde nuclear power plant - a BWR

- Boiling Water Reactors (BWR)

  A BWR is like a PWR without the steam generator. A boiling water reactor is cooled and moderated by water like a PWR, but at a lower pressure, which allows the water to boil inside the pressure vessel producing the steam that runs the turbines. Unlike a PWR, there is no primary and secondary loop. The thermal efficiency of these reactors can be higher, and they can be simpler, and even potentially more stable and safe. This is a thermal neutron reactor design, the newest of which are the Advanced Boiling Water Reactor and the Economic Simplified Boiling Water Reactor.

The CANDU Qinshan Nuclear Power Plant

- Pressurized Heavy Water Reactor (PHWR)

  A Canadian design, (known as CANDU) these reactors are heavy-water-cooled and -moderated Pressurized-Water reactors. Instead of using a single large pressure vessel as in a PWR, the fuel is contained in hundreds of pressure tubes. These reactors are fueled with natural uranium and are thermal neutron reactor designs. PHWRs can be refueled while at full power, which makes them very efficient in their use of uranium (it allows for precise flux control in the core). CANDU PHWR's have been built in Canada, Argentina, China, India (pre-NPT), Pakistan (pre-NPT), Romania, and South Korea. India also operates a number of PHWR's, often termed 'CANDU-derivatives', built after the Government of Canada halted nuclear dealings with India following the 1974 Smiling Buddha nuclear weapon test.

The Ignalina Nuclear Power Plant—a still operating RBMK

- Reaktor Bolshoy Moshchnosti Kanalniy (High Power Channel Reactor) (RBMK)

  A Soviet Union design, built to produce plutonium as well as power. RBMKs are water cooled with a graphite moderator. RBMKs are in some respects similar to CANDU in that they are refuelable during power operation and employ a pressure tube design instead of a PWR-style pressure vessel. However, unlike CANDU they are very unstable and too large to have containment buildings, making them dangerous in the case of an accident. A series of critical safety flaws have also been identified with the RBMK design, though some of these were corrected following the Chernobyl accident. RBMK reactors are generally considered one of the most dangerous reactor designs in use. The Chernobyl plant had four RBMK reactors.

The Torness nuclear power station—an AGR

- Gas Cooled Reactor (GCR) and Advanced Gas Cooled Reactor (AGR)

  These are generally graphite moderated and CO_2 cooled. They can have a high thermal efficiency compared with PWRs due to higher operating temperatures. There are a number of operating reactors of this design, mostly in the United Kingdom, where the concept was developed. Older designs (that is, Magnox stations) are either shut down or will be in the near future. However, the AGCRs have an anticipated life of a further 10 to 20 years. This is a thermal neutron reactor design. Decommissioning costs can be high due to large volume of reactor core.

- Liquid Metal Fast Breeder Reactor (LMFBR)

  This is a reactor design that is cooled by liquid metal, totally unmoderated, and produces more fuel than it consumes. They are said to "breed" fuel, because they produce fissionable fuel during operation because of neutron capture. These reactors can function much like a PWR in terms of efficiency, and do not require much high pressure containment, as the liquid metal does not need to be kept at high pressure, even at very high temperatures. Superphénix in France was a reactor of this type, as was Fermi-I in the United States. The Monju reactor in Japan suffered a sodium leak in 1995 and was approved for restart in 2008. All three use/used liquid sodium. These reactors are fast neutron, not thermal neutron designs. These reactors come in two types:

The Superphenix, one of the few FBRs

  - Lead cooled

    Using lead as the liquid metal provides excellent radiation shielding, and allows for operation at very high temperatures. Also, lead is (mostly) transparent to neutrons, so fewer neutrons are lost in the coolant, and the coolant does not become radioactive. Unlike sodium, lead is mostly inert, so there is less risk of explosion or accident, but such large quantities of lead may be problematic from toxicology and disposal points of view. Often a reactor of this type would use a lead-bismuth eutectic mixture. In this case, the bismuth would present some minor radiation problems, as it is not quite as transparent to neutrons, and can be transmuted to a radioactive isotope more readily than lead.

- Sodium cooled

Most LMFBRs are of this type. The sodium is relatively easy to obtain and work with, and it also manages to actually prevent corrosion on the various reactor parts immersed in it. However, sodium explodes violently when exposed to water, so care must be taken, but such explosions wouldn't be vastly more violent than (for example) a leak of superheated fluid from a SCWR or PWR. EBR-I, the first reactor to have a core meltdown, was of this type.

- Aqueous Homogeneous Reactor

Nuclear Fuel Cycle

Thermal reactors generally depend on refined and enriched uranium. Some nuclear reactors can operate with a mixture of plutonium and uranium. The process by which uranium ore is mined, processed, enriched, used, possibly reprocessed and disposed of is known as the nuclear fuel cycle.

Under 1 percent of the uranium found in nature is the easily fissionable U-235 isotope and as a result most reactor designs require enriched fuel. Enrichment involves increasing the percentage of U-235 and is usually done by means of gaseous diffusion or gas centrifuge. The enriched result is then converted into uranium dioxide powder, which is pressed and fired into pellet form. These pellets are stacked into tubes which are then sealed and called fuel rods. Many of these fuel rods are used in each nuclear reactor.

Most BWR and PWR commercial reactors use uranium enriched to about 4% U-235, and some commercial reactors with a high neutron economy do not require the fuel to be enriched at all (that is, they can use natural uranium). According to the International Atomic Energy Agency there are at least 100 research reactors in the world fueled by highly enriched (weapons-grade/90 percent enrichment uranium). Theft risk of this fuel (potentially used in the production of a nuclear weapon) has led to campaigns advocating conversion of this type of reactor to low-enrichment uranium (which poses less threat of proliferation).

It should be noted that fissionable U-235 and non-fissionable U-238 are both used in the fission process. U-235 is fissionable by thermal (that is, slow-moving) neutrons. A thermal neutron is one which is moving about the same speed as the atoms around it. Since all atoms vibrate proportionally to their absolute temperature, a thermal neutron has the best opportunity to fission U-235 when it is moving at this same vibrational speed. On the other hand, U-238 is more likely to capture a neutron when the neutron is moving very fast. This U-239 atom will soon decay into plutonium-239, which is another fuel. Pu-239 is a viable fuel and must be accounted for even when a highly enriched uranium fuel is used. Plutonium fissions will dominate the U-235 fissions in some reactors, especially after the initial loading of U-235 is spent. Plutonium is fissionable with both fast and thermal neutrons, which make it ideal for either nuclear reactors or nuclear bombs.

Most reactor designs in existence are thermal reactors and typically use water as a neutron moderator (moderator means that it slows down the neutron to a thermal speed) and as a coolant. But in a fast breeder reactor, some other kind of coolant is used which will not moderate or slow the neutrons down much. This enables fast neutrons to dominate, which can effectively be used to constantly replenish the fuel supply. By merely placing cheap unenriched uranium into such a core, the non-fissionable U-238 will be turned into Pu-239, "breeding" fuel.

Fueling of Nuclear Reactors

The amount of energy in the reservoir of nuclear fuel is frequently expressed in terms of "full-power days," which is the number of 24-hour periods (days) a reactor is scheduled for operation at full power output for the generation of heat energy. The number of full-power days in a reactor's operating cycle (between refueling outage times) is related to the amount of fissile uranium-235 (U-235) contained in the fuel assemblies at the beginning of the cycle. A higher percentage of U-235 in the core at the beginning of a cycle will permit the reactor to be run for a greater number of full-power days.

At the end of the operating cycle, the fuel in some of the assemblies is "spent" and is discharged and replaced with new (fresh) fuel assemblies, although in practice it is the buildup of reaction poisons in nuclear fuel that determines the lifetime of nuclear fuel in a reactor. Long before all possible fission has taken place, the buildup of long-lived neutron absorbing fission byproducts impedes the chain reaction. The fraction of the reactor's fuel core replaced during refueling is typically one-fourth for a boiling-water reactor and one-third for a pressurized-water reactor.

Not all reactors need to be shut down for refueling; for example, pebble bed reactors, RBMK reactors, molten salt reactors, Magnox, AGR and CANDU reactors allow fuel to be shifted through the reactor while it is running. In a CANDU reactor, this also allows individual fuel elements to be situated within the reactor core that are best suited to the amount of U-235 in the fuel element.

The amount of energy extracted from nuclear fuel is called its "burn up," which is expressed in terms of the heat energy produced per initial unit of fuel weight. Burn up is commonly expressed as megawatt days thermal per metric ton of initial heavy metal.

Thermal-neutron Reactor

Free neutrons can be classified according to their kinetic energy. This energy is usually given in electron volts (eV). The term temperature can also describe this energy representing thermal equilibrium between a neutron and a medium with a certain temperature.

Thermal Neutrons are neutrons in thermal equilibrium with a surrounding medium of temperature 290K (17 °C or 62 °F). Most probable energy at 17°C (62°F) for Maxwellian distribution is 0.025 eV (~2 km/s). This part of neutron's energy spectrum constitutes most important part of spectrum in thermal reactors.

Thermal neutrons have a different and often much larger effective neutron absorption cross-section (fission or radiative capture) for a given nuclide than fast neutrons. Therefore the criticality of a thermal reactor can be achieved with a much lower enrichment of nuclear fuel.

Moreover, thermal neutrons are in the 1/v region and the cross-section behaves according to the 1/v Law. In this region absorption cross-section increases as the velocity (kinetic energy) of the neutron decreases. Therefore the 1/v Law can be used to determine shift in absorbtion cross-section, if the neutron is in equilibrium with a surrounding medium. This phenomenon is due to the fact the nuclear force between the target nucleus and the neutron has a longer time to interact.

$\sigma_a \sim \frac{1}{\upsilon} \sim \frac{1}{\sqrt{E}} \sim \frac{1}{\sqrt{T}}$ Example of cross- sections in 1/v region:

The absorbtion cross-section for 238U at 20°C = 293K (~0.0253 eV) is:

$$\sigma_a(293\ K) = 2.68b.$$

The absorbtion cross-section for 238U at 1000°C = 1273K is equal to:

$$\sigma_a(1273\ K) = \sigma_a(293\ K).\sqrt{\frac{T_0}{T_1}} = 2.68.\sqrt{\frac{293}{1273}} = 1.286b$$

This cross-section reduction is caused only due to the shift of temperature of surrounding medium.

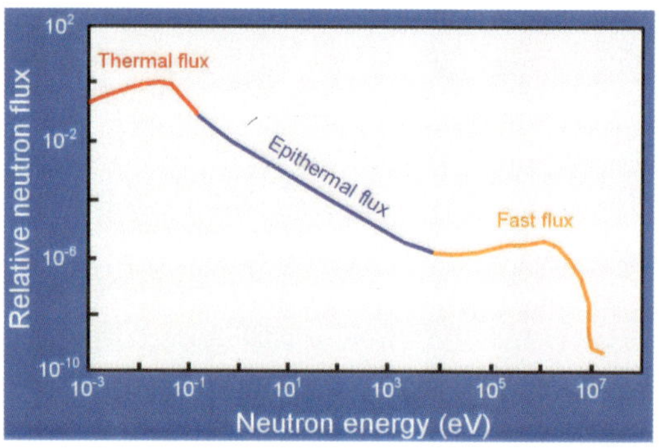

Distribution of kinetic energies of neutrons in the thermal reactor. The fission neutrons (fast flux) are immediately slowed down to the thermal energies via a process called neutron moderation.

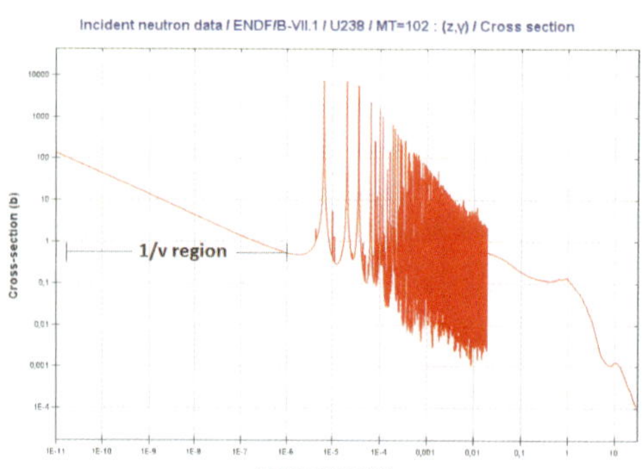

For thermal neutrons (in 1/v region), absorption cross sections increases as the velocity (kinetic energy) of the neutron decreases.

Thermal Neutron – Diffraction

Thermal neutrons are also widely used in neutron diffraction experiments. Neutron diffraction experiments use an elastic neutron scattering to determine the atomic (or magnetic) structure of a material. The neutron diffraction is based the fact that thermal or cold neutrons have the wavelengths similar to atomic spacings.

An examined sample (crystalline solids, gasses, liquids or amorphous materials) must be placed in a neutron beam of thermal (0.025 eV) or cold (neutrons in thermal equilibrium with very cold surroundings such as liquid deuterium) neutrons to obtain a diffraction pattern that provides information about the structure of the examined material. The neutron diffraction experiments are similar to X-ray diffraction experiments, but neutrons interact with matter differently. Photons (X-rays) interact primarily with the electrons surrounding (atomic electron cloud) a nucleus, but neutrons interact only with nuclei. Neither the electrons surrounding (atomic electron cloud) a nucleus nor the electric field caused by a positively charged nucleus affect a neutron's flight. Due to their different properties, both methods together (neutron diffraction and X-ray diffraction) can provide complementary information about the structure of the material.

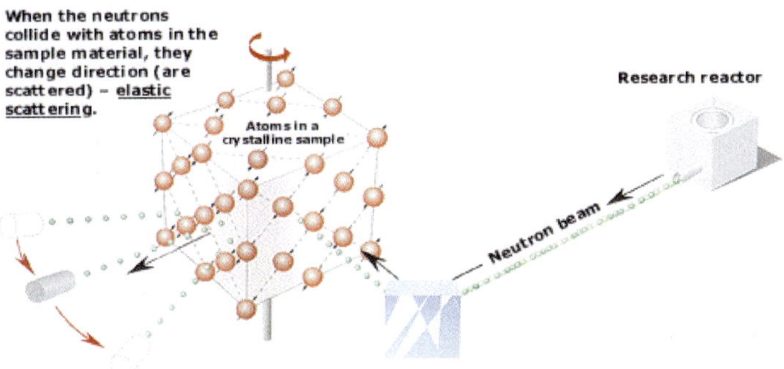

Simple scheme of neutron diffraction experiment.

Breeder Reactor

Breeder reactors are a type of nuclear reactor which produce *more* fissile materials than they consume. They are designed to extend the nuclear fuel supply for the generation of electricity, and have even been mistakenly called a potential renewable energy source. Breeder reactors certainly have the ability to make nuclear fuels quite sustainable though which was Dr. Cohen's main point. Concerns about nuclear weapons proliferation have been one large impediment to creating commercial breeder reactors.

Unlike normal reactors which only use uranium-235 as their fuel, which is only available in scarce concentrations of around 0.7% of natural uranium without enrichment, breeder reactors also make use of natural uranium-238 which is much more common. They can use approximately 70% of the uranium-238 for production of power, whereas normal reactors can only use around 1% of it. They can also use thorium-232 to breed uranium-233, another fissionable product.

The most common breeding is of plutonium-239, which is bred through the process seen in figure below.

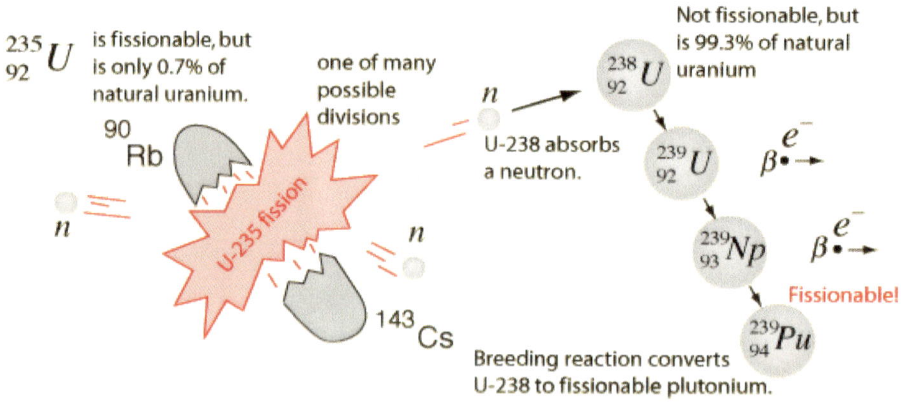

The breeding of ^{239}Pu in a breeder reactor. The neutrons are supplied by the decay of ^{235}U, which transmute ^{238}U to plutonium.

The first experimental breeder reactor (EBR-1) developed was in 1951 in Idaho, U.S.A. Subsequently Russia, Japan, Great Britain and France all developed experimental breeder reactors, however no nation has developed one suitable for high-capacity commercial use. So far, France has made the largest implementation of breeder reactors with their Super-Phenix fast breeder reactor.

Types

There are two categories of breeder reactors, based on the speed of the neutrons. Fast breeder reactors which use uranium-238 as fuel and thermal breeder reactors which use thorium-232 as fuel. Fast breeders do not require moderation since the neutrons need to be moving fast, whereas thermal breeders make us of moderation to achieve slower-moving neutrons.

Fast

The most promising type of breeder reactor is the Liquid Metal Fast Breeder Reactor (LMFBR), which operates by using liquid sodium as its coolant, and breeds plutonium from uranium-238. It works by using highly enriched uranium, between 15-20% uranium-235 content, surrounded or "blanketed" by natural uranium-238 in the reactor core. No moderator is used to slow down the neutrons, because fast neutrons transmute uranium-238 much more efficiently than slow neutrons.

Using water as a coolant would reduce the neutron abundance, since neutrons are absorbed by water. Therefore liquid sodium is used instead. This immediately raised concerns of safety when initially thought of, since sodium is a highly reactive element. It is important to keep the liquid sodium from contact with air or oxygen to avoid explosions, however they aren't any more dangerous than pressurized water reactors. This is because the sodium doesn't need to be pressurized to remain in a liquid state like water does; its boiling point is 892°C. This makes the bursting of pipes far less likely than in other water-reactors. Liquid sodium is also a very good choice because of its heat transfer capabilities, due to its high specific heat capacity.

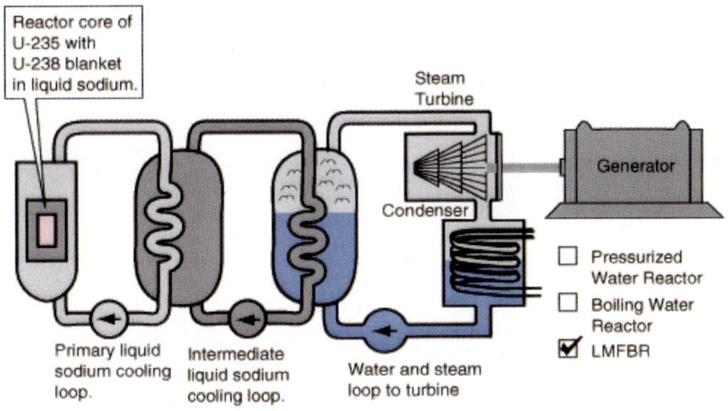

Liquid metal fast breeder reactor.

Other fast breeder types include supercritical water cooled reactors, molten salt reactors, and gas-cooled reactors. You can read more about fast breeder reactors here.

Thermal

The possibility to breed fissile material in slow neutron reactors is unique to thorium, as uranium cannot use thermal neutrons to do so. These reactors would need a fissile material to start the breeding just like fast breeder reactors, however the neutrons produced from this fissile material would need to be slowed down by a moderator.

The technology is much simpler than that of the liquid metal fast breeder; light water is used as the coolant to remove the heat produced by the continuous series of fission reactions rather than a liquid metal system.

Thorium hasn't been used in large scale reactors, however some reactors have used it successfully in the past. A light water breeder reactor in Shippingport, Pa. USA operated for 5 years, and by the end of its operation it had 1.4% more fissile fuel than it began with.

Breeding Ratio

An important concept for a breeder reactor is how much fissionable fuel is being produced compared to how much fuel is being used. This is known as the breeding ratio. For example for the breeding of plutonium, the ratio would be the amount of plutonium produced to the amount of uranium-235 used. In the liquid metal fast breeder reactor (LMFBR), the breeding ratio is 1.4, however the actual achieved ratio is around 1.2.

The number 1.4 is based off of the average number of neutrons given off by a fission reaction of uranium-235, which is 2.4. Only 1 neutron is needed for the fission chain reaction to be stable, so the remaining 1.4 neutrons (on average) could be used for the breeding of uranium-238.

Doubling Time

The amount of time for a breeder to produce enough material to fuel a second reactor is called its doubling time. Present goals for a breeder reactor's doubling time is 10 years, which means it would operate for 10 years producing energy and breeding fissionable material, after which this material could be used to produce the same energy output for another 10 years.

Breeding vs Burnup

All commercial light water reactors breed fuel, but they have breeding ratios that are very low (though still significant) compared to machines traditionally considered "breeders." In recent years, the commercial power industry has been emphasizing high-burnup fuels, which are typically enriched to higher percentages of U-235 than standard reactor fuels so that they last longer in the reactor core. As burnup increases, a higher percentage of the total power produced in a reactor is due to the fuel bred inside the reactor.

At a burnup of 30 gigawatt-days per metric ton of uranium (GWd/MTU), about thirty percent of the total energy released comes from bred plutonium. At 40 GWd/MTU, that percentage increases to about forty percent. This corresponds to a breeding ratio for these reactors of about 0.4 to 0.5. That is to say, about half of the fissile fuel in these reactors is bred there. Correspondingly, this effect extends the cycle life for such fuels to sometimes nearly twice what it would be otherwise. MOX fuel has a smaller breeding effect than U-235 fuel and is thus more challenging and slightly less economic to use due to a quicker drop off in reactivity through cycle life.

This is of interest largely because next-generation reactors such as the European Pressurized Reactor, AP1000 and ESBWR are designed to achieve very high burnup. This directly translates to higher breeding ratios. Current commercial power reactors have achieved breeding ratios of roughly 0.55, and next-generation designs like the AP1000 and EPR should have breeding ratios of 0.7 to 0.8, meaning that they produce 70 to 80 percent as much fuel as they consume, improving their fuel economy by roughly 15 percent compared to current high-burnup reactors.

Breeding of fissile fuel is a common feature in reactors, but in commercial reactors not optimized for this feature it is referred to as "enhanced burnup". Up to a third of all electricity produced in the current US reactor fleet comes from bred fuel, and the industry is working steadily to increase that percentage as time goes on.

Fuel Efficiency and Types of Nuclear Waste

Breeder reactors could, in principle, extract almost all of the energy contained in uranium or thorium, decreasing fuel requirements by a factor of 100 compared to widely used once-through light water reactors, which extract less than 1% of the energy in the uranium mined from the earth. The high fuel-efficiency of breeder reactors could greatly reduce concerns about fuel supply or energy used in mining. Adherents claim that with seawater uranium extraction, there would be enough fuel for breeder reactors to satisfy our energy needs for 5 billion years at 1983's total energy consumption rate, thus making nuclear energy effectively a renewable energy.

Nuclear waste became a greater concern by the 1990s. In broad terms, spent nuclear fuel has two main components. The first consists of fission products, the leftover fragments of fuel atoms after they have been split to release energy. Fission products come in dozens of elements and hundreds of isotopes, all of them lighter than uranium. The second main component of spent fuel is transuranics (atoms heavier than uranium), which are generated from uranium or heavier atoms in the fuel when they absorb neutrons but do not undergo fission. All transuranic isotopes fall within the actinide series on the periodic table, and so they are frequently referred to as the actinides.

The physical behavior of the fission products is markedly different from that of the transuranics.

In particular, fission products do not themselves undergo fission, and therefore cannot be used for nuclear weapons. Furthermore, only seven long-lived fission product isotopes have half-lives longer than a hundred years, which makes their geological storage or disposal less problematic than for transuranic materials.

With increased concerns about nuclear waste, breeding fuel cycles became interesting again because they can reduce actinide wastes, particularly plutonium and minor actinides. Breeder reactors are designed to fission the actinide wastes as fuel, and thus convert them to more fission products.

After "spent nuclear fuel" is removed from a light water reactor, it undergoes a complex decay profile as each nuclide decays at a different rate. Due to a physical oddity referenced below, there is a large gap in the decay half-lives of fission products compared to transuranic isotopes. If the transuranics are left in the spent fuel, after 1,000 to 100,000 years, the slow decay of these transuranics would generate most of the radioactivity in that spent fuel. Thus, removing the transuranics from the waste eliminates much of the long-term radioactivity of spent nuclear fuel.

Today's commercial light water reactors do breed some new fissile material, mostly in the form of plutonium. Because commercial reactors were never designed as breeders, they do not convert enough uranium-238 into plutonium to replace the uranium-235 consumed. Nonetheless, at least one-third of the power produced by commercial nuclear reactors comes from fission of plutonium generated within the fuel. Even with this level of plutonium consumption, light water reactors consume only part of the plutonium and minor actinides they produce, and nonfissile isotopes of plutonium build up, along with significant quantities of other minor actinides.

Fission Probabilities of Selected Actinides, Thermal vs. Fast Neutrons

Isotope	Thermal Fission Cross Section	Thermal Fission %	Fast Fission Cross Section	Fast Fission %
Th-232	nil	1 (non-fissile)	0.350 barn	3 (non-fissile)
U-232	76.66 barn	59	2.370 barn	95
U-233	531.2 barn	89	2.450 barn	93
U-235	584.4 barn	81	2.056 barn	80
U-238	11.77 microbarn	1 (non-fissile)	1.136 barn	11
Np-237	0.02249 barn	3 (non-fissile)	2.247 barn	27
Pu-238	17.89 barn	7	2.721 barn	70
Pu-239	747.4 barn	63	2.338 barn	85
Pu-240	58.77 barn	1 (non-fissile)	2.253 barn	55
Pu-241	1012 barn	75	2.298 barn	87
Pu-242	0.002557 barn	1 (non-fissile)	2.027 barn	53
Am-241	600.4 barn	1 (non-fissile)	0.2299 microbarn	21
Am-242m	6409 barn	75	2.550 barn	94
Am-243	0.1161 barn	1 (non-fissile)	2.140 barn	23
Cm-242	5.064 barn	1 (non-fissile)	2.907 barn	10
Cm-243	617.4 barn	78	2.500 barn	94
Cm-244	1.037 barn	4 (non-fissile)	0.08255 microbarn	33

Reprocessing

Fission of the nuclear fuel in any reactor produces neutron-absorbing fission products. Because of this unavoidable physical process, it is necessary to reprocess the fertile material from a breeder reactor to remove those neutron poisons. This step is required to fully utilize the ability to breed as much or more fuel than is consumed. All reprocessing can present a proliferation concern, since it extracts weapons-usable material from spent fuel. The most-common reprocessing technique, PUREX, presents a particular concern, since it was expressly designed to separate pure plutonium. Early proposals for the breeder-reactor fuel cycle posed an even greater proliferation concern because they would use PUREX to separate plutonium in a highly attractive isotopic form for use in nuclear weapons.

Several countries are developing reprocessing methods that do not separate the plutonium from the other actinides. For instance, the non-water-based pyrometallurgical electrowinning process, when used to reprocess fuel from an integral fast reactor, leaves large amounts of radioactive actinides in the reactor fuel. More-conventional water-based reprocessing systems include SANEX, UNEX, DIAMEX, COEX, and TRUEX, and proposals to combine PUREX with co-processes.

All these systems have modestly better proliferation resistance than PUREX, though their adoption rate is low.

In the thorium cycle, thorium-232 breeds by converting first to protactinium-233, which then decays to uranium-233. If the protactinium remains in the reactor, small amounts of uranium-232 are also produced, which has the strong gamma emitter thallium-208 in its decay chain. Similar to uranium-fueled designs, the longer the fuel and fertile material remain in the reactor, the more of these undesirable elements build up. In the envisioned commercial thorium reactors, high levels of uranium-232 would be allowed to accumulate, leading to extremely high gamma-radiation doses from any uranium derived from thorium. These gamma rays complicate the safe handling of a weapon and the design of its electronics; this explains why uranium-233 has never been pursued for weapons beyond proof-of-concept demonstrations.

While the thorium cycle may be proliferation-resistant with regard to uranium-233 extraction from fuel (because of the presence of uranium-232), it poses a proliferation risk from an alternate route of uranium-233 extraction, which involves chemically extracting protactinium-233 and allowing it to decay to pure uranium-233 outside of the reactor. This process could happen beyond the oversight of organizations such as the International Atomic Energy Agency (IAEA).

Waste Reduction

Actinides and fission products by half-life								
Actinides by decay chain				**Half-life range (y)**	**Fission products of ^{235}U by yield**			
4*n*	4*n*+1	4*n*+2	4*n*+3			4.5–7%	0.04–1.25%	<0.001%
^{228}Ra№				**4–6**	†		^{155}Euþ	
^{244}Cmƒ	^{241}Puƒ	^{250}Cf	^{227}Ac№	**10–29**		^{90}Sr	^{85}Kr	^{113m}Cdþ
^{232}Uƒ		^{238}Puƒ№	^{243}Cmƒ	**29–97**		^{137}Cs	^{151}Smþ	^{121m}Sn

^{248}Bk	^{249}Cf^{f}	^{242m}Am^{f}		**141–351**	No fission products have a half-life in the range of 100–210 k years ...			
	^{241}Am^{f}		^{251}Cf^{f}	**430–900**				
		^{226}Ra$^{№}$	^{247}Bk	**1.3 k – 1.6 k**				
^{240}Pu$^{f№}$	^{229}Th$^{№}$	^{246}Cm^{f}	^{243}Am^{f}	**4.7 k – 7.4 k**				
	^{245}Cm^{f}	^{250}Cm		**8.3 k – 8.5 k**				
			^{239}Pu$^{f№}$	**24.1 k**				
		^{230}Th$^{№}$	^{231}Pa$^{№}$	**32 k – 76 k**				
^{236}Np^{f}	^{233}U$^{f№}$	^{234}U$^{№}$		**150 k – 250 k**	‡	^{99}Tc$^{₡}$	^{126}Sn	
^{248}Cm		^{242}Pu^{f}		**327 k – 375 k**			^{79}Se$^{₡}$	
				1.53 M		^{93}Zr		
	^{237}Np$^{f№}$			**2.1 M – 6.5 M**		^{135}Cs$^{₡}$	^{107}Pd	
^{236}U$^{№}$			^{247}Cm^{f}	**15 M – 24 M**			^{129}I$^{₡}$	
^{244}Pu$^{№}$				**80 M**	... nor beyond 15.7 M years			
^{232}Th$^{№}$		^{238}U$^{№}$	^{235}U$^{f№}$	**0.7 G – 14.1 G**				

Legend for superscript symbols
₡ has thermal neutron capture cross section in the range of 8–50 barns
f fissile
m metastable isomer
№ naturally occurring radioactive material (NORM)
þ neutron poison (thermal neutron capture cross section greater than 3k barns)
† range 4–97 y: Medium-lived fission product
‡ over 200,000 y: Long-lived fission product

Nuclear waste became a greater concern by the 1990s. Breeding fuel cycles attracted renewed interest because of their potential to reduce actinide wastes, particularly plutonium and minor actinides. Since breeder reactors on a closed fuel cycle would use nearly all of the actinides fed into them as fuel, their fuel requirements would be reduced by a factor of about 100. The volume of waste they generate would be reduced by a factor of about 100 as well. While there is a huge reduction in the *volume* of waste from a breeder reactor, the *activity* of the waste is about the same as that produced by a light-water reactor.

In addition, the waste from a breeder reactor has a different decay behavior, because it is made up of different materials. Breeder reactor waste is mostly fission products, while light-water reactor waste has a large quantity of transuranics. After spent nuclear fuel has been removed from a light-water reactor for longer than 100,000 years, these transuranics would be the main source of radioactivity. Eliminating them would eliminate much of the long-term radioactivity from the spent fuel.

In principle, breeder fuel cycles can recycle and consume all actinides, leaving only fission products. Indicates, fission products have a peculiar ‘gap’ in their aggregate half-lives, such that no fission products have a half-life longer than 91 years and shorter than two hundred thousand years. As a result of this physical oddity, after several hundred years in storage, the activity of the radioactive waste from a Fast Breeder Reactor would quickly drop to the low level of the long-lived fission products. However, to obtain this benefit requires the highly efficient separation of transuranics from spent fuel. If the fuel reprocessing methods used leave a large fraction of the transuranics in the final waste stream, this advantage would be greatly reduced.

Both types of breeding cycles can reduce actinide wastes:

- The fast breeder reactor's fast neutrons can fission actinide nuclei with even numbers of both protons and neutrons. Such nuclei usually lack the low-speed "thermal neutron" resonances of fissile fuels used in LWRs.
- The thorium fuel cycle inherently produces lower levels of heavy actinides. The fertile material in the thorium fuel cycle has an atomic weight of 232, while the fertile material in the uranium fuel cycle has an atomic weight of 238. That mass difference means that thorium-232 requires six more neutron capture events per nucleus before the transuranic elements can be produced. In addition to this simple mass difference, the reactor gets two chances to fission the nuclei as the mass increases: First as the effective fuel nuclei U233, and as it absorbs two more neutrons, again as the fuel nuclei U235.

A reactor whose main purpose is to destroy actinides, rather than increasing fissile fuel-stocks, is sometimes known as a burner reactor. Both breeding and burning depend on good neutron economy, and many designs can do either. Breeding designs surround the core by a breeding blanket of fertile material. Waste burners surround the core with non-fertile wastes to be destroyed. Some designs add neutron reflectors or absorbers.

Subcritical–Critical–Supercritical Reactor

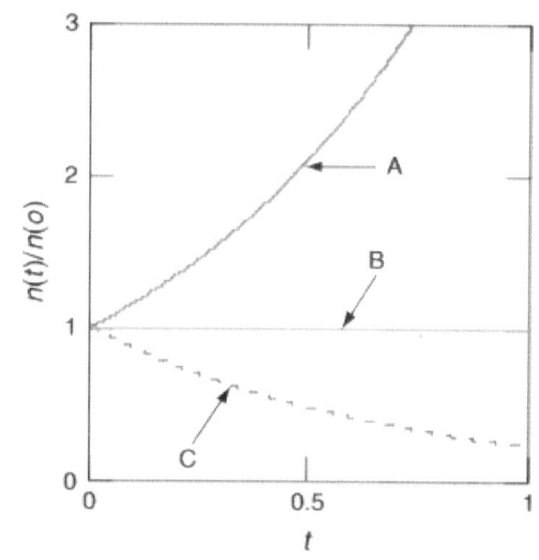

Reactor criticality. A – supercritical state; B – critical state; C – subcritical state

The basic classification of states of a reactor is according to the multiplication factor as eigenvalue which is a measure of the change in the fission neutron population from one neutron generation to the subsequent generation.

- $k_{eff} < 1$. If the multiplication factor for a multiplying system is less than 1.0, then the number of neutrons is decreasing in time (with the mean generation time) and the chain reaction will never be self-sustaining. This condition is known as the subcritical state.

- $k_{eff} = 1$. If the multiplication factor for a multiplying system is equal to 1.0, then there is no change in neutron population in time and the chain reaction will beself-sustaining. This condition is known as the critical state.
- $k_{eff} > 1$. If the multiplication factor for a multiplying system is greater than 1.0, then the multiplying system produces more neutrons than are needed to be self-sustaining. The number of neutrons is exponentially increasing in time (with the mean generation time). This condition is known as the supercritical state.

The definitions described above are fully applicable to a reactor at zero power level, that is at such power level in which all thermal considerations are not important to the chain reaction (let say from zero power to 1% of rated power).

Prompt Criticality

The basic classification of states of a reactor may be in some case insufficient and a finer classification is needed. The finer classification takes into account the two groups of neutrons that are produced in fission.

The prompt critical state is defined as:

- $k_{eff} > 1$; $\rho \geq \beta_{eff}$, where the reactivity of a reactor is higher than the effective delayed neutron fraction. In this case, the production of prompt neutrons alone is enough to balance neutron losses and increase the neutron population. The number of neutrons is exponentially increasing in time (as rapidly as the prompt neutron generation life-time ~10^{-5}s).

The prompt subcritical and delayed supercritical state is defined as:

- $k_{eff} > 1$; $0 < \rho < \beta_{eff}$, where the reactivity of a reactor is higher than zero and lower than the effective delayed neutron fraction. In this case, the production of prompt neutrons alone is insufficient to balance neutron losses and the delayed neutrons are needed in order to sustain the chain reaction. The neutron population increases, but much more slowly (as the mean generation lifetime with delayed neutrons~0.1 s).

The prompt subcritical and delayed critical state is defined as:

- $k_{eff} = 1$; $\rho = 0$, where the reactivity of a reactor is equal to zero. In this case, the production of prompt neutrons alone is insufficient to balance neutron losses and the delayed neutrons are needed in order to sustain the chain reaction. There is no change in neutron population in time and the chain reaction will be self-sustaining. This state is the same state as the critical state from basic classification.

The prompt subcritical and delayed subcritical state is defined as:

- $k_{eff} < 1$; $\rho < 0$, where the reactivity of a reactor is lower than zero. In this case, the production of all neutrons is insufficient to balance neutron losses and the chain reaction is not self-sustaining. If the reactor core contains external or internal neutron sources, the reactor is in the state that is usually referred to as the subcritical multiplication.

Research Reactor

Research reactor is a nuclear reactor that is a source of neutron and gamma radiation and is designed for various studies in numerous areas of science and technology.

Research reactors are used for studies in nuclear and neutron physics, solid-state physics, nuclear and radiation chemistry, the science of materials, biology, and medicine. The reactors are also used for testing the fuel elements of projected power reactors, as well as the materials used for the construction of reactors. Research reactors have been used to develop the method of activation analysis, which makes it possible to determine the composition of various materials without their destruction and to detect minute quantities (up to 10^{-6} u, g) of chemical elements. Research reactors are also used for the production of radioisotopes.

Research reactors have a core, which contains the fission material, whereas thermal reactors also contain neutron moderators (light or heavy water, graphite, beryllium). Heat removal is provided in the reactor core. The core is surrounded by a neutron reflector. The reactor is also surrounded by a biological shield, which may be pierced by channels to obtain neutron beams. In order to obtain a powerful flux of thermal neutrons without fast neutrons, a device called the thermal column is used. The column, consisting of a good moderator (most frequently graphite), is placed with one end directly at the core; the other end emerges on premises suitable for the performance of experiments. Special devices or channels are provided for loading the materials to be tested into the core. Figure shows the vertical cross section of a Soviet VVR-M research reactor, which is designed for studies in nuclear physics, radiochemistry, radiobiology, and other areas.

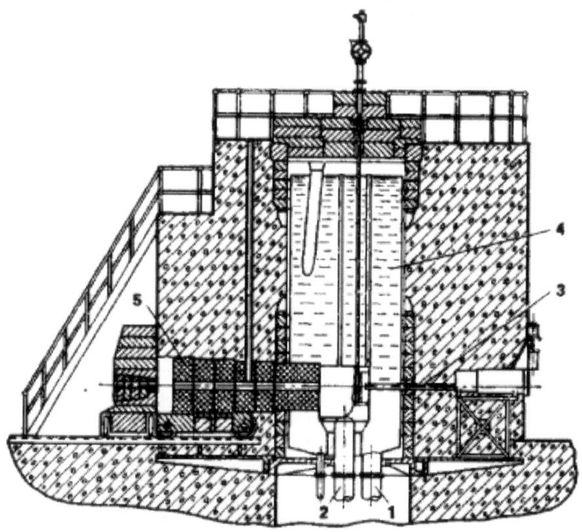

Figure. Cross section of the VVR-M research reactor: (1) water inlet, (2) water outlet, (3) experimental channel, (4) core tank, and (5) thermal column

According to the neutron spectrum in the core, research reactors, as well as nuclear reactors in general, are divided into fast reactors and thermal reactors. Most of the research reactors are thermal reactors, mainly of the heterogeneous type, meaning that the fuel elements alternate with the moderator in a certain order. Within the general range of 1012-1015 neutrons/(cm2.sec), research reactors are divided into those with low, medium, and high neutron flux in the core. Pulsed research reactors are designed to provide short-term increases of the neutron flux without the corresponding increase of the average reactor output and the corresponding complication of the heat removal system. For example, a Soviet pulsed fast reactor with an average power level of 3 kW, generates at the moment of the chain reaction burst, a neutron flux of 1.3 X 1018 neutrons/ (cm2.sec) at the pulse maximum, which corresponds to an instantaneous power of 23 MW. The excessive reactivity is obtained by a piece of uranium 235 pressed into a disk, which rotates between the fuel rods. The pulsed reactor is used in the study of the physics of solids and liquids and in neutron spectrometry.

The following reactor types are distinguished according to reactor core design: tank-type reactors (SM-2 in the USSR and ETR in the USA), tank-type reactors submerged in pools (BR-2 in Belgium), and channel-type reactors (RFT in the USSR).

The cores of tank-type research reactors are highly compact and therefore possess the best physical characteristics. The reactors submerged in pools of water (swimming-pool reactors) are the safest, since all manipulations with radioactive products are performed through a layer of water. The channel-type reactors are convenient in terms of the layout and replacement of the elements or samples under study. However, all three types of reactors have an essential disadvantage: access to the reactor core or the interchannel space is difficult, which complicates studies.

A fourth type of research reactor has been developed in the USSR. This reactor is of the channel type in which the core with the fuel channels and pipelines are submerged in a pool of water. This reactor combines the advantages of research reactors of the channel-type as well as the swimming-pool type. The MR research reactor, put into operation in Moscow in 1963, is of this type and is designed mainly for the testing of fuel elements and materials. This reactor achieves a thermal neutron flux of 8 X 1014 neutrons/(cm2.sec) in the central neutron trap, consisting of a cylinder of 100 mm in diameter filled with water, at a power output of 20 MW. The power level of this reactor was increased to 40 MW by 1968. The fuel channels with tubular fuel elements, in which the samples to be irradiated are installed, are cooled with pressurized water. Beryllium blocks are used as moderators. Up to 25 experimental channel loops may be loaded in the reactor. The control rod drives are mounted on a movable carriage in order to facilitate access to the reactor core during reloading. The reactor pool is connected via a lock system with the storage pool containing a gamma-irradiation installation, which uses the spent fuel assemblies as the source of radiation. Figure shows a cross section of the MR reactor with the storage pool. An even more powerful research reactor, the MIR, is in operation in Dimitrovgrad (USSR). This reactor makes it possible to achieve a neutron flux of 1.5 X 1015 neutrons/ (cm2.sec).

Several hundred research reactors of various types are in operation abroad. For example, the heavy-water tank-type reactor DIDO (Great Britain) is used for the production of isotopes and the testing of reactor materials. The graphite reactor PCTR (USA) is used in the study of physical constants. The swimming-pool reactor with light water FRI-1 (Federal Republic of Germany) is used for studies in neutron physics, chemistry, biology, and medicine.

Every research reactor is used for a variety of studies, but the experience gained when designing and running research reactors in the USSR and abroad indicates that it is more reasonable to build research reactors that are specialized for use in specific research areas.

Technical Aspects

Research reactors are simpler than power reactors and operate at lower temperatures. They need far less fuel, and far less fission products build up as the fuel is used. On the other hand, their fuel requires more highly enriched uranium, typically up to 20% U-235, although some use 93% U-235; while 20% enrichment is not generally considered usable in nuclear weapons, 93% is commonly referred to as "weapons grade". They also have a very high power density in the core, which requires special design features. Like power reactors, the core needs cooling, typically natural or forced convection with water, and a moderator is required to slow the neutron velocities and enhance fission. As neutron production is their main function, most research reactors benefit from reflectors to reduce neutron loss from the core.

Conversion to LEU

The International Atomic Energy Agency and the U.S. Department of Energy initiated a program in 1978 to develop the means to convert research reactors from using highly enriched uranium to the use of low enriched uranium, in support of its nonproliferation policy. By that time the U.S. had supplied research reactors and highly enriched uranium to 41 countries as part of its Atoms for Peace program. In 2004, the U.S. Department of Energy extended its Foreign Research Reactor Spent Nuclear Fuel Acceptance program until 2019.

Also in 2004, the Texas A&M reactor switched to LEU after decades using HEU. These changes are a part of an anti-terrorism initiative since 9/11 begun by the Bush Administration.

Designers and Constructors

While in the 1950s, 1960s and 1970s there were a number of companies that specialized in the design and construction of research reactors, the activity of this market cooled down afterwards, and many companies withdrew.

The market has consolidated today into a few companies that concentrate the key projects on a worldwide basis.

The most recent international tender (1999) for a research reactor was that organized by ANSTO for the design, construction and commissioning of the OPAL reactor. Four companies were prequalified: AECL, INVAP, Siemens and Technicatom. The project was awarded to INVAP that built the reactor. In recent years, AECL withdrew from this market, and Siemens and Technicatom activities were merged into AREVA.

Classes of Research Reactors

- Aqueous homogeneous reactor
- Argonaut class reactor
- DIDO class, six high-flux reactors worldwide
- TRIGA, a highly successful class with >50 installations worldwide

- SLOWPOKE reactor class, developed by AECL, Canada
- Miniature neutron source reactor, based on the SLOWPOKE design, developed by AECL, currently exported by China

Thorium-based Nuclear Power

Thorium (Th-232) is not itself fissile and so is not directly usable in a thermal neutron reactor. However, it is 'fertile' and upon absorbing a neutron will transmute to uranium-233 (U-233)[a], which is an excellent fissile fuel material[b]. In this regard it is similar to uranium-238 (which transmutes to plutonium-239). All thorium fuel concepts therefore require that Th-232 is first irradiated in a reactor to provide the necessary neutron dosing to produce protactinium-233. The Pa-233 that is produced can either be chemically separated from the parent thorium fuel and the decay product U-233 then recycled into new fuel, or the U-233 may be usable 'in-situ' in the same fuel form, especially in molten salt reactors (MSRs).

Thorium fuels therefore need a fissile material as a 'driver' so that a chain reaction (and thus supply of surplus neutrons) can be maintained. The only fissile driver options are U-233, U-235 or Pu-239. (None of these is easy to supply).

It is possible – but quite difficult – to design thorium fuels that produce more U-233 in thermal reactors than the fissile material they consume (this is referred to as having a fissile conversion ratio of more than 1.0 and is also called breeding). Thermal breeding with thorium requires that the neutron economy in the reactor has to be very good (ie, there must be low neutron loss through escape or parasitic absorption). The possibility to breed fissile material in slow neutron systems is a unique feature for thorium-based fuels and is not possible with uranium fuels.

Another distinct option for using thorium is as a 'fertile matrix' for fuels containing plutonium that serves as the fissile driver while being consumed (and even other transuranic elements like americium). Mixed thorium-plutonium oxide (Th-Pu MOX) fuel is an analog of current uranium-MOX fuel, but no new plutonium is produced from the thorium component, unlike for uranium fuels in U-Pu MOX fuel, and so the level of net consumption of plutonium is high. Production of all actinides is lower than with conventional fuel, and negative reactivity coefficient is enhanced compared with U-Pu MOX fuel.

In fresh thorium fuel, all of the fissions (thus power and neutrons) derive from the driver component. As the fuel operates the U-233 content gradually increases and it contributes more and more to the power output of the fuel. The ultimate energy output from U-233 (and hence indirectly thorium) depends on numerous fuel design parameters, including: fuel burn-up attained, fuel arrangement, neutron energy spectrum and neutron flux (affecting the intermediate product protactinium-233, which is a neutron absorber). The fission of a U-233 nucleus releases about the same amount of energy (200 MeV) as that of U-235.

An important principle in the design of thorium fuel systems is that of heterogeneous fuel arrangement in which a high fissile (and therefore higher power) fuel zone called the seed region is physically separated from the fertile (low or zero power) thorium part of the fuel – often called the blanket. Such an arrangement is far better for supplying surplus neutrons to thorium nuclei so they can convert to fissile U-233, in fact all thermal breeding fuel designs are heterogeneous. This principle applies to all the thorium-capable reactor systems.

Th-232 is fissionable with fast neutrons of over 1 MeV energy. It could therefore be used in fast molten salt and other Gen IV reactors with uranium or plutonium fuel to initiate fission. However, Th-232 fast fissions only one tenth as well as U-238, so there is no particular reason for using thorium in fast reactors, given the huge amount of depleted uranium awaiting use.

In Norway, Thor Energy is developing and testing two thorium-bearing fuels for use in existing nuclear power plants. Fuel rods containing thorium additive (Th-Add) and also thorium MOX (with Pu) fuel rods have been in a five-year irradiation trial since April 2013 at the Halden test reactor. The company is working towards obtaining regulatory approval for the commercial production and use of Th-Add fuel by 2017-18, and to market the fuel soon thereafter. In mid-2015 a second batch of Th-MOX fuel pellets will commence testing. Thor Energy and several utilities from North America and Europe are initiating feasibility studies to investigate the use of Th-Add fuel in commercial reactors. This fuel is promoted as a means to improve power profiles within commercial reactors.

Reactors Able to use Thorium

There are seven types of reactor into which thorium can be introduced as a nuclear fuel. The first five of these have all entered into operational service at some point. The last two are still conceptual:

- Heavy Water Reactors (PHWRs): These are well suited for thorium fuels due to their combination of: (i) excellent neutron economy (their low parasitic neutron absorption means more neutrons can be absorbed by thorium to produce useful U-233), (ii) slightly faster average neutron energy which favours conversion to U-233, (iii) flexible on-line refueling capability. Furthermore, heavy water reactors (especially CANDU) are well established and widely-deployed commercial technology for which there is extensive licensing experience.
- There is potential application to Enhanced Candu 6 (EC6) and ACR-1000 reactors fueled with 5% plutonium (reactor grade) plus thorium. In the closed fuel cycle, the driver fuel required for starting off is progressively replaced with recycled U-233, so that on reaching equilibrium 80% of the energy comes from thorium. Fissile drive fuel could be LEU, plutonium, or recycled uranium from LWR. Fleets of PHWRs with near-self-sufficient equilibrium thorium fuel cycles could be supported by a few fast breeder reactors to provide plutonium.
- High-Temperature Gas-Cooled Reactors (HTRs): These are well suited for thorium-based fuels in the form of robust 'TRISO' coated particles of thorium mixed with plutonium or enriched uranium, coated with pyrolytic carbon and silicon carbide layers which retain fission gases. The fuel particles are embedded in a graphite matrix that is very stable at high temperatures. Such fuels can be irradiated for very long periods and thus deeply burn their original fissile charge. Thorium fuels can be designed for both 'pebble bed' and 'prismatic' types of HTR reactors.

- Boiling (Light) Water Reactors (BWRs): BWR fuel assemblies can be flexibly designed in terms of rods with varying compositions (fissile content), and structural features enabling the fuel to experience more or less moderation (eg, half-length fuel rods). This design flexibility is very good for being able to come up with suitable heterogeneous arrangements and create well-optimised thorium fuels. So it is possible, for example, to design thorium-plutonium BWR fuels that are tailored for 'burning' surplus plutonium. And importantly, BWRs are a well-understood and licensed reactor type.

- Pressurised (Light) Water Reactors (PWRs): Viable thorium fuels can be designed for a PWR, though with less flexibility than for BWRs. Fuel needs to be in heterogeneous arrangements in order to achieve satisfactory fuel burn-up. It is not possible to design viable thorium-based PWR fuels that convert significant amounts of U-233. Even though PWRs are not the perfect reactor in which to use thorium, they are the industry workhorse and there is a lot of PWR licensing experience. They are a viable early-entry thorium platform.

- Fast Neutron Reactors (FNRs): Thorium can serve as a fuel component for reactors operating with a fast neutron spectrum – in which a wider range of heavy nuclides are fissionable and may potentially drive a thorium fuel. There is, however, no relative advantage in using thorium instead of depleted uranium (DU) as a fertile fuel matrix in these reactor systems due to a higher fast-fission rate for U-238 and the fission contribution from residual U-235 in this material. Also, there is a huge amount of surplus DU available for use when more FNRs are commercially available, so thorium has little or no competitive edge in these systems.

- Molten Salt Reactors (MSRs): These reactors are still at the design stage but are likely to be very well suited for using thorium as a fuel. The unique fluid fuel can incorporate thorium and uranium (U-233 and/or U-235) fluorides as part of a salt mixture that melts in the range 400-700°C, and this liquid serves as both heat transfer fluid and the matrix for the fissioning fuel. The fluid circulates through a core region and then through a chemical processing circuit that removes various fission products (poisons) and/or the valuable U-233. The level of moderation is given by the amount of graphite built into the core. Certain MSR designs[c] will be designed specifically for thorium fuels to produce useful amounts of U-233.

- Accelerator Driven Reactors (ADS): The sub-critical ADS system is an unconventional nuclear fission energy concept that is potentially 'thorium capable'. Spallation neutrons are produced[d] when high-energy protons from an accelerator strike a heavy target like lead. These neutrons are directed at a region containing a thorium fuel, eg, Th-plutonium which reacts to produce heat as in a conventional reactor. The system remains subcritical ie, unable to sustain a chain reaction without the proton beam. Difficulties lie with the reliability of high-energy accelerators and also with economics due to their high power consumption.

A key finding from thorium fuel studies to date is that it is not economically viable to use low-enriched uranium (LEU – with a U-235 content of up to 20%) as a fissile driver with thorium fuels, unless the fuel burn-up can be taken to very high levels – well beyond those currently attainable in LWRs with zirconium cladding.

With regard to proliferation significance, thorium-based power reactor fuels would be a poor source for fissile material usable in the illicit manufacture of an explosive device. U-233 contained in spent thorium fuel contains U-232 which decays to produce very radioactive daughter nuclides and these create a strong gamma radiation field. This confers proliferation resistance by creating significant handling problems and by greatly boosting the detectability (traceability) and ability to safeguard this material.

Prior Thorium-fuelled Electricity Generation

There have been several significant demonstrations of the use of thorium-based fuels to generate electricity in several reactor types. Many of these early trials were able to use high-enriched uranium (HEU) as the fissile 'driver' component, and this would not be considered today.

The 300 MWe Thorium High Temperature Reactor (THTR) at Hamm-Uentrop in Germany operated with thorium-HEU fuel between 1983 and 1989, when it was shut down due to technical problems. Over half of its 674,000 pebbles contained Th-HEU fuel particles (the rest comprised graphite moderator and some neutron absorbers). These were continuously moved through the reactor as it operated, and on average each fuel pebble passed six times through the core.

The 40 MWe Peach Bottom HTR in the USA was a demonstration thorium-fuelled reactor that ran from 1967-74.[2] It used a thorium-HEU fuel in the form of microspheres of mixed thorium-uranium carbide coated with pyrolytic carbon. These were embedded in annular graphite segments (not pebbles). This reactor produced 33 billion kWh over 1349 equivalent full-power days with a capacity factor of 74%.

The 330 MWe Fort St Vrain HTR in Colorado, USA, was a larger-scale commercial successor to the Peach Bottom reactor and ran from 1976-89. It also used thorium-HEU fuel in the form of microspheres of mixed thorium-uranium carbide coated with silicon oxide and pyrolytic carbon to retain fission products. These were embedded in graphite 'compacts' that were arranged in hexagonal columns ('prisms'). Almost 25 tonnes of thorium was used in fuel for the reactor, much of which attained a burn-up of about 170 GWd/t.

A unique thorium-fuelled light water breeder reactor operated from 1977 to 1982 at Shippingport in the USA[3] – it used uranium-233 as the fissile driver in special fuel assemblies that had movable 'seed' regions which allowed the level of neutron moderation to be gradually increased as the fuel aged[e]. The reactor core was housed in a reconfigured early PWR. It operated with a power output of 60 MWe (236 MWt) and an availability factor of 86% producing over 2.1 billion kWh. Post-operation inspections revealed that 1.39% more fissile fuel was present at the end of core life, proving that breeding had occurred. A 2007 NRC report quotes a breeding ratio of 1.01. Chemically reprocessing the fuel was not attempted.

Indian heavy water reactors (PHWRs) have for a long time used thorium-bearing fuel bundles for power flattening in some fuel channels – especially in initial cores when special reactivity control measures are needed.

Developing a Thorium-based Fuel Cycle

Thorium fuel cycles offer attractive features, including lower levels of waste generation, less transuranic elements in that waste, and providing a diversification option for nuclear fuel supply. Also, the use of thorium in most reactor types leads to extra safety margins. Despite these merits, the commercialization of thorium fuels faces some significant hurdles in terms of building an economic case to undertake the necessary development work.

A great deal of testing, analysis and licensing and qualification work is required before any thorium fuel can enter into service. This is expensive and will not eventuate without a clear business case and government support. Also, uranium is abundant and cheap and forms only a small part of the cost of nuclear electricity generation, so there are no real incentives for investment in a new fuel type that may save uranium resources.

Other impediments to the development of thorium fuel cycle are the higher cost of fuel fabrication and the cost of reprocessing to provide the fissile plutonium driver material. The high cost of fuel fabrication (for solid fuel) is due partly to the high level of radioactivity that builds up in U-233 chemically separated from the irradiated thorium fuel. Separated U-233 is always contaminated with traces of U-232 which decays (with a 69-year half-life) to daughter nuclides such as thallium-208 that are high-energy gamma emitters. Although this confers proliferation resistance to the fuel cycle by making U-233 hard to handle and easy to detect, it results in increased costs. There are similar problems in recycling thorium itself due to highly radioactive Th-228 (an alpha emitter with two-year half life) present. Some of these problems are overcome in the LFTR or other molten salt reactor and fuel cycle designs, rather than solid fuel.

Particularly in a molten salt reactor, the equilibrium fuel cycle is expected to have relatively low radiotoxicity, being fission products only plus short-lived Pa-233, without transuranics. These are continually removed in on-line reprocessing, though this is more complex than for the uranium-plutonium fuel cycle.

Nevertheless, the thorium fuel cycle offers energy security benefits in the long-term – due to its potential for being a self-sustaining fuel without the need for fast neutron reactors. It is therefore an important and potentially viable technology that seems able to contribute to building credible, long-term nuclear energy scenarios.

Possible Benefits

The World Nuclear Association explains some of the possible benefits:

The thorium fuel cycle offers enormous energy security benefits in the long-term – due to its potential for being a self-sustaining fuel without the need for fast neutron reactors. It is therefore an important and potentially viable technology that seems able to contribute to building credible, long-term nuclear energy scenarios.

Moir and Teller agree, noting that the possible advantages of thorium include "utilization of an abundant fuel, inaccessibility of that fuel to terrorists or for diversion to weapons use, together with good economics and safety feature " Thorium is considered the "most abundant, most readily available, cleanest, and safest energy source on Earth," adds science writer Richard Martin.

- Thorium is three times as abundant as uranium and nearly as abundant as lead and gallium in the Earth's crust. The Thorium Energy Alliance estimates "there is enough thorium in the United States alone to power the country at its current energy level for over 1,000 years." "America has buried tons as a by-product of rare earth metals mining," notes Evans-Pritchard. Almost all thorium is fertile Th-232, compared to uranium that is composed of 99.3% fertile U-238 and 0.7% more valuable fissile U-235.

- It is difficult to make a practical nuclear bomb from a thorium reactor's byproducts. According to Alvin Radkowsky, designer of the world's first full-scale atomic electric power plant, "a thorium reactor's plutonium production rate would be less than 2 percent of that of a standard reactor, and the plutonium's isotopic content would make it unsuitable for a nuclear detonation." Several uranium-233 bombs have been tested, but the presence of uranium-232 tended to "poison" the uranium-233 in two ways: intense radiation from the uranium-232 made the material difficult to handle, and the uranium-232 led to possible pre-detonation. Separating the uranium-232 from the uranium-233 proved very difficult, although newer laser techniques could facilitate that process.
- There is much less nuclear waste—up to two orders of magnitude less, state Moir and Teller, eliminating the need for large-scale or long-term storage; "Chinese scientists claim that hazardous waste will be a thousand times less than with uranium." The radioactivity of the resulting waste also drops down to safe levels after just a one or a few hundred years, compared to tens of thousands of years needed for current nuclear waste to cool off.
- According to Moir and Teller, "once started up [it] needs no other fuel except thorium because it makes most or all of its own fuel." This only applies to breeding reactors, that produce at least as much fissile material as they consume. Other reactors require additional fissile material, such as uranium-235 or plutonium.
- Thorium fuel cycle is a potential way to produce long term nuclear energy with low radio-toxicity waste. In addition, the transition to thorium could be done through the incineration of weapons grade plutonium (WPu) or civilian plutonium.
- Since all natural thorium can be used as fuel no expensive fuel enrichment is needed. However the same is true for U-238 as fertile fuel in the uranium-plutonium cycle.
- Comparing the amount of thorium needed with coal, Nobel laureate Carlo Rubbia of CERN, (European Organization for Nuclear Research), estimates that one ton of thorium can produce as much energy as 200 tons of uranium, or 3,500,000 tons of coal.
- Liquid fluoride thorium reactors are designed to be meltdown proof. A plug at the bottom of the reactor melts in the event of a power failure or if temperatures exceed a set limit, draining the fuel into an underground tank for safe storage.
- Mining thorium is safer and more efficient than mining uranium. Thorium's ore monazite generally contains higher concentrations of thorium than the percentage of uranium found in its respective ore. This makes thorium a more cost efficient and less environmentally damaging fuel source. Thorium mining is also easier and less dangerous than uranium mining, as the mine is an open pit which requires no ventilation, unlike underground uranium mines, where radon levels can be potentially harmful.

Summarizing some of the potential benefits, Martin offers his general opinion: "Thorium could provide a clean and effectively limitless source of power while allaying all public concern—weapons proliferation, radioactive pollution, toxic waste, and fuel that is both costly and complicated to process. From an economics viewpoint, UK business editor has suggested that "Obama could kill fossil fuels overnight with a nuclear dash for thorium," suggesting a "new Manhattan Project," and adding, "If it works, Manhattan II could restore American optimism and strategic leadership at a stroke" Moir and Teller estimated in 2004 that the cost for their recommended prototype would be "well under $1 billion with operation costs likely on the order of $100 million per year," and as a result a "large-scale nuclear power plan" usable by many countries could be set up within a decade.

A report by the Bellona Foundation in 2013 concluded that the economics are quite speculative. Thorium nuclear reactors are unlikely to produce cheaper energy, but the management of spent fuel is likely to be cheaper than for uranium nuclear reactors.

Possible Disadvantages

Some experts note possible specific disadvantages of thorium nuclear power:

- Breeding in a thermal neutron spectrum is slow and requires extensive reprocessing. The feasibility of reprocessing is still open.
- Significant and expensive testing, analysis and licensing work is first required, requiring business and government support. In a 2012 report on the use of thorium fuel with existing water-cooled reactors, the Bulletin of the Atomic Scientists suggested that it would "require too great an investment and provide no clear payoff", and that "from the utilities' point of view, the only legitimate driver capable of motivating pursuit of thorium is economics".
- There is a higher cost of fuel fabrication and reprocessing than in plants using traditional solid fuel rods.
- Thorium, when being irradiated for use in reactors, will make uranium-232, which is very dangerous due to the gamma rays it emits. This irradiation process may be altered slightly by removing protactinium-233. The irradiation would then make uranium-233 in lieu of uranium-232, which can be used in nuclear weapons to make thorium into a dual purpose fuel.

Gaseous Fission Reactor

The Gaseous Core Reactor (GCR) concept has the potential to surpass all existing and current plans for advanced reactors in the areas of sustainability, proliferation resistance, and power conversion efficiency. The fuel in a GCR (as its name implies) is in a gaseous state throughout the lifecycle of the core. Because of its gaseous state, the fuel can be pumped into the system allowing continuous online fueling, giving operators greater control over the nuclear environment in the core. The unique ability to rapidly change the neutron environment through online fueling opens up a range of abilities to manage fission products, manage actinide production, and limit proliferation dangerous materials.

The GCR design is being studied as a viable alternative to solid-fuel power reactors, because of its promise in the areas of waste products and high burnup capabilities. Currently there are no regulations differentiating solid or gas fuel reactors for core limitations, therefore the GCR must abide by the current regulations placed on existing reactors. The GCR design is being studied under the assumption that these current regulations and restrictions do not change. The GCR design would prove even more effective if limitations on the reactor were less severe than on solid-fuel reactors.

Features of the Gas Core Reactor

The GCR design can be characterized as an externally moderated, circulating fuel reactor with a thermal neutron spectrum. The fuel will typically be enriched uranium tetra-fluoride (UF4) and will remain with current licensing limitations for the purpose of this study.

The GCR features a unique list of traits that promise to make it an exceptionally safe and environmentally suitable reactor. The GCR design features a very low fuel inventory, requiring nearly 2 orders of magnitude lower inventory than a Light Water Reactor (LWR). Because the fuel is constantly being fed into the core excess fuel will be required on site, so that the core can remain critical, however this can be kept secure while not in use. The system's criticality is a direct function of the core pressure, therefore any pressure leak will immediately cause a loss in reactivity, thus protecting the workers and the public from criticality accidents. The reactor will be kept isolated from the workers, and operated remotely, limiting the radiation dose to plant workers compared to current operating reactors. The GCR design also shows promise of being able to achieve extremely high burnups, thus limiting the amount of nuclear waste that must be stored in containment.

Neutronically and electrically the GCR design promises to set the upper limit in performance for power reactors. Plant operators can directly change the neutronics in the core through a variety of methods. Rotating BeO columns, mass flow rate, and pressure can be adjusted to control core reactivity and neutron energy spectrum, allowing the users to achieve optimum neutronics. The system is designed to use a minimum of a combined cycle (Brayton and Rankine), as well as possibly including a Magneto-Hydro-Dynamic (MDH) third tier system. The electrical efficiencies are expected to achieve upwards of 60% for the combined cycle design and 70% when including the MHD topping cycle.

Mechanically the system is very simple compared to current nuclear power plants. There will be no control or fuel rods, or any other mechanical structures in the core: simply a cylinder containing the gaseous fuel, and an outside cylinder containing the reflector material. This drastically reduces the risk of accidents caused from mechanical limitations. There will be no cladding as the fuel is gaseous, further eliminating any mechanical or material concerns. The MHD system will present a mechanical challenge, however the system with just the combined cycle is no different than current combinedcycle systems used throughout the world, and therefore imposes no additional danger to mechanical limits.

The backend of the cycle will function like existing power plants, there will be filters to remove unwanted products (in this case fission products will most likely be chemically separated) and the fuel will be pressurized to preset specifications. The backend of the cycle will also include a feeding valve where new fuel is mixed with the existing fuel, thus removing the current shutdown period required by current nuclear power plants to reload the fuel.

Theory of Operation

The vapor core reactor (VCR), also called a gas core reactor (GCR), has been studied for some time. It would have a gas or vapor core composed of UF_4 with some ^4He added to increase the electrical conductivity, the vapor core may also have tiny UF_4 droplets in it. It has both terrestrial and space based applications. Since the space concept doesn't necessarily have to be economical in the traditional sense, it allows the enrichment to exceed what would be acceptable for a terrestrial system. It also allows for a higher ratio of UF_4 to helium, which in the terrestrial version would be kept just high enough to ensure criticality in order to increase the efficiency of direct conversion. The terrestrial version is designed for a vapor core inlet temperature of about 1,500 K and exit temperature of 2,500 K and a UF_4 to helium ratio of around 20% to 60%. It is thought that the outlet temperature could be raised to that of the 8,000 K to 15,000 K range where the exhaust would be a fission-generated non-equilibrium electron gas, which would be of much more importance for a rocket design. The space based concept would be cut off at the end of the MHD channel.

Reasoning for He-4 Addition

He-4 may be used in increase the ability of the design to extract energy and be controlled. A few sentences from Anghaie et al. sheds light on the reasoning:

> "The power density in the MHD duct is proportional to the product of electrical conductivity, velocity squared and magnetic field squared $\sigma v^2 B^2$. Therefore, the enthalpy extraction is very sensitive to the MHD input-output fluid conditions. The vapor core reactor provides a hotter-than-most fluid with potential for adequate thermal equilibrium conductivity and duct velocities. Considering the product $v^2 \times B^2$, it is apparent that a light working fluid should dominate the thermal properties and the UF_4 fraction should be small. Additional electrical conductivity enhancement might be needed from thermal ionization of suitable seed materials, and from non-equilibrium ionization by fission fragments and other ionizing radiation produced by the fissioning process."

Spacecraft

The spacecraft variant of the gaseous fission reactor is called the gas core reactor rocket. There are two approaches: the open and closed cycle. In the open cycle, the propellant, most likely hydrogen, is fed to the reactor, heated up by the nuclear reaction in the reactor, and exits out the other end. Unfortunately, the propellant will be contaminated by fuel and fission products, and although the problem can be mitigated by engineering the hydrodynamics within the reactor, it renders the rocket design completely unsuitable for use in atmosphere.

One might attempt to circumvent the problem by confining the fission fuel magnetically, in a manner similar to the fusion fuel in a tokamak. Unfortunately it is not likely that this arrangement will actually work to contain the fuel, since the ratio of ionization to particle momentum is not favourable. Whereas a tokamak would generally work to contain singly ionized deuterium or tritium with a mass of two or three daltons, the uranium vapour would be at most triply ionized with a mass of

235 dalton (unit). Since the force imparted by a magnetic field is proportional to the charge on the particle, and the acceleration is proportional to the force divided by the mass of the particle, the magnets required to contain uranium gas would be impractically large; most such designs have focused on fuel cycles that do not depend upon retaining the fuel in the reactor.

In the closed cycle, the reaction is entirely shielded from the propellant. The reaction is contained in a quartz vessel and the propellant merely flows outside of it, being heated in an indirect fashion. The closed cycle avoids contamination because the propellant can't enter the reactor itself, but the solution carries a significant penalty to the rocket's Isp.

Energy Production

For energy production purposes, one might use a container located inside a solenoid. The container is filled with gaseous uranium hexafluoride, where the uranium is enriched, to a level just short of criticality. Afterward, the uranium hexafluoride is compressed by external means, thus initiating a nuclear chain reaction and a great amount of heat, which in turn causes an expansion of the uranium hexafluoride. Since the UF_6 is contained within the vessel, it can't escape and thus compresses elsewhere. The result is a plasma wave moving in the container, and the solenoid converts some of its energy into electricity at an efficiency level of about 20%. In addition, the container must be cooled, and one can extract energy from the coolant by passing it through a heat exchanger and turbine system as in an ordinary thermal power plant.

However, there are enormous problems with corrosion during this arrangement, as the uranium hexafluoride is chemically very reactive.

Inhour Equation

If the reactivity is constant, the model of point kinetics equations contains a set (1 + 6) of linear ordinary differential equations with constant coefficient and can be solved analytically. Solution of six-group point kinetics equations with Laplace transformation leads to the relation between the reactivity and the reactor period. This relation is known as the inhour equation (which comes from inverse hour, when it was used as a unit of reactivity that corresponded to e-fold neutron density change during one hour) may be derived.

General Form

The point kinetics equations may be solved for the case of an initially critical reactor without external source in which the properties are changed at $t = 0$ in such a way as to introduce a step reactivity ρ_0 which is then constant over time. The system of coupled first-order differential equations can be solved with Laplace transformation or by trying the solution $n(t) = A.\exp(s.t)$ (equation for the neutron flux) and $C_i(t) = C_{i,0}.\exp(s.t)$ (equations for the density of precursors).

Substitution of these assumed exponential solutions in the equation for precursors gives the relation between the coefficients of the neutron density and the precursors.

$$\frac{dC_i}{dt} = \frac{\beta_i}{\wedge} n - \lambda_i C_i$$

$$C_i(t) = C_{i,0} \exp(s.t)$$

$$\downarrow$$

$$C_{i,0} = \frac{1}{\wedge} \frac{\beta_i}{s + \lambda_i} n_0$$

The subsequent substitution in the equation for neutron density yields an equation for s, which after some manipulation can be written as:

$$\rho_0 = s\left(\wedge + \sum_i^6 \frac{\beta_i}{s + \lambda_i}\right)$$

$$s_0 = \frac{1}{\tau_e}$$

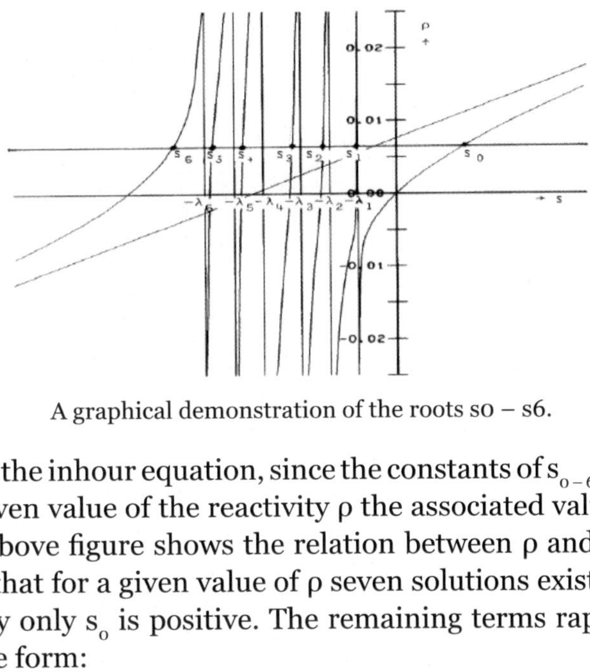

A graphical demonstration of the roots so – s6.

This equation is known as the inhour equation, since the constants of s_{0-6} was originally determined in inverse hours. For a given value of the reactivity ρ the associated values of s_{0-6} are determined with this equation. The above figure shows the relation between ρ and roots s graphically. From this figure it can be seen that for a given value of ρ seven solutions exist for s. The figure indicates that for positive reactivity only s_0 is positive. The remaining terms rapidly die away, yielding an asymptotic solution in the form:

$$n(t) = \sum_{i=0}^{6} A_i \exp(s_i t)$$

$$\downarrow$$

$$n(t) = A_0 \exp(s_0 t)$$

where $s_0 = 1/\tau_e$ is the stable reactor period or asymptotic period of reactor. This root, s_0, is positive for $\rho > 0$ and negative for $\rho < 0$, therefore this root describes the reactor response, which is lasting after the transition phenomena have died out. The figure also shows that a negative reactivity leads to a negative period: All of the s_i are negative, but the root s_0 will die away more slowly than the others. Thus the solution $n(t) = A_0 \exp(s_0 t)$ is valid for positive as well as negative reactivity insertions.

To determine the reactivity required to produce a given period a plot of ρ vs. τ_e must be constructed using the delayed neutron data for a particular fissionable isotope or mix of isotopes, and for a given prompt generation time. To determine the stable reactor period, which results from a given reactivity insertion, it is convenient to use the following form of inhour equation.

Reactor Period Form of Inhour Equation

$$\tau_e = \frac{l}{\rho} + \frac{\beta_{eff} - \rho}{\lambda_{eff} \cdot \rho + \rho}$$

where:

l = prompt neutron lifetime

β_{eff} = effective delayed neutron fraction

λ_{eff} = effective delayed neutron precursor decay constant

τ_e = reactor period

ρ = reactivity

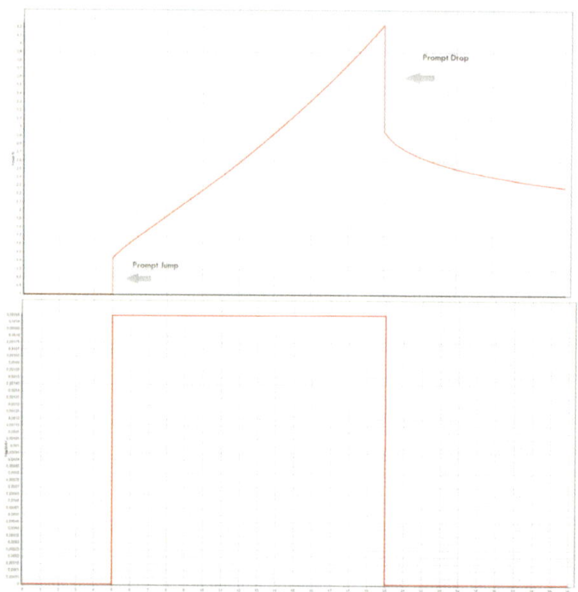

The presence of delayed neutrons brings many interesting phenomena. This chart shows the typical response of reactor on constant reactivity insertion.

The first term in this formula is the prompt term and it causes that the positive reactivity insertion is followed immediately by a immediate power increase called the prompt jump. This power increase occurs because the rate of production of prompt neutrons changes immediately as the reactivity is inserted. After the prompt jump, the rate of change of power cannot increase any more rapidly than the built-in time delay the precursor half-lives allow. Therefore the second term in this formula is called the delayed term. The presence of delayed neutrons causes the power rise to be controllable and the reactor can be controlled by control rods or another reactivity control mechanism.

Special Cases of Inhour Equation

$\rho \ll \beta$ - low reactivity

Assume that ρ is very small. This causes the first root, s_o, of the reactivity to be also very small. In this case, $s_o \ll \lambda_i$ and therefore, we can ignore the term s_o in the denominator of the inhour equation. The inhour equation then takes form:

$$\rho_0 = s\left(\wedge + \sum_i^6 \frac{\beta_i}{s+\lambda_i}\right) \xrightarrow{s<<\lambda_i} s\left(\wedge + \sum_i^6 \frac{\beta_i}{\lambda_i}\right) \rightarrow \tau_e = \frac{l}{\rho_0}\left(\wedge + \sum \tau_i \beta_i\right) = \frac{l_d}{k_{eff}-1}$$

$$\tau_e = \frac{l_d}{k_{eff}-1}$$

As can be seen, this special case results in the same period as in the case of simple point kinetics equation, which also uses the mean generation time with delayed neutrons (l_d):

$$\frac{d}{dt}n(t) = \frac{(k-1)}{l_d}n(t)$$

$$n(t) = n(0)\exp\left[\frac{(k-1)}{l_d}t\right]$$

Thus for small reactivities—positive or negative—the reactor period is governed almost completely by the delayed neutron properties. Despite the fact the amount of delayed neutrons is only on the order of tenths of percent of the total amount, the timescale in seconds (τ_i) plays the extremely important role.

The assumption is $s_o \ll \lambda_i$. The largest value of $\tau_i = 1/\lambda_i = 80s$, therefore the this formula is valid for periods, τ_e, higher than 80s. On the other hand, this formula is useful and accurate enough for most purposes for reactivities up to about $\rho = 0.0005 = 50pcm$.

Mean generation time with delayed neutrons (l_d):

$$l_d = (1-\beta).l_p + \sum l_i \,.\, \beta_i \Rightarrow l_d = (1-\beta).l_p + \sum \tau_i \,.\, \beta_i$$

where

- $(1-\beta)$ is the fraction of all neutrons emitted as prompt neutrons
- l_p is the prompt neutron lifetime
- τ_i is the mean precursor lifetime, the inverse value of the decay constant $\tau_i = 1/\lambda_i$

- The weighted delayed generation time is given by $\tau = \sum \tau_i \cdot \beta_i / \beta = 13.05$ s
- Therefore the weighted decay constant $\lambda = 1 / \tau \approx 0.08\ s^{-1}$

The number, 0.08 s^{-1}, is relatively high and have a dominating effect of reactor time response, although delayed neutrons are a small fraction of all neutrons in the core. This is best illustrated by calculating a weighted mean generation time with delayed neutrons:

$$l_d = (1 - \beta).l_p + \sum \tau_i \cdot \beta_i = (1 - 0.0065).\ 2 \times 10^{-5} + 0.085 = 0.00001987 + 0.085 \approx 0.085$$

In short, the mean generation time with delayed neutrons is about ~0.1 s, rather than ~10^{-5} as in section Prompt Neutron Lifetime, where the delayed neutrons were omitted.

Example:

Let us consider that the mean generation time with delayed neutrons is ~0.085 and k (k_∞ – neutron multiplication factor) will be step increased by only 0.01% (i.e. 10pcm or ~1.5 cents), that is k_∞=1.0000 will increase to k_∞=1.0001.

It must be noted such reactivity insertion (10pcm) is very small in case of LWRs (e.g. one step by control rods). The reactivity insertions of the order of one pcm are for LWRs practically unrealizable. In this case the reactor period will be:

$$T = l_d / (k_\infty - 1) = 0.085 / (1.0001 - 1) = 850s$$

This is a very long period. In ~14 minutes the neutron flux (and power) in the reactor would increase by a factor of e = 2.718. This is completely different dimension of the response on reactivity insertion in comparison with the case without presence of delayed neutrons, where the reactor period was 1 second.

ρ ≥ β - High Reactivity - Prompt Criticality

Assume that ρ is higher than β. In this case, the production of prompt neutrons alone is enough to balance neutron losses and increase the neutron population. The number of neutrons is exponentially increasing in time (as rapidly as the prompt neutron generation lifetime ~10^{-5}s).

This causes the first root, s_0, of the reactivity becomes large compared to each λ_i. In this case, $\lambda_i \ll s_0$ and therefore, we can ignore the term λ_i in the denominator of the inhour equation. The inhour equation then takes form:

$$\rho_0 = s\left(\wedge + \sum_i^6 \frac{\beta_i}{s+\lambda_i}\right) \xrightarrow{\lambda_i << s} s\left(\wedge + \sum_i^6 \frac{\beta_i}{s}\right) \rightarrow \rho_0 = s\wedge + \beta \rightarrow$$

$$\tau_e = \frac{l}{k_{eff} - 1}$$

As can be seen, this special case results in the same period as in the case of simple point kinetics equation without delayed neutrons.

Obviously, reaching the reactivity of β (e.g. 650pcm) completely changes the response of a reactor, therefore this situation can only be accidental. As prompt criticality is reached, the distinction between the prompt jump and the reactor period vanishes, for now the prompt neutron lifetime rather than the delayed neutron half-lives largely determines the rate of exponential increase. Reactors with such a kinetics would be very difficult to control by mechanical means such as the movement of control rods. In normal operation, the reactivity of a reactor must remain far below the prompt criticality threshold with sufficient margin.

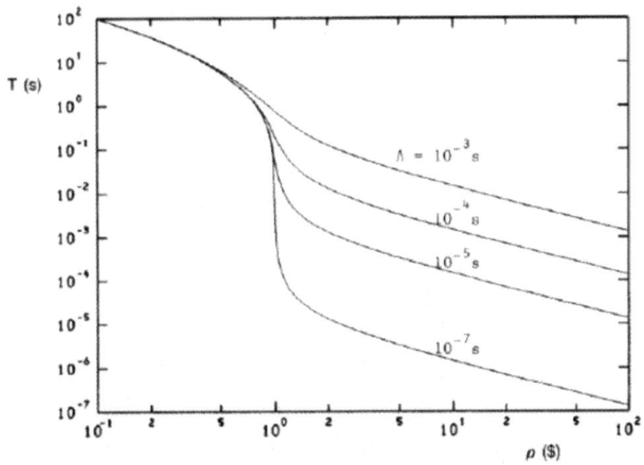

reactor period vs. reactivity

In design basis accidents (DBA), such a kinetics can occur. For example under RIA conditions (Reactivity-Initiated Accidents) reactors should withstand a jump-like insertion of relatively large (~1 $ or even more) positive reactivity (e.g. in case of control rod ejection) and the PNL (prompt neutron lifetime) plays here the key role. The longer prompt neutron lifetimes can substantially improve kinetic response of reactor (the longer prompt neutron lifetime gives simply slower power increase). Therefore the PNL should be verified in a reload safety evaluation (RSE) process. Management of this accident is based on reactivity feedbacks, especially on the doppler temperature coefficient (DTC). This coefficient is of the highest importance in the reactor stability. The doppler temperature coefficient is generally considered to be even more important than the moderator temperature coefficient (MTC). Especially in case of a control rod ejection the doppler temperature coefficient will be the first and the most important feedback, that will compensate the inserted positive reactivity. The time for heat to be transferred to the moderator is usually measured in seconds, while the fuel temperature coefficient is effective almost instantaneously. Therefore this coefficient is also called the prompt temperature coefficient because it causes an immediate response on changes in fuel temperature.

Shutdown

A reactor is considered to be shut down when it is subcritical and sufficient shutdown reactivity exists so there is no immediate probability of secondary criticality. Despite the fact the chain reaction is quickly broken down, all of the fissioning in the core is not stopped. There are always neutrons in the core and therefore a residual fissioning occurs. On the other hand, the amount of heat being generated due to the residual fissions (initiated by source neutrons) which are not stopped and the decay heat is much less than that which can be removed by the plant systems.

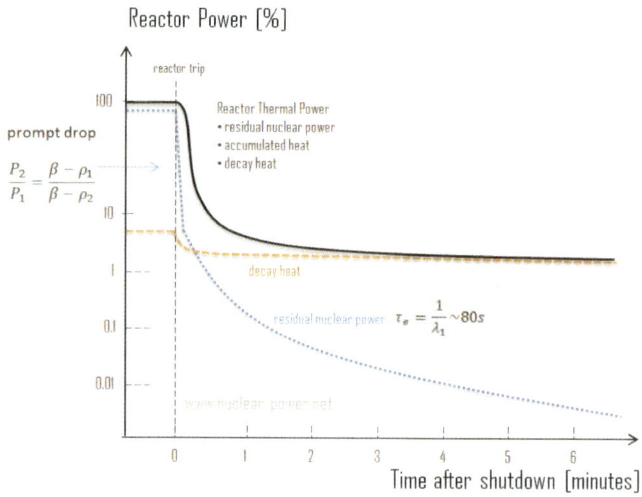

In general, there are two types of reactor shutdown:

- Emergency Shutdown – SCRAM. A reactor "SCRAM" (or "reactor trip") is the rapid insertion or fall of the control rods into the core to stop the fission chain reaction. It is a key safety function of each nuclear reactor and usually does not need any power supply, since control rods fall only by gravitational force.
- Gradual Shutdown. Usually there is no need to stop the chain reaction promptly. Shutdown is normally accomplished by gradual insertion of some (or all) of the control rods, or by introduction of soluble neutron poison (boric acid in case of PWRs) into the reactor coolant. Reactor thermal power is in the first instance decreased from 100% of rated power to about 10 – 30% of rated thermal power. At this power level the turbine is disconnected from the electrical grid, while all the steam production is taken over by the turbine bypass system (TBS). The reactor is now at zero electrical load and the thermal power can be decreased under 1% of rated power (to the Startup Mode) and the reactor dynamics changes to zero-power reactor dynamics. Further introduction of negative reactivity causes controlled exponential decrease in neutron power and after reaching shutdown margin (e.g. k_{eff} = 0.98) the reactor is in safe shutdown state.

Emergency Shutdown – SCRAM

A reactor "SCRAM" (or "reactor trip") is the rapid insertion or fall of the control rods into the core to stop the fission chain reaction. At PWRs, all control rods are usually inserted within two to four seconds. Control rods are an important safety and control system of nuclear reactors. Their prompt action and prompt response of the reactor is indispensable. Control rods are also used for maintaining the desired state of chain reaction within a nuclear reactor (i.e. subcritical state, critical state, supercritical state). Reactor Trip System (RTS), which actuates a reactor trip, is a part of a Reactor Protection System (RPS).

In a pressurized water reactor, the control rods are inserted (dropped) from the top of the reactor vessel into the core. In a boiling water reactor, the control rods are inserted from the bottom of the reactor vessel into the core.

SCRAM from Hot Zero Power – Dynamics

Assume a reactor trip from hot zero power (from "zero power criticality"). The power decrease, in this case, does not influence the criticality (k_{eff}). In this case, the reactivity of the multiplying system becomes quickly negative and higher than β (i.e.that ρ is negative and ρ ≪ β).

The rate at which the reactor fission rate decays immediately following shutdown is similar for all reactors provided a large amount of negative reactivity is inserted. After a large negative reactivity addition the neutron level undergoes a rapid decrease of about two decades (prompt drop) until it is at the level of production of delayed neutrons.

But even if it were possible to insert an infinite negative reactivity, the neutron flux would not immediately fall to zero. Prompt neutrons will be absorbed almost immediately. It is consistent with the prompt drop formula. Therefore the resulting neutron flux will be:

$$\frac{n_2}{n_1} = \frac{\beta - \rho_1}{\beta - \rho_2} \xrightarrow{\rho_2 \to \infty} n_2 = \beta . n_1$$

It is obvious, the neutron flux cannot drop below the value $\beta n_{1.}$ The real values are much higher. The integral worth of all control and emergency rods (PWRs) is for example -9000pcm. It is equal to ρ = -9000/600 = -15*β* = *-0.09*. (*β*= *600pcm* = *0.006*)

For this negative reactivity the prompt drop is equal to:

n_2/n_1 = 0.006/(0.006+0.09)=0.063

which is about ten times higher than in case of an infinite negative reactivity insertion.

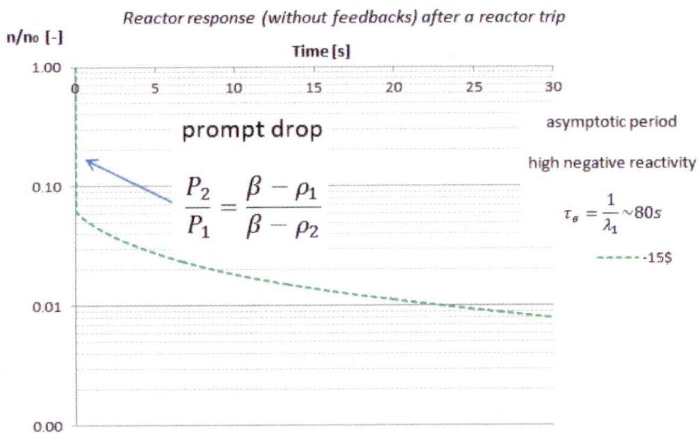

The neutron flux then continues to fall according to stable period. The first root of reactivity equation occurs at $s_0 = -\lambda_1$, the decay constant of the long-lived precursors group. The shortest negative stable period is then $_{e = -1/\lambda 1}$ = -80s. The neutron flux cannot be reduced more rapidly than this period. On the other hand, the prompt drop causes an immediate drop to about 6% of rated power and within few tens of seconds the thermal power which originates from nuclear fission is below the thermal power which originates from decay heat.

There is an exception to this behavior and it is a SCRAM of heavy water reactor. In a heavy water reactor, the photoneutron source is extremely large after shutdown due to the amount of deuterium in the moderator and the large number of high energy gammas from short-lived fission product decay. This high gamma flux from short-lived fission products will decrease rapidly after shutdown. The photoneutron source is large enough to have a significant impact on neutron population immediately after shutdown. The photoneutron source has the result of flux levels decreasing more slowly so that a heavy water reactor will have a significantly larger negative reactor period after a shutdown.

SCRAM from Hot Full Power – Dynamics

For power reactors at power conditions the reactor can behave differently as a result of the presence of reactivity feedbacks. Power reactors are initially started up from hot standby mode (subcritical state at 0% of rated power) to power operation mode (100% of rated power) by withdrawing control rods and by boron dilution from the primary coolant. During the reactor startup and up to about 1% of rated power, the reactor kinetics is exponential as in zero power reactor. This is due to the fact all temperature reactivity effects are minimal.

During further power increase from about 1% up to 100% of rated power, the temperature reactivity effects play very important role. As the neutron population increases, the fuel and the moderator increase its temperature, which results in decrease in reactivity of the reactor (almost all reactors are designed to have the temperature coefficients negative). The total amount of feedback reactivity that must be offset by control rod withdrawal or boron dilution during the power increase is known as the power defect. The power defects for PWRs, graphite-moderated reactors, and sodium-cooled fast reactors are:

- about 2500pcm for PWRs,
- about 800pcm for graphite-moderated reactors
- about 500pcm for sodium-cooled fast reactors

This exactly but in the opposite direction acts after a reactor trip (SCRAM) from the Hot Full Power state (HFP). It is logical, as power defects act against power increase, they act also against power decrease. When reactor power is decreased quickly, as in the case of reactor trip, power defect causes a positive reactivity insertion, and the initial rod insertion must be sufficient to make the reactor safe subcritical. Similarly as in the HZP state, the integral worth of all control and emergency rods (PWRs) is for example -9000pcm. It is equal to $\rho = -9000/600 = -15\beta = -0.09$. ($\beta = 600pcm = 0.006$). Also in this case a prompt drop occurs. The prompt drop is quicker than the deaccumulation of heat from fuel pellets. For this negative reactivity the prompt drop is equal to:

$$n_2/n_1 = 0.006/(0.006+0.09)=0.063$$

but the subsequent power decrease is strongly influenced by changes in core (fuel and moderator) temperatures. After reaching stable temperature, the neutron flux may continue to fall (when subcritical) according to stable period. It is obvious, if the power defect for PWRs is about 2500pcm (about 4 β_{eff}), the control rods must weigh more than 2500pcm to achieve the subcritical condition. To ensure the safe subcritical condition, the control rods must weigh more than 2500pcm plus value of SDM (SHUTDOWN MARGIN). The total weigh of control rods is design specific, but, for example, it may reach about 6000 to 9000pcm. To ensure that the control rods can safe shut down the reactor, they must be maintained above a minimum rod height (rods insertion limits) specified in the technical specifications.

A distinction should be made between indicated reactor power level measured by excore neutron detectors after shutdown and the actual thermal power level. The indicated reactor power level, usually known as the nuclear power, is the power produced directly from fission in the reactor core, but the actual thermal power drops more slowly due to decay heat production as previously discussed. Decay heat, although approximately 5 to 6% of the steady state reactor power prior to shutdown, diminishes to less than 1% of the pre-shutdown power level after about one hour.

Types of Shutdown State

There are three shutdown states:

1. Hot shutdown state.
2. Cold shutdown state.
3. Guaranteed shutdown state.

Hot Shutdown State

In this state, the reactor is shutdown from the normal operation delivering the state steady state power. The temperatures of coolant and moderator are at their normal levels. The positive reactivity to be compensated is the power coefficient. Shutdown system takes care of this task.

Cold Shutdown State

The reactor shutdown from normal power, with the coolant and moderator being at room temperatures is called the cold shutdown. The positive reactivity coefficients are due to the low temperature of coolant and moderator. These are the reactivity coefficients that need to be compensated during shutdown.

Guaranteed Shutdown State

It is a sub-critical state of a nuclear reactor. This state provides assurance that there cannot be any increase in nuclear fission reaction. This being a highly sub-critical cold shut down state with large amount of poisons in the moderator, withdrawal of a control rod would not lead to increase in flux levels.

References

- Helmreich, J.E. Gathering Rare Ores: The Diplomacy of Uranium Acquisition, 1943–1954, Princeton UP, 1986: ch. 10 ISBN 0-7837-9349-9
- Jaganathan, Venkatachari (11 May 2011). "India's new fast-breeder on track, nuclear power from September next". Chennai: Hindustan Times. Archived from the original on 13 May 2013. Retrieved 25 March 2012
- Nuclear-reactor: energyeducation.ca, Retrieved 21 March 2018

- G. Glazov, V. N. Leonov, V. V. Orlov, A. G. Sila-Novitskii, V. S. Smirnov, A. I. Filin, and V. S. Tsikunov (2007). "Brest Reactor and Plant-site Nuclear Fuel Cycle" (PDF). Atomic Energy. 103, No. 1: 501–508. doi:10.1007/s10512-007-0080-5
- FSUE "State Scientific Center of Russian Federation Research Institute of Atomic Reactors". "Experimental fast reactor BOR-60". Retrieved 15 June 2012
- Subcritical-reactor, reactor-criticality, nuclear-fission-chain-reaction, nuclear-power, reactor-physics: nuclear-power.net, Retrieved 31 March 2018
- Frank von Hippel; et al. (February 2010). Fast Breeder Reactor Programs: History and Status (PDF). International Panel on Fissile Materials. ISBN 978-0-9819275-6-5. Retrieved 28 April 2014
- S. R. Pillai, M. V. Ramana (2014). "Breeder reactors: A possible connection between metal corrosion and sodium leaks". Bulletin of the Atomic Scientists. 70 (3). doi:10.1177/0096340214531178. Retrieved 15 February 2015
- Non power reactor: encyclopedia2.thefreedictionary.com, Retrieved 23 May 2018
- Srikanth (27 November 2011). "80% of work on fast breeder reactor at Kalpakkam over". Kalpakkam: The Hindu. Retrieved 25 March 2012
- Thorium, information-library, current and future-generation: world-nuclear.org, Retrieved 30 June 2018

2

Types of Nuclear Reactors

Nuclear reactors are used for the generation of electric power and for the propulsion of ships. These are classified according to the type of nuclear reaction sustained or the type of moderator material or the coolant used. They can divided into light-water reactors, heavy water reactors, graphite-moderated reactors and molten salt reactors, etc. This chapter closely examines the different kinds of nuclear power reactors for providing a comprehensive understanding.

Light-water Reactor

The light water reactor is a type of thermal- neutron reactor that utilizes normal water as opposed to heavy water, a form of water that contains a larger amount of the hydrogen isotope deuterium. Normal water that contains the hydrogen-1 isotope called protium. Light water reactors use water as both and a coolant method and a neutron moderator that reduces the speed of fast moving neutrons. Light water reactors produce heat by controlled nuclear fission. There are several different vital parts of light water reactors that make the generation of nuclear energy possible. The nuclear reactions take place in the nuclear reactor core, where the nuclear fuel components are contained. The core is made of nuclear fuel rods that are pencil thin and are about 3.7 m long. These rods are grouped by the hundreds in bundles called fuel assemblies and are filled with pellets of uranium or uranium oxide. Control rods are lowered into the core and are filled with pellets of substances such as hafnium or cadmium that are responsible for capturing neutrons. The neutrons that are absorbed by the control rods cannot take part in the chain reaction, however, when the control rods are raised out of the core, more neutrons strike the fissile U-235 (or Pu-239) nuclei in nearby fuel rods, and the chain reaction intensifies as the control rods are out of the way. All of these components are enclosed in a water-filled steel pressure vessel, called the reactor vessel. The heat generated by controlled nuclear fission turns the water into steam, which drives the power- generating turbines. After the steam flows through the turbines, the steam turns back into water in the condenser.

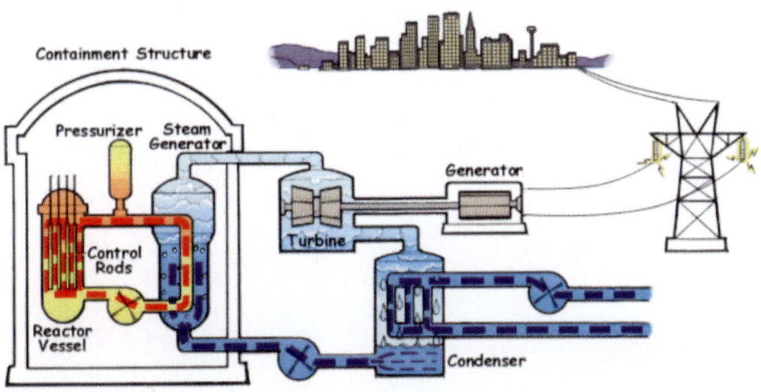

Pressure Water Reactor.

In general, LWR's are divided into two categories:

- Pressurized water reactors (PWR) – are characterized by high pressure primary circuit (to keep the water in liquid state)
- Boiling water reactors (BWR) – are characterized by controlled boiling in the primary circuit.

Light Water as the Moderator

To be an effective moderator, the probability of elastic reaction between neutron and the nucleus must be high. In terms of cross-sections, the elastic scattering cross section of a moderator's nucleus must be high. Therefore high elastic scattering cross-section is important, but does not describe comprehensively capabilities of moderators. In order to describe capabilities of a material to slow down neutrons, three new material variables must are defined:

- The Average Logarithmic Energy Decrement (ξ)
- The Macroscopic Slowing Down Power (MSDP)
- The Moderating Ratio (MR)

Light water has the highest ξ and σ_s among the moderators (resulting in the highest MSDP) shown in the table, but its moderating ratio is low due to its relatively higher absorption cross section.

Element	A	ξ	Collisions 2 MeV→1eV	MR
H_2O	-	0.920	16	71
D_2O	-	0.509	29	5670
Be	9	0.207	70	143
C	12	0.158	92	192
238U	238	0.008	1812	0.0092

Table of moderating ratios for some materials.

Pressurized Water Reactor

The Watts Bar Nuclear Generating Station in Tennessee uses PWRs in its operation.

Nuclear submarines make use of the high power-to-weight ratio of PWRs in their operation.

The pressurized water reactor (PWR) is a type of nuclear reactor used to the generate electricity and propel nuclear submarines and naval vessels. They make use of light water (ordinary water, as opposed to heavy water) as their coolant and neutron moderator. It is one of three types of light water reactors, with the others being the boiling water reactor and the supercritical water cooled reactor.

It was originally designed for the U.S. Navy, however, it quickly grew to become the most widely used reactor in nuclear power plants; with over 270 in operation around the world as of 2015. This makes them by far the most dominantly used reactor in the world, with the second most (the boiling water reactor) having only 80 in operation. Construction of PWRs diminished greatly after the Three Mile Island nuclear accident, mainly as a result of public support becoming weaker.

Their use on naval vessels and nuclear ships are of extreme importance to various militaries around the world.Nuclear power has a huge advantage over fuels like gasoline or diesel as it allows ships to run for very long periods without the need to refuel. PWRs make a good reactor for these ships since they have a high specific power (high power for their mass) due to their use of high pressure. This allows the reactors to be fairly compact, especially with the use of highly enriched uranium.

Characteristics

Fuel

Pressurized water reactors must use enriched uranium as their nuclear fuel, because of their use of light water. This is because light water would absorb too many neutrons if natural uranium was used, so the fuel content of fissile Uranium-235 must be increased. This is done through a uranium enrichment process, in which the concentration of Uranium-235 is increased from 0.7% to around 4%.

The enriched uranium is packed into fuel rods which are assembled into a fuel bundle, as seen in figure. There are about 200-300 rods in each bundle for a PWR, with a large reactor containing 150-250 bundles in their core. This corresponds to 80-100 tonnes of uranium.

The bundles are arranged vertically in fuel tubes within the core. As the fuel is "burned" in the reactor, its density gradually increases, resulting in small voids to develop inside the fuel tube. These void spaces can cause a problem because high pressures could cause stress to the tubes, increasing the likelihood of a rupture. To avoid this problem, the tubes are pressurized with helium at about 3.4 MPa. As fission gas products accumulate over the fuel's lifetime, the pressure gradually balances with the high pressure of the core.

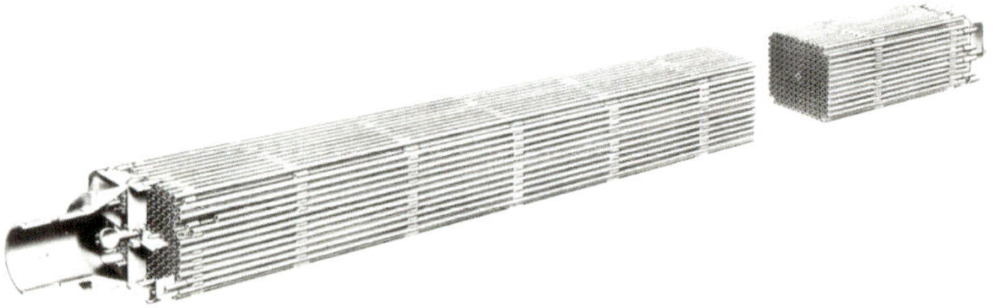

A nuclear fuel bundle for a PWR.

Coolant and Moderator

As mentioned before, light water is used as the coolant and moderator for a pressurized water reactor. Light water is *much* more abundant than heavy water, as it makes up 99.99% of natural water.

Light water does not make as good of a moderator as heavy water or graphite as a result of its relatively high absorption of neutrons. However, its use as a moderator makes for an important safety feature; if there is a loss of coolant accident (LOCA), there will also be a loss of moderator causing the nuclear chain reaction to stop. Also if the moderating water overheats and becomes steam inside the bottom reactor core, there will be less moderator and therefore the chain reaction will stop.

Pressure, Temperatures and Water Flow

The inverted U-tube bundle in the steam generator of a PWR.

As the name implies, the water in the reactor is pressurized. This is due to the fact that as the pressure gets higher, the boiling point of water increases with it. This means that at high pressures the water can operate at extremely high temperatures without boiling to steam. This is important for the reactor as higher pressures allow for greater power output and higher thermal efficiency. The pressure is maintained by the "pressurizer", which acts to stabilize pressure changes caused by changes in electrical load.

Water enters the reactor at 290°C, and by the time it exits it is at around 325°C. In order for it to remain a liquid at these temperatures, the pressure must be 15 MPa, or about 150 times atmospheric pressure. By keeping the water in liquid form, the control rod system is simplified as they are able to be placed in from the top, rather than from the bottom like in a boiling water reactor. Therefore, if the power is lost in the plant, the electromagnetic system holding the rods will give out, and gravity will cause the rods to fall into the core, stopping the reaction.

The hot water flowing from the reactor flows through inverted U-tubes which acts as a heat exchanger, heating up a secondary loop of water in what is called a "steam generator". This secondary loop is at a lower pressure so it is able to boil to steam, which then passes through turbines in order to generate electricity. Large reactors have up to *4 steam generators*, each of which may be larger than the reactor itself.

The basic operation of a PWR can be seen below.

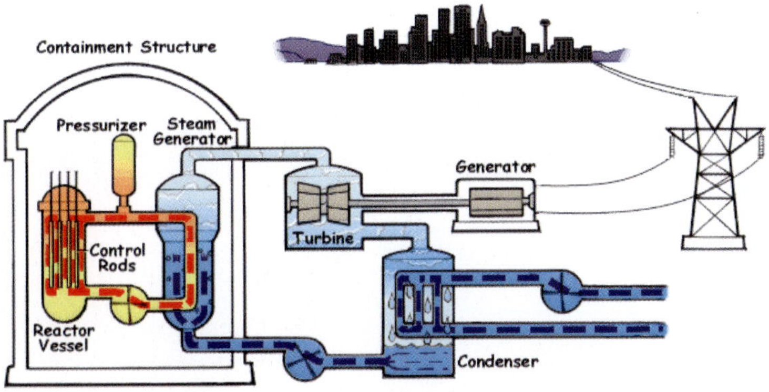

The basic cycle and water flow of a PWR.

Advantages

- PWR reactors are very stable due to their tendency to produce less power as temperatures increase; this makes the reactor easier to operate from a stability standpoint.
- PWR turbine cycle loop is separate from the primary loop, so the water in the secondary loop is not contaminated by radioactive materials.
- PWRs can passively scram the reactor in the event that offsite power is lost to immediately stop the primary nuclear reaction. The control rods are held by electromagnets and fall by gravity when current is lost; full insertion safely shuts down the primary nuclear reaction.
- PWR technology is favoured by nations seeking to develop a nuclear navy; the compact reactors fit well in nuclear submarines and other nuclear ships.

Disadvantages

- The coolant water must be highly pressurized to remain liquid at high temperatures. This requires high strength piping and a heavy pressure vessel and hence increases construction costs. The higher pressure can increase the consequences of a loss-of-coolant accident. The reactor pressure vessel is manufactured from ductile steel but, as the plant is operated, neutron flux from the reactor causes this steel to become less ductile. Eventually the ductility of the steel will reach limits determined by the applicable boiler and pressure vessel standards, and the pressure vessel must be repaired or replaced. This might not be practical or economic, and so determines the life of the plant.

- Additional high pressure components such as reactor coolant pumps, pressurizer, steam generators, etc. are also needed. This also increases the capital cost and complexity of a PWR power plant.
- The high temperature water coolant with boric acid dissolved in it is corrosive to carbon steel (but not stainless steel); this can cause radioactive corrosion products to circulate in the primary coolant loop. This not only limits the lifetime of the reactor, but the systems that filter out the corrosion products and adjust the boric acid concentration add significantly to the overall cost of the reactor and to radiation exposure. In one instance, this has resulted in severe corrosion to control rod drive mechanisms when the boric acid solution leaked through the seal between the mechanism itself and the primary system.
- Natural uranium is only 0.7% uranium-235, the isotope necessary for thermal reactors. This makes it necessary to enrich the uranium fuel, which significantly increases the costs of fuel production.
- Because water acts as a neutron moderator, it is not possible to build a fast neutron reactor with a PWR design. A reduced moderation water reactor may however achieve a breeding ratio greater than unity, though this reactor design has disadvantages of its own.

Boiling Water Reactor

Boiling water reactors (BWR) are a type of nuclear reactor that use light water (ordinary water, as opposed to heavy water) as their coolant and neutron moderator. They are the second most used reactor for nuclear power generation in the world, next to the pressurized water reactor (PWR)—with 75 in operation as of 2018.

A boiling water reactor is different from a pressurized water reactor (PWR) in the sense that in a BWR the reactor core heats the water, which transforms into steam, driving the steam turbine. In a PWR, the reactor core warms the water to just below its boiling point. Then, the hot water is exchanged with heat within a water system under low pressure. It is then turned into steam that serves as the driving force for the turbine.

Boiling water reactors are preferred in the industries because this type of light water nuclear reactor does not utilize boric acid to regulate the burn up of fission, resulting in lowered possibility of corrosion in the piping and reactor vessel. In a PWR, corrosion is a major problem, so accurate and prompt monitoring of corrosion should be observed.

A BWR also presents less possibility of rupture and core damage since it has fewer welds as well as large-diameter pipes and tubes to generate steam. BWRs are also compact and known for their strong power, making them a favored tool for low-cost, simple and safety-focused electricity generation.

Characteristics

Fuel

Boiling water reactors must use enriched uranium as their nuclear fuel, due to their use of light water. This is because light water absorbs too many neutrons to be used with natural uranium, so the fuel content of fissile Uranium-235 must be increased. This is done through uranium enrichment—which increases the concentration of Uranium-235 from 0.7% to around 4%.

The enriched uranium is packed into fuel rods, which are assembled into a fuel bundle, as seen in figure. There are about 90-100 rods in each bundle, with up to 750 bundles in a reactor. This corresponds to nearly 140 tonnes of uranium.

Coolant and Moderator

As mentioned before, light water is used as the coolant and moderator for a boiling water reactor. Light water is *much* more abundant than heavy water, as it makes up 99.99% of natural water.

The use of light water as a coolant is beneficial as it can cool down the fuel at the same time as it heats up to steam. The steam moves through the turbines which spin a generator to generate electricity. This simplifies the design of boiling water reactors, as there is *only one primary loop* needed which results in a lower cost of development.

Light water does not make as good of a moderator as heavy water or graphite due to its relatively high absorption of neutrons. However, its use as a moderator makes for an interesting and important safety feature; if there is a "loss of coolant accident: (LOCA), the moderator will also decline, causing the nuclear chain reaction to stop. In addition, if the moderating water overheats and turns to steam inside the bottom reactor core, there will be less moderator and therefore the chain reaction will stop.

Pressures

Boiling water reactors must operate at fairly high pressures in order for the water to perform adequately; at high pressures the water can remain in liquid form at temperatures much higher than its normal 100°C boiling point. Therefore, boiling water reactors operate at around 7 MPa—around 70 times higher than atmospheric pressure.

Although very high, these pressures are still less than half of the pressures needed for pressurized water reactors. Since the pressures in boiling water reactors are less than those of pressurized ones, the specific power is less. This means that the boiling water reactor needs to be larger in order to provide the same power as a pressurized reactor, which results in higher costs. However, the walls do not need to be as thick as in PWRs, so the costs are essentially the same.

Water Flow

Since the BWR has only one primary loop, as mentioned earlier, the very *water that flows through the reactor core is the same one used to spin the turbines*. In contrast, pressurized water reactors do not convert the hot water in the core to steam, therefore, two loops are necessary. This is the main difference between BWRs and PWRs.

Less water needs to flow through a BWR than through a PWR, however, there are complications because the water becomes radioactive in its circulation. This means that there must be extra shielding in the non-reactor parts of the power plant.

Control rods are placed in the bottom of the reactor, rather than the top as in PWRs. This is because in a boiling water reactor there is more steam at the top and placement of control rods in the steam portion would have a lesser effect than their placement in the water portion of the reactor. There is higher reactivity of the fuel in the water portion due to its moderation, so the control rods are needed to change the reactivity as deemed fit by the operators.

Seen below is the flow of water through a boiling water reactor power plant, and its production of electricity.

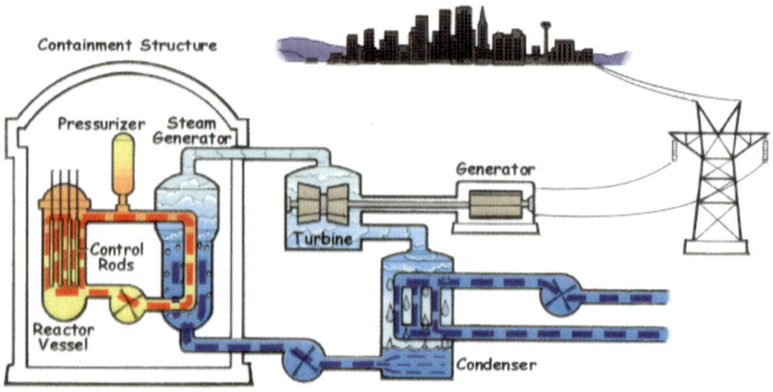

A boiling water nuclear reactor operates on a direct cycle, where the water from the core is also used as steam in the turbine section.

Components

Condensate and Feedwater

Steam exiting the turbine flows into condensers located underneath the low-pressure turbines, where the steam is cooled and returned to the liquid state (condensate). The condensate is then pumped through feedwater heaters that raise its temperature using extraction steam from various turbine stages. Feedwater from the feedwater heaters enters the reactor pressure vessel (RPV) through nozzles high on the vessel, well above the top of the nuclear fuel assemblies (these nuclear fuel assemblies constitute the "core") but below the water level.

The feedwater enters into the downcomer or annulus region and combines with water exiting the moisture separators. The feedwater subcools the saturated water from the moisture separators. This water now flows down the downcomer or annulus region, which is separated from the core by a tall shroud. The water then goes through either jet pumps or internal recirculation pumps that provide additional pumping power (hydraulic head). The water now makes a 180-degree turn and moves up through the lower core plate into the nuclear core, where the fuel elements heat the water. Water exiting the fuel channels at the top guide is saturated with a steam quality of about 15%. Typical core flow may be 45,000,000 kg/h (100,000,000 lb/h) with 6,500,000 kg/h (14,500,000 lb/h) steam flow. However, core-average void fraction is a significantly higher fraction (~40%). These sort of values may be found in each plant's publicly available Technical Specifications, Final Safety Analysis Report, or Core Operating Limits Report.

The heating from the core creates a thermal head that assists the recirculation pumps in recirculating the water inside of the RPV. A BWR can be designed with no recirculation pumps and rely entirely on the thermal head to recirculate the water inside of the RPV. The forced recirculation head from the recirculation pumps is very useful in controlling power, however, and allows achieving higher power levels that would not otherwise be possible. The thermal power level is easily varied by simply increasing or decreasing the forced recirculation flow through the recirculation pumps.

The two-phase fluid (water and steam) above the core enters the riser area, which is the upper region contained inside of the shroud. The height of this region may be increased to increase the thermal natural recirculation pumping head. At the top of the riser area is the moisture separator. By swirling the two-phase flow in cyclone separators, the steam is separated and rises upwards towards the steam dryer while the water remains behind and flows horizontally out into the downcomer or annulus region. In the downcomer or annulus region, it combines with the feedwater flow and the cycle repeats.

The saturated steam that rises above the separator is dried by a chevron dryer structure. The "wet" steam goes through a tortuous path where the water droplets are slowed down and directed out into the downcomer or annulus region. The "dry" steam then exits the RPV through four main steam lines and goes to the turbine.

Control Systems

Reactor power is controlled via two methods: by inserting or withdrawing control rods (control blades) and by changing the water flow through the reactor core.

Positioning (withdrawing or inserting) control rods is the normal method for controlling power when starting up a BWR. As control rods are withdrawn, neutron absorption decreases in the control material and increases in the fuel, so reactor power increases. As control rods are inserted, neutron absorption increases in the control material and decreases in the fuel, so reactor power decreases. Differently from the PWR, in a BWR the control rods (boron carbide plates) are inserted from below to give a more homogeneous distribution of the power: in the upper side the density of the water is lower due to vapour formation, making the neutron moderation less efficient and the fission probability lower. In normal operation, the control rods are only used to keep a homogeneous power distribution in the reactor and compensate the consumption of the fuel, while the power is controlled through the water flow. Some early BWRs and the proposed ESBWR (Economic Simplified BWR made by General Electric Hitachi) designs use only natural circulation with control rod positioning to control power from zero to 100% because they do not have reactor recirculation systems.

Changing (increasing or decreasing) the flow of water through the core is the normal and convenient method for controlling power from approximately 30% to 100% reactor power. When operating on the so-called "100% rod line," power may be varied from approximately 30% to 100% of rated power by changing the reactor recirculation system flow by varying the speed of the recirculation pumps or modulating flow control valves. As flow of water through the core is increased, steam bubbles ("voids") are more quickly removed from the core, the amount of liquid water in the core increases, neutron moderation increases, more neutrons are slowed down to be absorbed by the fuel, and reactor power increases. As flow of water through the core is decreased, steam voids remain longer in the core, the amount of liquid water in the core decreases, neutron moderation decreases, fewer neutrons are slowed down to be absorbed by the fuel, and reactor power decreases.

Reactor pressure in a BWR is controlled by the main turbine or main steam bypass valves. Unlike a PWR, where the turbine steam demand is set manually by the operators, in a BWR, the turbine valves will modulate to maintain reactor pressure at a set point. Under this control mode, the turbine will automatically follow reactor power changes. When the turbine is offline or trips, the main steam bypass/dump valves will open to direct steam directly to the condenser. These bypass valves will automatically or manually modulate as necessary to maintain reactor pressure and control the reactor's heatup and cooldown rates while steaming is still in progress.

Reactor water level is controlled by the main feedwater system. From about 0.5% power to 100% power, feedwater will automatically control the water level in the reactor. At low power conditions, the feedwater controller acts as a simple PID control by watching reactor water level. At high power conditions, the controller is switched to a "Three-Element" control mode, where the controller looks at the current water level in the reactor, as well as the amount of water going in and the amount of steam leaving the reactor. By using the water injection and steam flow rates, the feed water control system can rapidly anticipate water level deviations and respond to maintain water level within a few inches of set point. If one of the two feedwater pumps fails during operation, the feedwater system will command the recirculation system to rapidly reduce core flow, effectively reducing reactor power from 100% to 50% in a few seconds. At this power level a single feedwater pump can maintain the core water level. If all feedwater is lost, the reactor will scram and the Emergency Core Cooling System is used to restore reactor water level.

Steam Turbines

Steam produced in the reactor core passes through steam separators and dryer plates above the core and then directly to the turbine, which is part of the reactor circuit. Because the water around the core of a reactor is always contaminated with traces of radionuclides, the turbine must be shielded during normal operation, and radiological protection must be provided during maintenance. The increased cost related to operation and maintenance of a BWR tends to balance the savings due to the simpler design and greater thermal efficiency of a BWR when compared with a PWR. Most of the radioactivity in the water is very short-lived (mostly N-16, with a 7-second half-life), so the turbine hall can be entered soon after the reactor is shut down.

BWR steam turbines employ a high-pressure turbine designed to handle saturated steam, and multiple low-pressure turbines. The high-pressure turbine receives steam directly from the reactor. The high-pressure turbine exhaust passes through a steam reheater which superheats the steam to over 400 degrees F for the low-pressure turbines to use. The exhaust of the low-pressure turbines is sent to the main condenser. The steam reheaters take some of the reactor's steam and use it as a heating source to reheat what comes out of the high-pressure turbine exhaust. While the reheaters take steam away from the turbine, the net result is that the reheaters improve the thermodynamic efficiency of the plant.

Reactor Core

A modern BWR fuel assembly comprises 74 to 100 fuel rods, and there are up to approximately 800 assemblies in a reactor core, holding up to approximately 140 short tons of low-enriched uranium. The number of fuel assemblies in a specific reactor is based on considerations of desired reactor power output, reactor core size and reactor power density.

Safety Systems

A modern reactor has many safety systems that are designed with a defence in depth philosophy, which is a design philosophy that is integrated throughout construction and commissioning.

A BWR is similar to a pressurized water reactor (PWR) in that the reactor will continue to produce heat even after the fission reactions have stopped, which could make a core damage incident possible. This heat is produced by the radioactive decay of fission products and materials that have been activated by neutron absorption. BWRs contain multiple safety systems for cooling the core after emergency shut down.

Refueling Systems

The reactor fuel rods are occasionally replaced by removing them from the top of the containment vessel. A typical fuel cycle lasts 18–24 months, with about one third of fuel assemblies being replaced during a refueling outage. The remaining fuel assemblies are shuffled to new core locations to maximize the efficiency and power produced in the next fuel cycle.

Because they are hot both radioactively and thermally, this is done via cranes and under water. For this reason the spent fuel storage pools are above the reactor in typical installations. They are shielded by water several times their height, and stored in rigid arrays in which their geometry is controlled to avoid criticality. In the Fukushima reactor incident this became problematic because water was lost from one or more spent fuel pools and the earthquake could have altered the geometry. The fact that the fuel rods' cladding is a zirconium alloy was also problematic since this element can react with steam at extreme temperatures to produce hydrogen, which can ignite with oxygen in the air. Normally the fuel rods are kept sufficiently cool in the reactor and spent fuel pools that this is not a concern, and the cladding remains intact for the life of the rod.

Advantages

- The reactor vessel and associated components operate at a substantially lower pressure of about 70–75 bars (1,020–1,090 psi) compared to about 155 bars (2,250 psi) in a PWR.
- Pressure vessel is subject to significantly less irradiation compared to a PWR, and so does not become as brittle with age.
- Operates at a lower nuclear fuel temperature.
- Fewer components due to no steam generators and no pressurizer vessel. (Older BWRs have external recirculation loops, but even this piping is eliminated in modern BWRs, such as the ABWR.) This also makes BWRs simpler to operate.

- Lower risk (probability) of a rupture causing loss of coolant compared to a PWR, and lower risk of core damage should such a rupture occur. This is due to fewer pipes, fewer large diameter pipes, fewer welds and no steam generator tubes.
- NRC assessments of limiting fault potentials indicate if such a fault occurred, the average BWR would be less likely to sustain core damage than the average PWR due to the robustness and redundancy of the Emergency Core Cooling System (ECCS).
- Measuring the water level in the pressure vessel is the same for both normal and emergency operations, which results in easy and intuitive assessment of emergency conditions.
- Can operate at lower core power density levels using natural circulation without forced flow.
- A BWR may be designed to operate using only natural circulation so that recirculation pumps are eliminated entirely. (The new ESBWR design uses natural circulation.)
- BWRs do not use boric acid to control fission burn-up to avoid the production of tritium (contamination of the turbines), leading to less possibility of corrosion within the reactor vessel and piping. (Corrosion from boric acid must be carefully monitored in PWRs; it has been demonstrated that reactor vessel head corrosion can occur if the reactor vessel head is not properly maintained. Since BWRs do not utilize boric acid, these contingencies are eliminated.)
- The power control by reduction of the moderator density (vapour bubbles in the water) instead of by addition of neutron absorbers (boric acid in PWR) leads to breeding of U-238 by fast neutrons, producing fissile Pu-239.
- BWRs generally have *N*-2 redundancy on their major safety-related systems, which normally consist of four "trains" of components. This generally means that up to two of the four components of a safety system can fail and the system will still perform if called upon.
- Due to their single major vendor (GE/Hitachi), the current fleet of BWRs have predictable, uniform designs that, while not completely standardized, generally are very similar to one another. The ABWR/ESBWR designs are completely standardized. Lack of standardization remains a problem with PWRs, as, at least in the United States, there are three design families represented among the current PWR fleet (Combustion Engineering, Westinghouse, and Babcock & Wilcox), within these families, there are quite divergent designs. Still, some countries could reach a high level of standardisation with PWRs, like France.
    - Additional families of PWRs are being introduced. For example, Mitsubishi's APWR, Areva's US-EPR, and Westinghouse's AP1000/AP600 will add diversity and complexity to an already diverse crowd, and possibly cause customers seeking stability and predictability to seek other designs, such as the BWR.
- BWRs are overrepresented in imports, when the importing nation does not have a nuclear navy (PWRs are favored by nuclear naval states due to their compact, high-power design

used on nuclear-powered vessels; since naval reactors are generally not exported, they cause national skill to be developed in PWR design, construction, and operation). This may be due to the fact that BWRs are ideally suited for peaceful uses like power generation, process/industrial/district heating, and desalinization, due to low cost, simplicity, and safety focus, which come at the expense of larger size and slightly lower thermal efficiency.

- Sweden is standardized mainly on BWRs.
- Mexico's two reactors are BWRs.
- Japan experimented with both PWRs and BWRs, but most builds as of late have been of BWRs, specifically ABWRs.
- In the CEGB open competition in the early 1960s for a standard design for UK 2nd-generation power reactors, the PWR didn't even make it to the final round, which was a showdown between the BWR (preferred for its easily understood design as well as for being predictable and "boring") and the AGR, a uniquely British design; the indigenous design won, possibly on technical merits, possibly due to the proximity of a general election. In the 1980s the CEGB built a PWR, Sizewell B.

Disadvantages

- BWRs require more complex calculations for managing consumption of nuclear fuel during operation due to "two phase (water and steam) fluid flow" in the upper part of the core. This also requires more instrumentation in the reactor core.
- Larger pressure vessel than for a PWR of similar power, with correspondingly higher cost, in particular for older models that still use a main steam generator and associated piping.
- Contamination of the turbine by short-lived activation products. This means that shielding and access control around the steam turbine are required during normal operations due to the radiation levels arising from the steam entering directly from the reactor core. This is a moderately minor concern, as most of the radiation flux is due to Nitrogen-16 (activation of oxygen in the water), which has a half-life of 7 seconds, allowing the turbine chamber to be entered into within minutes of shutdown.
- Though the present fleet of BWRs is said to be less likely to suffer core damage from the "1 in 100,000 reactor-year" limiting fault than the present fleet of PWRs (due to increased ECCS robustness and redundancy), there have been concerns raised about the pressure containment ability of the as-built, unmodified Mark I containment – that such may be insufficient to contain pressures generated by a limiting fault combined with complete ECCS failure that results in extremely severe core damage. In this double failure scenario, assumed to be extremely unlikely prior to the Fukushima I nuclear accidents, an unmodified Mark I containment can allow some degree of radioactive release to occur. This is supposed to be mitigated by the modification of the Mark I containment; namely, the addition of an outgas stack system that, if containment pressure exceeds critical setpoints, is supposed to allow the orderly discharge of pressurizing gases after the gases pass through activated carbon filters designed to trap radionuclides.

- Control rods are inserted from below for current BWR designs. There are two available hydraulic power sources that can drive the control rods into the core for a BWR under emergency conditions. There is a dedicated high pressure hydraulic accumulator and also the pressure inside of the reactor pressure vessel available to each control rod. Either the dedicated accumulator (one per rod) or reactor pressure is capable of fully inserting each rod. Most other reactor types use top entry control rods that are held up in the withdrawn position by electromagnets, causing them to fall into the reactor by gravity if power is lost.

Technical and Background Information

Start-up ("Going Critical")

Reactor start up (criticality) is achieved by withdrawing control rods from the core to raise core reactivity to a level where it is evident that the nuclear chain reaction is self-sustaining. This is known as "going critical". Control rod withdrawal is performed slowly, as to carefully monitor core conditions as the reactor approaches criticality. When the reactor is observed to become slightly super-critical, that is, reactor power is increasing on its own, the reactor is declared critical.

Rod motion is performed using rod drive control systems. Newer BWRs such as the ABWR and ESBWR as well as all German and Swedish BWRs use the Fine Motion Control Rod Drive system, which allows multiple rods to be controlled with very smooth motions. This allows a reactor operator to evenly increase the core's reactivity until the reactor is critical. Older BWR designs use a manual control system, which is usually limited to controlling one or four control rods at a time, and only through a series of notched positions with fixed intervals between these positions. Due to the limitations of the manual control system it is possible while starting-up that the core can be placed into a condition where a single control rod can cause a large uneven reactivity change which can potentially challenge the fuel's thermal design margins. As a result, GE developed a set of rules in 1977 called BPWS (Banked Position Withdrawal Sequence) which help minimize the worth of any single control rod and prevent fuel damage in the case of a control rod drop accident. BPWS separates control rods into four groups, A1, A2, B1, and B2. Then, either all of the A control rods or B control rods are pulled full out in a defined sequence to create a "checkboard" pattern. Next the opposing group (B or A) is pulled in a defined sequence to positions 02, then 04, 08, 16, and finally full out (48), until the reactor enters the power operation range where thermal limits are no longer bounding. By following a BPWS compliant start-up sequence, the manual control system can be used to evenly and safely raise the entire core to critical, and prevent any fuel rods from exceeding 280 cal/gm energy release during any postulated event which could potentially damage the fuel.

Thermal Margins

Several calculated/measured quantities are tracked while operating a BWR:

- Maximum Fraction Limiting Critical Power Ratio, or MFLCPR;
- Fraction Limiting Linear Heat Generation Rate, or FLLHGR;

- Average Planar Linear Heat Generation Rate, or APLHGR;
- Pre-Conditioning Interim Operating Management Recommendation, or PCIOMR;

MFLCPR, FLLHGR, and APLHGR must be kept less than 1.0 during normal operation; administrative controls are in place to assure some margin of error and margin of safety to these licensed limits. Typical computer simulations divide the reactor core into 24–25 axial planes; relevant quantities (margins, burnup, power, void history) are tracked for each "node" in the reactor core (764 fuel assemblies x 25 nodes/assembly = 19100 nodal calculations/quantity).

Maximum Fraction Limiting Critical Power Ratio (MFLCPR)

Specifically, MFLCPR represents how close the leading fuel bundle is to "dry-out" (or "departure from nucleate boiling" for a PWR). Transition boiling is the unstable transient region where nucleate boiling tends toward film boiling. A water drop dancing on a hot frying pan is an example of film boiling. During film boiling a volume of insulating vapor separates the heated surface from the cooling fluid; this causes the temperature of the heated surface to increase drastically to once again reach equilibrium heat transfer with the cooling fluid. In other words, steam semi-insulates the heated surface and surface temperature rises to allow heat to get to the cooling fluid (through convection and radiative heat transfer).

MFLCPR is monitored with an empirical correlation that is formulated by vendors of BWR fuel (GE, Westinghouse, AREVA-NP). The vendors have test rigs where they simulate nuclear heat with resistive heating and determine experimentally what conditions of coolant flow, fuel assembly power, and reactor pressure will be in/out of the transition boiling region for a particular fuel design. In essence, the vendors make a model of the fuel assembly but power it with resistive heaters. These mock fuel assemblies are put into a test stand where data points are taken at specific powers, flows, pressures. Nuclear fuel could be damaged by film boiling; this would cause the fuel cladding to overheat and fail. Experimental data is conservatively applied to BWR fuel to ensure that the transition to film boiling does not occur during normal or transient operation. Typical SLMCPR/MCPRSL (Safety Limit MCPR) licensing limit for a BWR core is substantiated by a calculation that proves that 99.9% of fuel rods in a BWR core will not enter the transition to film boiling during normal operation or anticipated operational occurrences. Since the BWR is boiling water, and steam does not transfer heat as well as liquid water, MFLCPR typically occurs at the top of a fuel assembly, where steam volume is the highest.

Fraction Limiting Linear Heat Generation Rate (FLLHGR)

FLLHGR (FDLRX, MFLPD) is a limit on fuel rod power in the reactor core. For new fuel, this limit is typically around 13 kW/ft (43 kW/m) of fuel rod. This limit ensures that the centerline temperature of the fuel pellets in the rods will not exceed the melting point of the fuel material (uranium/gadolinium oxides) in the event of the worst possible plant transient/scram anticipated to occur. To illustrate the response of LHGR in transient imagine the rapid closure of the valves that admit steam to the turbines at full power. This causes the immediate cessation of steam flow and an immediate rise in BWR pressure. This rise in pressure effectively subcools the reactor coolant instantaneously; the voids (vapor) collapse into solid water. When the voids collapse in the reactor, the fission reaction is encouraged (more thermal neutrons); power increases drastically (120%) until it is terminated by the automatic insertion of the control rods. So, when the reactor is isolated from the turbine rapidly, pressure in the vessel rises rapidly, which collapses the water vapor, which causes a power excursion which is terminated by the Reactor Protection System. If a fuel pin was operating at 13.0 kW/ft prior to the transient, the void collapse would cause its power to rise. The FLLHGR limit is in place to ensure that the highest powered fuel rod will not melt if its power was rapidly increased following a pressurization transient. Abiding by the LHGR limit precludes melting of fuel in a pressurization transient.

Average Planar Linear Heat Generation Rate (APLHGR)

APLHGR, being an average of the Linear Heat Generation Rate (LHGR), a measure of the decay heat present in the fuel bundles, is a margin of safety associated with the potential for fuel failure to occur during a LBLOCA (large-break loss-of-coolant accident – a massive pipe rupture leading to catastrophic loss of coolant pressure within the reactor, considered the most threatening "design basis accident" in probabilistic risk assessment and nuclear safety and security), which is anticipated to lead to the temporary exposure of the core; this core drying-out event is termed core "uncovery", for the core loses its heat-removing cover of coolant, in the case of a BWR, light water. If the core is uncovered for too long, fuel failure can occur; for the purpose of design, fuel failure is assumed to occur when the temperature of the uncovered fuel reaches a critical temperature (1100 °C, 2200 °F). BWR designs incorporate failsafe protection systems to rapidly cool and make safe the uncovered fuel prior to it reaching this temperature; these failsafe systems are known as the Emergency Core Cooling System. The ECCS is designed to rapidly flood the reactor pressure vessel, spray water on the core itself, and sufficiently cool the reactor fuel in this event. However, like any system, the ECCS has limits, in this case, to its cooling capacity, and there is a possibility that fuel could be designed that produces so much decay heat that the ECCS would be overwhelmed and could not cool it down successfully.

So as to prevent this from happening, it is required that the decay heat stored in the fuel assemblies at any one time does not overwhelm the ECCS. As such, the measure of decay heat generation known as LHGR was developed by GE's engineers, and from this measure, APLHGR is derived. APLHGR is monitored to ensure that the reactor is not operated at an average power level that would defeat the primary containment systems. When a refueled core is licensed to operate, the fuel vendor/licensee simulate events with computer models. Their approach is to simulate worst case events when the reactor is in its most vulnerable state.

APLHGR is commonly pronounced as "Apple Hugger" in the industry.

Pre-Conditioning Interim Operating Management Recommendation (PCIOMR)

PCIOMR is a set of rules and limits to prevent cladding damage due to pellet-clad interaction. During the first nuclear heatup, nuclear fuel pellets can crack. The jagged edges of the pellet can rub and interact with the inner cladding wall. During power increases in the fuel pellet, the ceramic fuel material expands faster than the fuel cladding, and the jagged edges of the fuel pellet begin to press into the cladding, potentially causing a perforation. To prevent this from occurring, two corrective actions were taken. The first is the inclusion of a thin barrier layer against the inner walls of the fuel cladding which are resistant to perforation due to pellet-clad interactions, and the second is a set of rules created under PCIOMR.

The PCIOMR rules require initial "conditioning" of new fuel. This means, for the first nuclear heatup of each fuel element, that local bundle power must be ramped very slowly to prevent cracking of the fuel pellets and limit the differences in the rates of thermal expansion of the fuel. PCIOMR rules also limit the maximum local power change (in kW/ft*hr), prevent pulling control rods below the tips of adjacent control rods, and require control rod sequences to be analyzed against core modelling software to prevent pellet-clad interactions. PCIOMR analysis look at local power peaks and xenon transients which could be caused by control rod position changes or rapid power changes to ensure that local power rates never exceed maximum ratings.

Advanced Boiling Water Reactor

Construction of ABWR at Lungmen Nuclear Power Plant in New Taipei City, Taiwan.

The advanced boiling water reactor (ABWR) is a Generation III boiling water reactor. The ABWR is currently offered by GE Hitachi Nuclear Energy (GEH) and Toshiba. The ABWR generates electrical power by using steam to power a turbine connected to a generator; the steam is boiled from water using heat generated by fission reactions within nuclear fuel.

Boiling water reactors (BWRs) are the second most common form of light water reactor with a direct cycle design that uses fewer large steam supply components than the pressurized water reactor (PWR), which employs an indirect cycle. The ABWR is the present state of the art in boiling water reactors, and is the first Generation III reactor design to be fully built, with several reactors complete and operating. The first reactors were built on time and under budget in Japan, with others under construction there and in Taiwan. ABWRs were on order in the United States, including two reactors at the South Texas Project site.

The standard ABWR plant design has a net electrical output of about 1.35 GW (3926 MW of thermal power).

Overview of the Design

Pressure vessel from the ABWR. 1: Reactor core 2: Control rods 3: Internal Water Pump
4: Steam pipeline to the Turbine generator 5: Cooling water flow to the core

The ABWR represents an evolutionary route for the BWR family, with numerous changes and improvements to previous BWR designs.

Major areas of improvement include:

- The addition of reactor internal pumps (RIP) mounted on the bottom of the reactor pressure vessel (RPV) - 10 in total - which achieve improved performance while eliminating large recirculation pumps in containment and associated large-diameter and complex piping interfaces with the RPV (e.g. the recirculation loop found in earlier BWR models). Only the RIP motor is located outside of the RPV in the ABWR. According to the Tier 1 Design Control Document (which is the officially certified Nuclear Regulatory Commission document generally describing the design of the plant), each RIP has a nominal capacity of 6912 m^3/h.

- The control rod adjustment capabilities have been supplemented with the addition of an electro-hydraulic Fine Motion Control Rod Drive (FMCRD), allowing for fine position adjustment using an electrical motor, while not losing the reliability or redundancy of traditional hydraulic systems which are designed to accomplish rapid shutdown in 2.80 s from receipt of an initiating signal, or ARI (alternate rod insertion) in a greater but still insignificant time period. The FMCRD also improves defense-in-depth in the event of primary hydraulic and ARI contingencies.

- A fully digital Reactor Protection System (RPS) (with redundant digital backups as well as redundant manual backups) ensures a high level of reliability and simplification for safety condition detection and response. This system initiates rapid hydraulic insertion of control rods for shutdown (known as SCRAM by nuclear engineers) when needed. Two-out-of-four per parameter rapid shutdown logic ensures that nuisance rapid shutdowns are not triggered by single instrument failures. RPS can also trigger ARI, FMCRD rod run-in to shut down the nuclear chain reaction. The standby liquid control system (SLCS) actuation is provided as diverse logic in the unlikely event of an Anticipated Transient Without Scram.
- Fully digital reactor controls (with redundant digital backup and redundant manual backups) allow the control room to easily and rapidly control plant operations and processes. Separate redundant safety and non-safety related digital multiplexing buses allow for reliability and diversity of instrumentation and control.
    - In particular, the reactor is automated for startup (i.e., initiate the nuclear chain reaction and ascent to power) and for standard shutdown using automatic systems only. Of course, human operators remain essential to reactor control and supervision, but much of the busy-work of bringing the reactor to power and descending from power can be automated at operator discretion.
- The Emergency Core Cooling System (ECCS) has been improved in many areas, providing a very high level of defense-in-depth against accidents, contingencies, and incidents.
    - The overall system has been divided up into 3 divisions; each division is capable - by itself - of reacting to the maximally contingent Limiting Fault/Design Basis Accident (DBA) and terminating the accident prior to core uncovery, even in the event of loss of offsite power and loss of proper feedwater. Previous BWRs had 2 divisions, and uncovery (but no core damage) was predicted to occur for a short time in the event of a severe accident, prior to ECCS response.
    - Eighteen SORVs (safety overpressure relief valves), eight of which are part of the ADS (automatic depressurization system), ensure that RPV overpressure events are quickly mitigated, and that if necessary, that the reactor can be depressurized rapidly to a level where low pressure core flooder (LPCF, the high-capacity mode of the residual heat removal system, which replaces the LPCI and LPCS in previous BWR models) can be used.
    - Further, LPCF can inject against much higher RPV pressures, providing an increased level of safety in the event of intermediate-sized breaks, which could be small enough to result in slow natural depressurization but could be large enough to result in high pressure corespray/coolant injection systems' capacities for response being overwhelmed by the size of the break.
    - Though the Class 1E (safety-related) power bus is still powered by 3 highly-reliable emergency diesel generators that are safety related, an additional Plant Investment Protection power bus using a combustion gas turbine is located on-site to generate electricity to provide defense-in-depth against station blackout contingencies as well as to power important but non-safety critical systems in the event of a loss of offsite power.

- Though one division of the ECCS does not have high pressure flood (HPCF) capacities, there exists a steam-driven, safety-rated reactor core isolation cooling (RCIC) turbopump that is high-pressure rated and has extensive battery backup for its instrumentation and control systems, ensuring cooling is maintained even in the event of a full station blackout with failure of all 3 emergency diesel generators, the combustion gas turbine, primary battery backup, and the diesel firewater pumps.

- There exists an extremely thick basaltic reinforced concrete pad under the RPV that will both catch and hold any heated core melt that might fall on that pad in extraordinarily contingent situations. In addition, there are several fusible links within the wall separating the wetwell from the lower drywell that flood the pad using the wetwell's water supply, ensuring cooling of that area even with the failure of standard mitigation systems.

- The containment has been significantly improved over the conventional Mark I type. Like the conventional Mark I type, it is of the pressure suppression type, designed to handle evolved steam in the event of a transient, incident, or accident by routing the steam using pipes that go into a pool of water enclosed in the wetwell (or torus in the case of the Mark I), the low temperature of which will condense the steam back into liquid water. This will keep containment pressure low. Notably, the typical ABWR containment has numerous hardened layers between the interior of the primary containment and the outer shield wall, and is cubical in shape. One major enhancement is that the reactor has a standard safe shutdown earthquake acceleration of .3G; further, it is designed to withstand a tornado with >320 mph wind speed. Seismic hardening is possible in earthquake-prone areas and has been done at the Lungmen facility in Taiwan which has been hardened up 0.4 g in any direction.

- The ABWR is designed for a lifetime of at least 60 years. The comparatively simple design of the ABWR also means that no expensive steam generators need to be replaced either, decreasing total cost of operation.

- According to GEH's Probabilistic Risk Assessment, a core damage event would occur no more often than once in six million years as the core damage frequency (CDF) of the ABWR is 1.6×10^{-7}, second in lowest CDF probability to the ESBWR.

The RPV and Nuclear Steam Supply System (NSSS) have significant improvements, such as the substitution of RIPs, eliminating conventional external recirculation piping loops and pumps in the containment that in turn drive jet pumps producing forced flow in the RPV. RIPs provide significant improvements related to reliability, performance and maintenance, including a reduction in occupational radiation exposure related to containment activities during maintenance outages. These pumps are powered by wet-rotor motors with the housings connected to the bottom of the RPV and eliminating large diameter external recirculation pipes that are possible leakage paths. The 10 internal recirculation pumps are located at the bottom of the annulus downcomer region (i.e., between the core shroud and the inside surface of the RPV). Consequently, internal recirculation pumps eliminate all of the jet pumps in the RPV, all of the large external recirculation loop pumps and piping, the isolation valves and the large diameter nozzles that penetrated the RPV and needed to suction water from and return it to the RPV. This design therefore reduces the worst leak below the core region to effectively equivalent to a 2-inch-diameter (51 mm) leak. The conventional BWR3-BWR6 product line has an analogous potential leak of 24 or more inches in diameter. A major benefit of this design is that it greatly reduces the flow capacity required of the ECCS.

The first reactors to use internal recirculation pumps were designed by ASEA-Atom (now Westinghouse Electric Company by way of mergers and buyouts, which is owned by Toshiba) and built in Sweden. These plants have operated very successfully for many years.

The internal pumps reduce the required pumping power for the same flow to about half that required with the jet pump system with external recirculation loops. Thus, in addition to the safety and cost improvements due to eliminating the piping, the overall plant thermal efficiency is increased. Eliminating the external recirculation piping also reduces occupational radiation exposure to personnel during maintenance.

An operational feature in the ABWR design is electric fine motion control rod drives, first used in the BWRs of AEG (later Kraftwerk Union AG, now AREVA). Older BWRs use a hydraulic locking piston system to move the control rods in six-inch increments. The electric fine motion control rod design greatly enhances positive actual control rod position and similarly reduces the risk of a control rod drive accident to the point that no velocity limiter is required at the base of the cruciform control rod blades.

ABWR-II Design

A number of design variants have been considered, with power outputs varying from 600 to 1800 MWe. The most developed design variant is the ABWR-II, started in 1991, an enlarged 1718 MWe ABWR, intended to make nuclear power generation more competitive in the late 2010s. None of these designs have been deployed.

The new designs hoped to achieve 20% reductions in operating costs, 30% reduction in capital costs, and tight planned construction schedule of 30 months. The design would allow for more flexibility in choices of nuclear fuels.

Supercritical Water Reactor

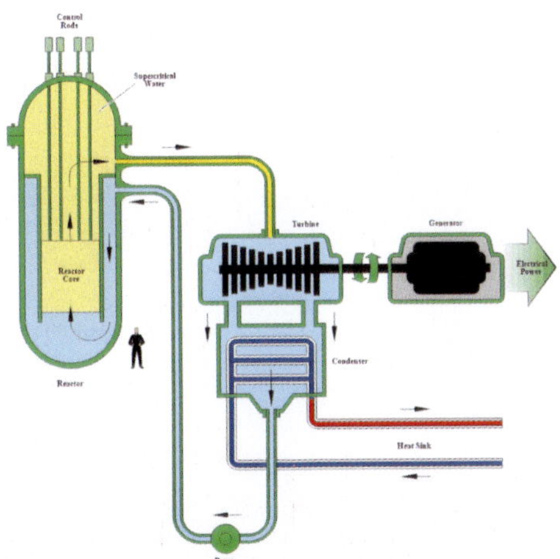

Concept of the SCWR

The supercritical water reactor (SCWR) is a concept of Generation IV reactor, that is operated at supercritical pressure (i.e. greater than 22.1 MPa). The term supercritical in this context refers to the thermodynamic critical point of water (T_{CR} = 374 °C; p_{CR} = 22.1 MPa), and must not be confused with the criticality of the reactor core, that describes changes in the neutron population in the reactor core.

The supercritical water reactor may be operated as a thermal reactor or as a fast-neutron reactor, depending on the core design. The concept of the supercritical water reactor may be based on classical pressure vessel as in commercial PWRs or on pressure tubes as in CANDU reactors. The pressure-vessel design of supercritical water reactors is developed largely in the EU, US, Japan, Korea, and China, while the pressure-channel design is developed largely in Canada and in Russia. The pressure-vessel design allows using a traditional high-pressure circuit layout. The pressure-channel design allows the key features of passive accident and decay heat removal by radiation and convection from the distributed channels even with no active cooling and fuel melting and use of multi-pass reactor flows making reheating and superheating possible.

For both pressure vessel and pressure-tube designs, a once through steam cycle has been envisaged, omitting any coolant recirculation inside the reactor. It is similar as in boiling water reactors, steam will be supplied directly to the steam turbine and the feed water from the steam cycle will be supplied back to the core.

As well as the supercritical water reactor may use light water or heavy water as neutron moderator. As can be seen, there are many SCWR designs, but all SCWRs have a key feature, that is the use of water beyond the thermodynamic critical point as primary coolant. Since this feature allows to increase the peak temperature, the supercritical water reactors are considered a promising advancement for nuclear power plants because of its high thermal efficiency (~45 % vs. ~33 % for current LWRs).

Critical Point of Water

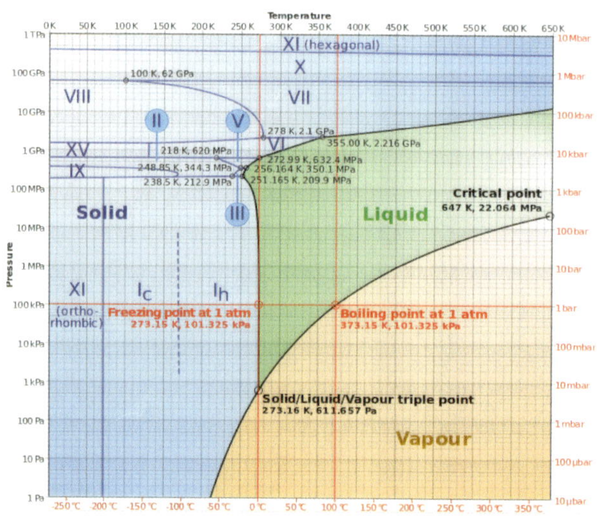

Phase diagram of water.

In thermodynamics, a critical point (or critical state) is the end point of a phase equilibrium curve. The phase diagram of water is a pressure-temperature diagram for water that shows how all three phases (solid, liquid, and vapor) may coexist together in thermal equilibrium. Along the vaporization line, the liquid and vapor phases are in equilibrium, along the fusion line, the solid and liquid phases are in equilibrium and along the sublimation line, the solid and vapor phases are in equilibrium. The only point at which all three phases may exist in equilibrium is the triple point. The vaporization line ends at the critical point because there is no distinct change from the liquid phase to the vapor phase above the critical point.

Above the critical point, there is no constant-temperature vaporization process. At the critical point the saturated-liquid and saturated-vapor states are identical. The temperature, pressure, and specific volume at the critical point are called the critical temperature, critical pressure, and critical volume. For water, these parameters are the following:

- P_{cr} = 22.09 MPa
- T_{cr} = 374.14 °C (or 647.3 K)
- v_{cr} = 0.003155 m3/kg
- $u_f = u_g$ = 2014 kJ/kg
- $h_f = h_g$ = 2084 kJ/kg
- $s_f = s_g$ =4.406 kJ/kg K

A supercritical phase (e.g. water at a pressure above the critical pressure) does not separate into two phases when it is cooled at constant pressure (along a horizontal line above the critical point in phase diagram). Instead, its properties change gradually and continuously from those we ordinarily associate with a gas (low density, large compressibility) to those of a liquid (high density, small compressibility) without a phase change.

As can be seen from steam tables, as we approach the critical point, the differences in physical properties (such as density and enthalpy) between the liquid and vapor phases become smaller. For example, just below the critical point at the pressure of 21.8 MPa and the temperature of 373°C the specific enthalpy of saturated liquid is 1970 kJ/kg, while the specific enthalpy of saturated vapor is 2230 kJ/kg. The specific heat of vaporization is only 260 kJ/kg. Exactly at the critical point these differences become zero, and at this point the distinction between liquid and vapor disappears. The heat of vaporization also becomes zero at the critical point.

Near the critical point, the physical properties of the liquid and the vapor change dramatically. For example, liquid water under normal conditions has a low thermal expansion coefficient, is nearly incompressible, is an excellent solvent for electrolytes, and has a high dielectric constant. But near the critical point, all these properties change into the exact opposite: water becomes compressible, has a significant thermal expansion coefficient, has a low dielectric constant, is a bad solvent for electrolytes. At pressures greater than the critical pressure the physical properties also undergo a fast transition, but without singularities. With further increase in pressure the transition becomes smoother. This transition region is called pseudocritical region, at a given pressure the temperature where the specific heat has its maximum is called pseudocritical point.

For nearly all familiar materials the critical pressures are much greater than atmospheric pressure, therefore we don't observe this behavior in everyday life. But in energy engineering, there are many applications for supercritical water. One of ways to increase thermal efficiency of thermodynamic cycles (e.g. Rankine cycle) is to increase the peak temperature and pressure in the boiler. High-pressure steam boilers in thermal power plants regularly run at pressures and temperatures well above the critical point.

In nuclear engineering, the supercritical water reactor is considered a promising advancement for nuclear power plants because of its high thermal efficiency (~45 % vs. ~33 % for current LWRs). This concept of reactor operates at supercritical pressure (i.e. greater than 22.1 MPa) and belongs to Generation IV reactor designs.

Properties of Supercritical Water

A supercritical fluid is a fluid that is at pressures higher than its thermodynamic critical values. At the critical and supercritical pressures a fluid is considered as a single-phase substance in spite of the fact that all thermophysical properties undergo significant changes within the critical and pseudocritical regions.

Pseudocritical line – pseudocritical points. Pseudocritical line consist of pseudocritical points, which are points at a pressure above the critical pressure and at a temperature ($T_{pc} > T_{cr}$) corresponding to the maximum value of the specific heat at this particular pressure.

Supercritical fluid properties - 25MPa

Temperature - °C -	Density - kg/m3 -	Specific Enthalpy - kJ/kg
250	820,9	1087
260	807,1	1134
270	792,5	1182
280	777,0	1231
290	760,6	1280
300	743,0	1331
310	724,1	1384
320	703,5	1439
330	680,7	1496
340	655,1	1558
350	625,5	1624
360	589,3	1699
370	540,5	1790
380	450,8	1936
390	215,2	2396
400	166,5	2579
410	145,3	2687
420	131,9	2769
430	122,4	2838
440	115,0	2897
450	109,0	2951
460	104,0	2999
470	99,7	3045
480	96,0	3087
490	92,7	3128
500	89,7	3166
520	84,7	3238
540	80,4	3307
560	76,8	3371
580	73,6	3433
600	70,7	3494

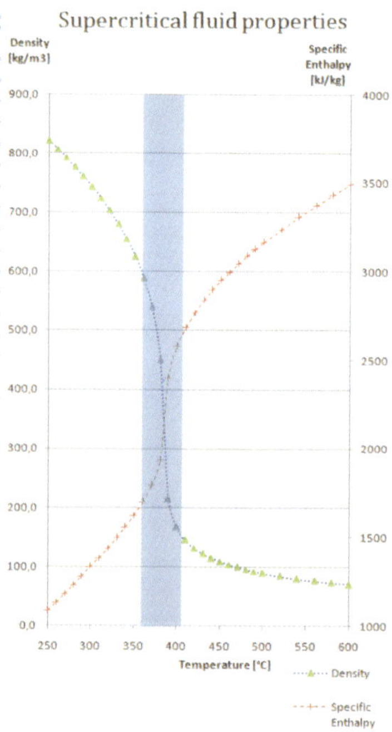

At pressures above the critical pressure, properties of water in the reactor change gradually and continuously from those we ordinarily associate with a liquid (high density, small compressibility) to those of a gas (low density, large compressibility) without a phase change. There is no change in the phase of water in the core. On the other hand, physical properties such as density, specific heat, specific enthalpy undergo significant changes, especially in the temperature range of the pseudocritical region (for 25 MPa between 372°C and 392°C). For example,

- The density of supercritical water at the inlet and at the outlet is about 777 kg/m3 (for 25MPa and 280°C) and 90 kg/m3 (for 25MPa and 500°C).
- The specific enthalpy of supercritical water at the inlet and at the outlet is about 1230 kJ/kg (for 25MPa and 280°C) and 3165 kJ/kg (for 25MPa and 500°C).

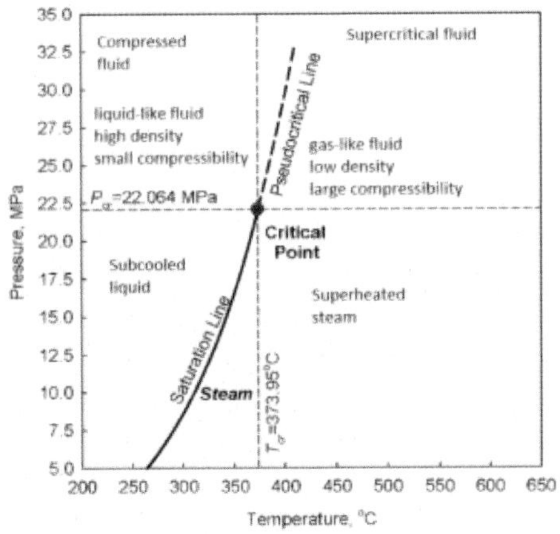

Following figures shows the behaviour of thermophysical properties of water near the critical (22.1MPa) and pseudocritical (25MPa) points. Near the critical point these property changes are dramatic. In the vicinity of the pseudocritical point at 25 MPa, these property changes become less significant. At 25 MPa the most significant property changes occur within ±25°C around pseudo-critical point (389.4°C), this region is known as the pseudocritical region. For convenience, below the pseudocritical point fluid properties are considered to show liquid-like behaviour and above the pseudocritical point they are considered to show gas-like behaviour.

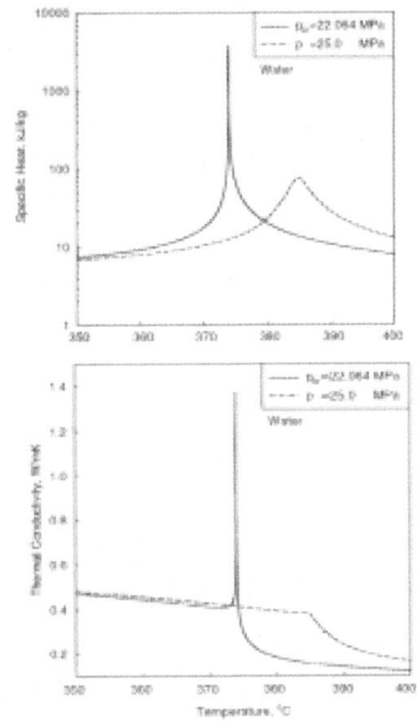

Characteristics of SCWRs

SCWRs operate at pressures above the critical pressure, properties of water in the reactor change gradually and continuously from those we ordinarily associate with a liquid (high density, small compressibility) to those of a gas (low density, large compressibility) without a phase change. There is no change in the phase of water in the core. On the other hand, physical properties such as density, specific heat, specific enthalpy undergo significant changes, especially in the temperature range of the pseudocritical region (for 25 MPa between 372°C and 392°C). For example:

- The density of supercritical water at the inlet and at the outlet is about 777 kg/m³ (for 25MPa and 280°C) and 90 kg/m³ (for 25MPa and 500°C).
- The specific enthalpy of supercritical water at the inlet and at the outlet is about 1230 kJ/kg (for 25MPa and 280°C) and 3165 kJ/kg (for 25MPa and 500°C).

From a neutronics point of view, the density of water is the most important factor. There is a significant change of neutron spectrum, which changes with the axial coordinate of the core. It is caused by the greater moderation in places with greater density. For this reason, a coupled neutronics – thermal hydraulics calculation is inevitable to obtain neutron flux distribution within the core. In each nuclear reactor, there is a direct proportionality between the neutron flux and the reactor thermal power.

Since the coolant (moderator) undergo significant change in density, the SCWR can also be designed as a fast neutron reactor. This property depends on certain reactor design.

Advantages and Challenges of SCWRs

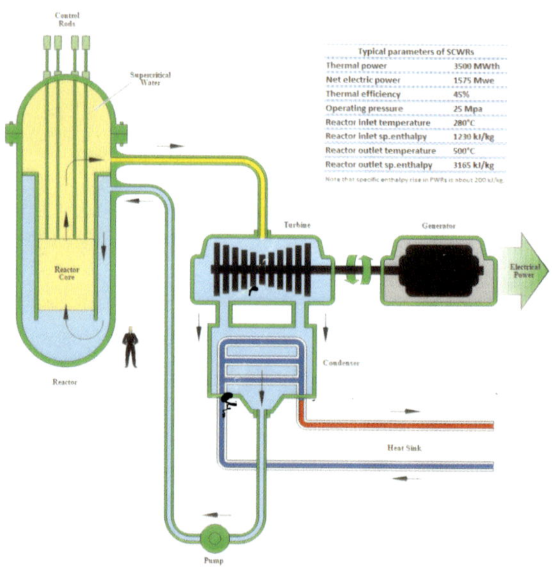

As can be inferred from its name, the SCWR's key feature is the use of water beyond the thermodynamic critical point (T_{CR} = 374 °C; p_{CR} = 22.1 MPa) as primary coolant. SCWR designs have unique features that offer many advantages compared to current light water reactors (LWRs).

- High thermal efficiency. Since SCWRs supply supercritical steam at pressures and temperatures, that are much higher than in conventional LWRs, they will reach higher thermodynamic efficiency of the power plant (~45 % vs. ~33 % for current LWRs).
- Lower coolant mass flow rate. The reactor coolant flow rate of SCWR is much smaller than

that of BWR and PWR because the enthalpy rise in the core is much larger, which results in low capacity components of the primary system. Thus, the coolant pumps, pipes and other supporting equipment become smaller. Moreover, the only pumps driving the coolant under normal operating conditions are the feed water pumps and the condensate extraction pumps.

- The higher steam enthalpy allows to decrease the size of the turbine system and thus to lower the capital costs of the conventional island.
- Smaller coolant inventory. The coolant inventory is smaller which reduces the size of the containment with pressure suppression pools.
- Since supercritical water does not undergo a phase change and it exists in a single thermodynamic phase, the boiling crisis (i.e. departure from nucleate boiling – DNB or dry out) cannot occur.
- With a direct cycle of supercritical coolant, components like steam dryers, separators and steam generators are omitted entirely, reducing the number of major components and eliminating their associated costs.
- Many of the components (e.g. turbine) are readily developed and available from supercritical fossil fired power plants.

The combination of advanced water-cooled reactor technology and advanced supercritical fossil technology is expected to result in a reactor concept that can be used to generate base-load electricity very economically and efficiently. This feature also makes the SCWR a very attractive concept for utilities, especially those that have experience with both water-cooled reactors and supercritical fossil plants.

On the other hand numerous challenges must be solved before the first power plants of this type can be built:

- The cladding materials have to withstand higher temperatures (600-650°C in normal operating conditions) than in today's LWRs.
- The heat transfer phenomena from cladding surface into supercritical fluid must be understood thoroughly. Especially in case of within transients with depressurization from supercritical to subcritical conditions.
- Due to the smaller coolant inventory, the dry out of the core progresses faster in case of a loss of coolant accident (LOCA).

Heavy-water Reactors

The most efficient moderator is heavy water. Heavy water has lower ξ and σ_s, but it has the highest moderating ratio owing to its lowest neutron absorption cross-section. Therefore heavy water is commonly used as a moderator in pressurized heavy water reactors (PHWR). Most of PHWR are heavy water cooled and moderated pressurized water reactors. Instead of using a single large reactor vessel as in a PWR or BWR, the nuclear core is contained in hundreds of pressure tubes. PHWRs generally use natural uranium (0.7% U-235) oxide or slightly enriched uranium as fuel, hence needs a more efficient moderator, in this case heavy water (D_2O).

Pressurized Heavy-water Reactor

The pressurized heavy water reactor utilizes natural uranium as the fuel and heavy water as both moderator and coolant. The use of heavy water as moderator facilitates the use of natural uranium as the fuel. The cycle diagram of coolant in PHWR is similar to that in a PWR with the difference being the use of heavy water as primary coolant. A schematic diagram of working of a PHWR plant is shown in Figure .

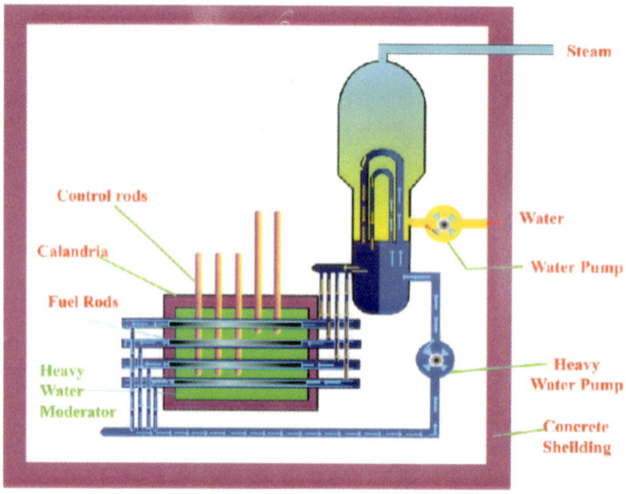

Fig. Schematic diagram of Pressurized Heavy Water Reactor

PHWR Core

CANDU (Canada Deuterium Uranium) type reactors are the common form of PHWR. The core configuration of CANDU is different from that of a PWR. UO2 pellets (12.2 mm diameter and 16.4 mm long) are stacked in zircaloy tube to form fuel rods of 13.1 mm diameter and 490 mm long. These rods are arranged in concentric circles with a pitch of about 14.6 mm to form an assembly of 102 mm diameter and 495 mm long that houses 37 fuel rods. A number of such fuel assemblies are inserted into pressure tubes of calandria.

Calandria is a low-pressure horizontal vessel, consisting of circular channels arranged in a regular fashion and immersed in moderator (heavy water). The pressure tubes (containing fuel assemblies) are loaded in the channels of calandria with the coolant (heavy water) flowing around the zircaloy tubes inside the pressure tubes. This extracts the heat released during fission and transfers the same to the secondary coolant (light water) that boils to generate steam.

The Indian PHWR (220 MWe) consists of a calandria that contains heavy water (moderator) at near ambient pressure and temperature. About 306 pressure tubes made of zircaloy are inserted in the calandria. Each pressure tube contains 12 fuel bundles. Each fuel bundle has 19 fuel rods and is 495 mm long weighing about 16 kg. A schematic diagram of arrangement of fuel pellets in zircaloy tubes to form fuel bundle, their loading inside pressure tubes and the placement of pressure tubes in calandria are shown in Figure.

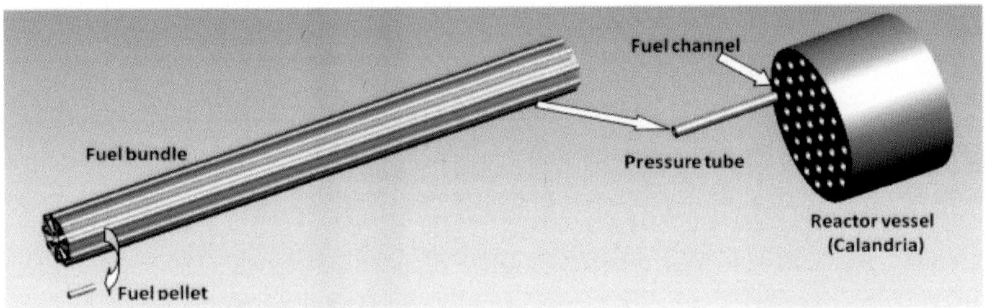

Fig. Arrangement of fuel pellets in fuel bundles and pressure tubes

The power cycle adopted in PHWRs is the two-coolant cycle (indirect cycle). Heavy water acts as both coolant and moderator, similar to the roles of light water in pressurized water reactor. However, the heavy water used as coolant alone is at high pressures. The moderator (also heavy water) is at lower pressure. Another difference between coolant system of PWR and PHWR is the use of different coolant material for primary and secondary circuits. The primary coolant is heavy water, while the secondary coolant is light water. The annular region between the coolant tube and pressure tube is filled with an annulus gas like carbon dioxide. The primary coolant passing thorough the coolant tubes removes the heat generated by fission from the fuel rods. In the steam generator, heavy water transfer heat to light water that undergoes evaporation to produce saturated steam. The steam then expands in a turbine generator producing electricity. The spent steam is condensed in condenser and returned to the steam generator for extraction of heat from heavy water.

The calandria has provisions for accommodating liquid poisons in tubes at specific locations in the calandria. The moderator also acts as neutron reflectors. Therefore, control of neutron flux and hence the reactor power can be achieved by controlling the level of moderator. In nutshell, calandria represents the most important assembly of PHWRs as fuel, coolant, moderator, control rods etc. are all present in it. One of the important aspects of CANDU reactors is the facility for online refueling without shutting down the reactor. The spent-fuel bundles can be replaced by the bundles of fresh fuel during the normal operation of the reactor.

There are two primary coolant loops with one pump used for each loop. Four steam generators (@ two steam generators per loop) are used. The primary coolant pumps themselves maintain the primary coolant pressure and hence PHWRs do not require pressurizer. The operating conditions in a typical PHWR are given in table.

Table. Operating conditions of a typical PHWR

Parameter	Value
Primary coolant pressure (MPa)	10
Primary coolant inlet temperature (°C)	267
Primary coolant outlet temperature (°C)	310
Secondary coolant inlet temperature (°C)	187
Secondary coolant outlet temperature (°C)	260
Steam generator pressure (MPa)	4.7

Purpose of using Heavy Water

The key to maintaining a nuclear reaction within a nuclear reactor is to use the neutrons released during fission to stimulate fission in other nuclei. With careful control over the geometry and reaction rates, this can lead to a self-sustaining chain reaction, a state known as "criticality".

Natural uranium consists of a mixture of various isotopes, primarily ^{238}U and a much smaller amount (about 0.72% by weight) of ^{235}U. ^{238}U can only be fissioned by neutrons that are relatively energetic, about 1 MeV or above. No amount of ^{238}U can be made "critical" since it will tend to parasitically absorb more neutrons than it releases by the fission process. ^{235}U, on the other hand, can support a self-sustained chain reaction, but due to the low natural abundance of ^{235}U, natural uranium cannot achieve criticality by itself.

The "trick" to making a working reactor fuelled by natural or low enriched Uranium is to slow enough of the neutrons to the point where their probability of causing nuclear fission in ^{235}U increases to a level that permits a sustained chain reaction in the uranium as a whole. This requires the use of a neutron moderator, which absorbs some of the neutrons' kinetic energy, slowing them down to an energy comparable to the thermal energy of the moderator nuclei themselves (leading to the terminology of "thermal neutrons" and "thermal reactors"). During this slowing-down process it is beneficial to physically separate the neutrons from the uranium, since ^{238}U nuclei have an enormous parasitic affinity for neutrons in this intermediate energy range (a reaction known as "resonance" absorption). This is a fundamental reason for designing reactors with discrete solid fuel separated by moderator, rather than employing a more homogeneous mixture of the two materials.

Water makes an excellent moderator; the hydrogen atoms in the water molecules are very close in mass to a single neutron, and the collisions thus have a very efficient momentum transfer, similar conceptually to the collision of two billiard balls. However, despite being a good moderator, water is relatively effective at absorbing neutrons. Using water as a moderator will absorb so many neutrons that there will be too few left to react with the small amount of ^{235}U in the fuel, thus precluding criticality in natural uranium. Instead, in order to fuel a light-water reactor, first the amount of ^{235}U in the uranium must be increased, producing enriched uranium, which generally contains between 3% and 5% ^{235}U by weight (the waste from this process is known as depleted uranium, consisting primarily of ^{238}U). In this enriched form there *is* enough ^{235}U to react with the water-moderated neutrons to maintain criticality.

One complication of this approach is the need for uranium enrichment facilities, which are generally expensive to build and operate. They also present a nuclear proliferation concern; the same systems used to enrich the ^{235}U can also be used to produce much more "pure" weapons-grade material (90% or more ^{235}U), suitable for producing a nuclear weapon. This is not a trivial exercise by any means, but feasible enough that enrichment facilities present a significant nuclear proliferation risk.

An alternative solution to the problem is to use a moderator that does *not* absorb neutrons as readily as water. In this case potentially all of the neutrons being released can be moderated and used in reactions with the ^{235}U, in which case there *is* enough ^{235}U in natural uranium to sustain criticality. One such moderator is heavy water, or deuterium-oxide. Although it reacts dynamically with the neutrons in a fashion similar to light water (albeit with less energy transfer on average, given that heavy hydrogen, or deuterium, is about twice the mass of hydrogen), it already has the extra neutron that light water would normally tend to absorb.

Advantages and Disadvantages

The use of heavy water as the moderator is the key to the PHWR (pressurized heavy water reactor) system, enabling the use of natural uranium as the fuel (in the form of ceramic UO_2), which means that it can be operated without expensive uranium enrichment facilities. The mechanical arrangement of the PHWR, which places most of the moderator at lower temperatures, is particularly efficient because the resulting thermal neutrons are "more thermal" than in traditional designs, where the moderator normally is much hotter. These features mean that a PHWR can use natural uranium and other fuels, and does so more efficiently than light water reactors (LWRs).

Pressurised heavy-water reactors do have some drawbacks. Heavy water generally costs hundreds of dollars per kilogram, though this is a trade-off against reduced fuel costs. The reduced energy content of natural uranium as compared to enriched uranium necessitates more frequent replacement of fuel; this is normally accomplished by use of an on-power refuelling system. The increased rate of fuel movement through the reactor also results in higher volumes of spent fuel than in LWRs employing enriched uranium. However, since unenriched uranium fuel accumulates a lower density of fission products than enriched uranium fuel, it generates less heat, allowing more compact storage.

Nuclear Proliferation

Heavy-water reactors may pose a greater risk of nuclear proliferation verses comparable light-water reactors due the low neutron absorption properties of heavy water, discovered in 1937 by Hans von Halban and Otto Frisch. Occasionally, when an atom of U-238 is exposed to neutron radiation, its nucleus will capture a neutron, changing it to U-239. The U-239 then rapidly undergoes two β^- decays — an emission of an electron and an anti-neutrino, — the first β^- decay transforming the U-239 into neptunium-239, and the second β^- decay transforming the Np-239 into Pu-239. Although this process takes place with other moderators such as ultra-pure graphite or beryllium, heavy water is by far the best.

Plutonium-239 is a fissile material suitable for use in nuclear weapons. As a result, if the fuel of a heavy-water reactor is changed frequently, significant amounts of weapons-grade plutonium can be chemically extracted from the irradiated natural uranium fuel by nuclear reprocessing.

In addition, the use of heavy water as a moderator results in the production of small amounts of tritium when the deuterium nuclei in the heavy water absorb neutrons, a very inefficient reaction. Tritium is essential for the production of boosted fission weapons, which in turn enable the easier production of thermonuclear weapons, including neutron bombs. It is unclear whether it is possible to use this method to produce tritium on a practical scale.

The proliferation risk of heavy-water reactors was demonstrated when India produced the plutonium for Operation Smiling Buddha, its first nuclear weapon test, by extraction from the spent fuel of a heavy-water research reactor known as the CIRUS reactor.

Graphite-moderated Reactor

In a graphite-moderated reactor, un-enriched uranium acts as a nuclear fuel and carbon acts as a neutron moderator. The first nuclear reactor that used graphite as a moderator was the Chicago Pile-1. Many types of graphite-moderated nuclear reactors are used for commercial generation of power, such as water-cooled and gas-cooled reactors, etc.

Water-cooled, Graphite-moderated Reactor

Water-cooled, graphite-moderated reactor (RBMK) (Реактор Большой Мощности Канальный – High power channel reactor in Russian) For moderator in this type of reactors is used graphite and for coolant – light water. The fuel is low enriched uranium compound, situated in channels in graphite blocks. Circulation pumps drive the water trough the channels it heats up and starts to boil and the steam-water mixture leaves the reactor core with a temperature of about 290 °C. From the reactor it is directed to the separators (a device that separates the steam from the water) and then to the turbine, which drives the generator for electricity production. The coolant is then pumped back to the reactor.

After the Chernobyl accident, a number of measures have been taken to enhance the safety of the RBMK reactors, which are still in operation. The only country besides the Russian federation that owns RBMK reactors is Lithuania. There are two reactors, one of them is shut down and the other is going to be in operation till 2009.

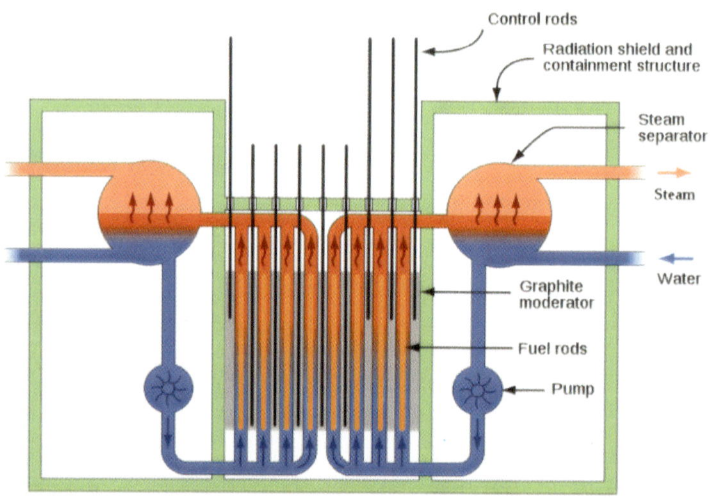

Gas-cooled Reactor

A schematic diagram of a gas-cooled reactor is shown in figure. This is a type of nuclear reactor that uses a gas as the coolant. Mostly CO2 is used. Graphite blocks are used as moderator, within which channels are made for housing fuel rods. Control rods are inserted into the graphite blocks. Channels are established between the graphite blocks for the flow of coolant. Natural uranium is used as the fuel while cladding is made of a magnesium alloy called magnox.This reactor derives its name from the alloy used for cladding, 'magnox'. The coolant gas is supplied by a gas circulator and enters the core from bottom. Gas flows through the coolant channels between the graphite blocks. As the gas moves up through the core, it gets heated up and leaves the top of the core at high temperature.

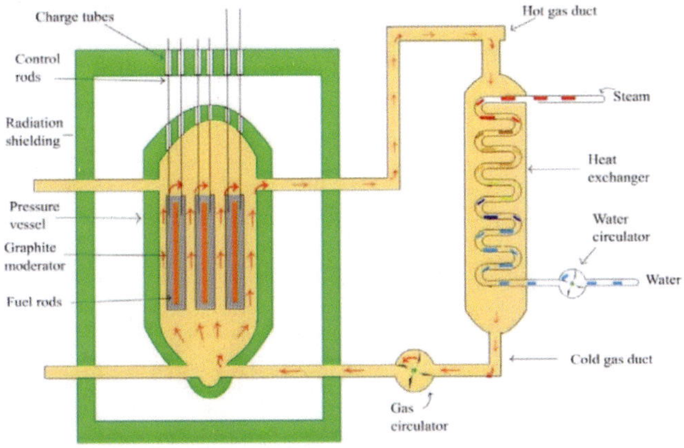

This high temperature gas exchanges heat with water in a heat exchanger, resulting in the production of steam, which runs the turbine. The spent steam is condensed and returned back to the heat exchanger, while the gas returns to the reactor. The heat exchanger is located outside the pressure vessel and the containment. Magnox reactor finds extensive use in United Kingdom.

Types

Gas-cooled reactor types include:

- Gas-cooled reactor (graphite moderated, CO_2 cooled)
    - Magnox (British design, 28 built, 1956-2015)
    - UNGG reactor (French design, 10 built, 1956-1994)
    - Advanced Gas-cooled Reactor (Magnox successor, 15 built, 1962-today)
- Heavy water gas cooled reactor (heavy water moderated, CO_2 cooled)
    - Brennilis Nuclear Power Plant (1967-1985)
    - KS 150 (1972-1979)
- High & Very-high temperature reactor (graphite moderated, He cooled)
    - Prismatic block reactor
        - Dragon reactor (1964-1975)
        - Peach Bottom Atomic Power Station (1967-1974)
        - Fort Saint Vrain Generating Station (1979-1989)
        - High-temperature engineering test reactor (1999-today)

        - Gas Turbine Modular Helium Reactor (General Atomics design)
        - Steam Cycle High-Temperature Gas-Cooled Reactor (Areva SMR design)
    - Pebble bed reactor
        - AVR reactor (1966-1988)
        - THTR-300 (1983-1989)
        - HTR-10 (2003-today)
        - HTR-PM (under construction)
        - Pebble bed modular reactor (design)
- Gas-cooled fast reactor (No moderator, He cooled)
    - Energy Multiplier Module (General Atomics design)

Advanced Gas-cooled Reactor

An advanced gas-cooled reactor (AGR) is a British design of nuclear reactor. AGRs are using graphite as the neutron moderator and carbon dioxide as coolant. AGRs were developed from the Magnox type reactor.These are the second generation of British gas-cooled reactors. AGRs are operating at a higher gas temperature for improved thermal efficiency, thus requires stainless steel fuel cladding to withstand the higher temperature. Because the stainless steel fuel cladding has a higher neutron capture cross section than Magnox fuel (magnesium non-oxidising alloy), low enriched uranium fuel is needed.

The fuel is uranium oxide pellets, enriched to 2.5-3.5%, in stainless steel tubes. The carbon dioxide circulates through the core, reaching 650°C (for improved thermal efficiency) and then past steam generator tubes outside it, but still inside the concrete and steel pressure vessel (hence 'integral' design). Control rods penetrate the moderator and a secondary shutdown system involves injecting nitrogen to the coolant.

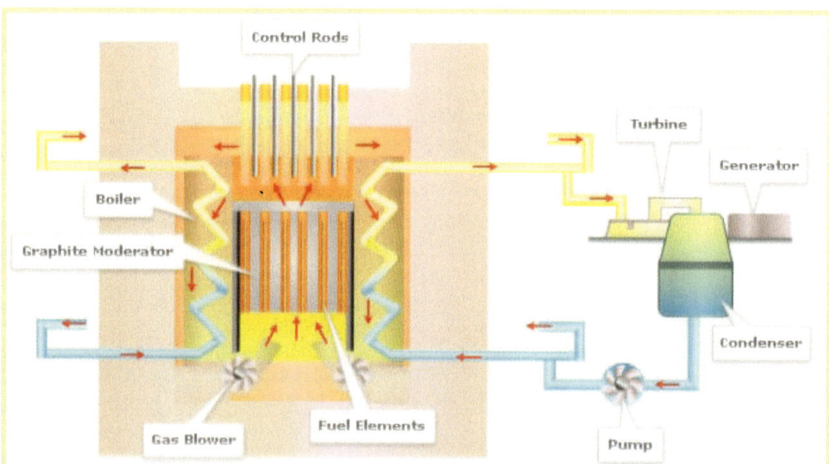

AGR – Advanced Gas-cooled Reactor

AGR design

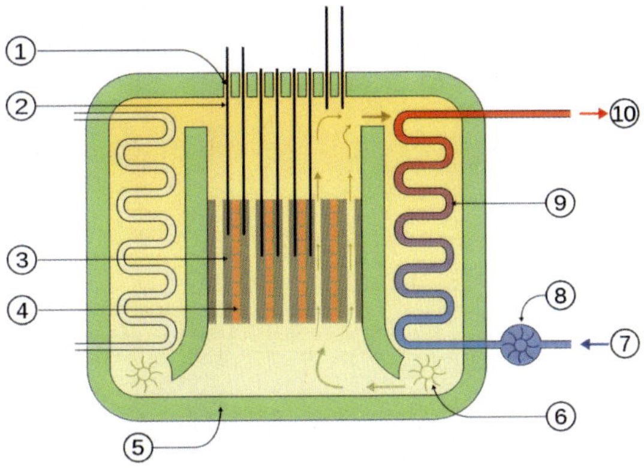

Schematic diagram of the Advanced Gas-cooled Reactor.

Note that the heat exchanger is contained within the steel-reinforced concrete combined pressure vessel and radiation shield.

1. Charge tubes
2. Control rods
3. Graphite moderator
4. Fuel assemblies
5. Concrete pressure vessel and radiation shielding
6. Gas circulator
7. Water
8. Water circulator
9. Heat exchanger
10. Steam

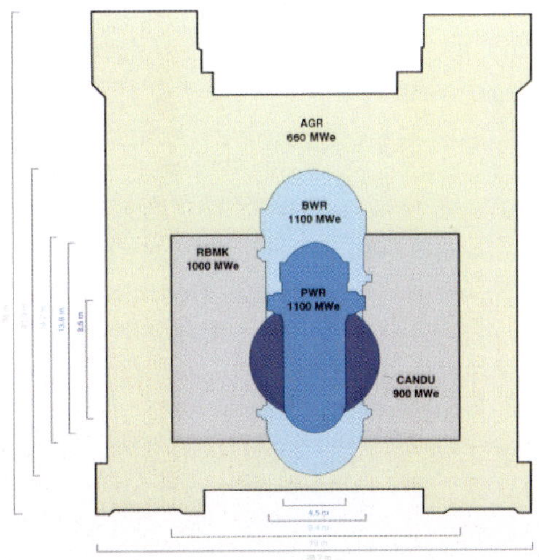

AGR reactor size compared to other technologies

The design of the AGR was such that the final steam conditions at the boiler stop valve were identical to that of conventional coal-fired power stations, thus the same design of turbo-generator plant could be used. The mean temperature of the hot coolant leaving the reactor core was designed to be 648 °C. In order to obtain these high temperatures, yet ensure useful graphite core life (graphite oxidises readily in CO_2 at high temperature) a re-entrant flow of coolant at the lower boiler outlet temperature of 278 °C is utilised to cool the graphite, ensuring that the graphite core temperatures do not vary too much from those seen in a Magnox station. The superheater outlet temperature and pressure were designed to be 2,485 psi (170 bar) and 543 °C.

The fuel is uranium dioxide pellets, enriched to 2.5-3.5%, in stainless steel tubes. The original design concept of the AGR was to use a beryllium based cladding. When this proved unsuitable due to brittle fracture, the enrichment level of the fuel was raised to allow for the higher neutron capture losses of stainless steel cladding. This significantly increased the cost of the power produced by an AGR. The carbon dioxide coolant circulates through the core, reaching 640 °C (1,184 °F) and a pressure of around 40 bar (580 psi), and then passes through boiler (steam generator) assemblies outside the core but still within the steel-lined, reinforced concrete pressure vessel. Control rods penetrate the graphite moderator and a secondary system involves injecting nitrogen into the coolant to hold the reactor temperature down. A tertiary shutdown system which operates by injecting boron balls into the reactor is included in case the reactor has to be depressurized with insufficient control rods lowered. This would mean that nitrogen pressure cannot be maintained.

The AGR was designed to have a high thermal efficiency (electricity generated/heat generated ratio) of about 41%, which is better than modern pressurized water reactors which have a typical thermal efficiency of 34%. This is due to the higher coolant outlet temperature of about 640 °C (1,184 °F) practical with gas cooling, compared to about 325 °C (617 °F) for PWRs. However the reactor core has to be larger for the same power output, and the fuel burnup ratio at discharge is lower so the fuel is used less efficiently, countering the thermal efficiency advantage.

Like the Magnox, CANDU and RBMK reactors, and in contrast to the light water reactors, AGRs are designed to be refuelled without being shut down first (see Online refuelling). This on-load refuelling was an important part of the economic case for choosing the AGR over other reactor types, and in 1965 allowed the Central Electricity Generating Board (CEGB) and the government to claim that the AGR would produce electricity cheaper than the best coal-fired power stations. However fuel assembly vibration problems arose during on-load refuelling at full power, so in 1988 full power refuelling was suspended until the mid-1990s, when further trials led to a fuel rod becoming stuck in a reactor core. Only refuelling at part load or when shut down is now undertaken at AGRs.

The pre-stressed concrete pressure vessel contains the reactor core and the boilers. To minimise the number of penetrations into the vessel (and thus reduce the number of possible breach sites) the boilers are of the once through design where all boiling and superheating is carried out within the boiler tubes. This necessitates the use of ultra pure water to minimise the buildup of salts in the evaporator and subsequent corrosion problems.

The AGR was intended to be a superior British alternative to American light water reactor designs. It was promoted as a development of the operationally (if not economically) successful Magnox design, and was chosen from a multitude of competing British alternatives - the helium cooled High Temperature Reactor, the Steam Generating Heavy Water Reactor and the Fast Breeder Reactor

- as well as the American light water pressurised and boiling water reactors (PWR and BWR) and Canadian CANDU designs. The CEGB conducted a detailed economic appraisal of the competing designs and concluded that the AGR proposed for Dungeness B would generate the cheapest electricity, cheaper than any of the rival designs and the best coal-fired stations.

Pebble-bed Reactor

Pebble Bed Reactors could allow nuclear plants to support the goal of reducing global climate change in an energy hungry world. They are small, modular, inherently safe, use a demonstrated nuclear technology and can be competitive with fossil fuels. Pebble bed reactors are helium cooled reactors that use small tennis ball size fuel balls consisting of only 9 grams of uranium per pebble to provide a low power density reactor. The low power density and large graphite core provide inherent safety features such that the peak temperature reached even under the complete loss of coolant accident without any active emergency core cooling system is significantly below the temperature that the fuel melts. This feature should enhance public confidence in this nuclear technology. With advanced modularity principles, it is expected that this type of design and assembly could lower the cost of new nuclear plants removing a major impediment to deployment.

A pebble bed reactor is graphically illustrated in Figure. The reactor core contains approximately 360,000 uranium fuelled pebbles about the size of tennis balls. Each pebble contains about 9 gm of low enriched uranium in 10,000–15,000 (depending on the design) tiny grains of sand-like microsphere coated particles each with its own a hard silicon carbide shell. These microspheres are embedded in a graphite matrix material as shown in figure. The unique feature of pebble bed reactors is the online refuelling capability in which the pebbles are recirculated with checks on integrity and consumption of uranium. This system allows new fuel to be inserted during operation and used fuel to be discharged and stored on site for the life of the plant. It is projected that each pebble will pass through the reactor 6–10 times before discharge in a three year period on average. With online refuelling, capability outages are determined by turbine generator maintenance, which is expected to require six year maintenance intervals.

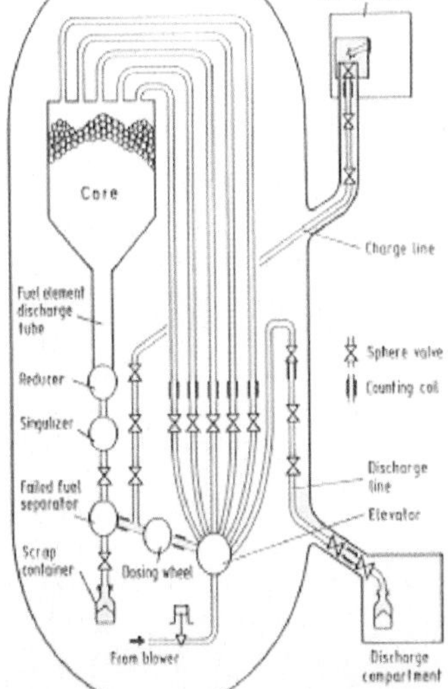

Schematic of a pebble bed reactor

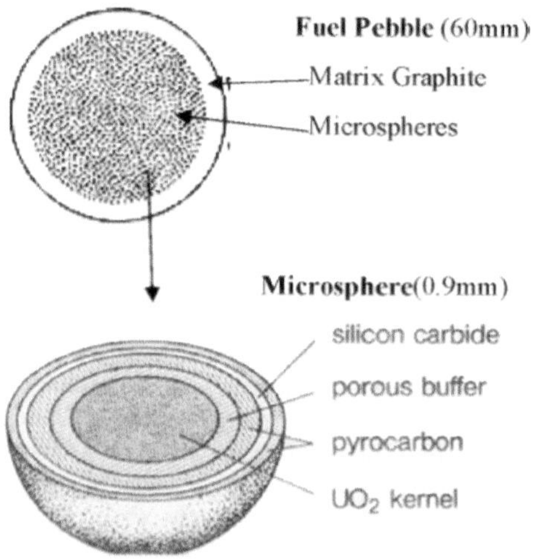

Fuel pebble and microsphere

Safety Features

Pebble-bed reactors have an advantage over conventional light-water reactors in operating at higher temperatures. A technical advantage is that some designs are throttled by temperature, not by control rods. The reactor can be simpler because it does not need to operate well at the varying neutron profiles caused by partially withdrawn control rods.

Pebble-bed reactors are also capable of using fuel pebbles made from different fuels in the same basic design of reactor (though perhaps not at the same time). Proponents claim that some kinds of pebble-bed reactors should be able to use thorium, plutonium and natural unenriched uranium, as well as the customary enriched uranium. There is a project in progress to develop pebbles and reactors that use MOX fuel, that mixes uranium with plutonium from either reprocessed fuel rods or decommissioned nuclear weapons.

In most stationary pebble-bed reactor designs, fuel replacement is continuous. Instead of shutting down for weeks to replace fuel rods, pebbles are placed in a bin-shaped reactor. A pebble is recycled from the bottom to the top about ten times over a few years, and tested each time it is removed. When it is expended, it is removed to the nuclear-waste area, and a new pebble inserted.

When the nuclear fuel increases in temperature, the rapid motion of the atoms in the fuel causes an effect known as Doppler broadening. The fuel then sees a wider range of relative neutron speeds. Uranium-238, which forms the bulk of the uranium in the reactor, is much more likely to absorb fast or epithermal neutrons at higher temperatures. This reduces the number of neutrons available to cause fission, and reduces the power of the reactor. Doppler broadening therefore creates a negative feedback: as fuel temperature increases, reactor power decreases. All reactors have reactivity feedback mechanisms, but the pebble-bed reactor is designed so that this effect is very strong. Also, it is automatic and does not depend on any kind of machinery or moving parts. If the rate of fission increases, temperature will increase and Doppler broadening will occur, decreasing the rate of fission. This creates passive cooling.

Because of this, and because the pebble-bed reactor is designed for higher temperatures, the reactor will passively reduce to a safe power-level in an accident scenario. This is the main passive safety feature of the pebble-bed reactor, and it distinguishes the pebble-bed design (as well as most other very-high-temperature reactors) from conventional light-water reactors, which require active safety controls.

The reactor is cooled by an inert, fireproof gas, so it cannot have a steam explosion as a light-water reactor can. The coolant has no phase transitions—it starts as a gas and remains a gas. Similarly, the moderator is solid carbon; it does not act as a coolant, move, or have phase transitions (i.e., between liquid and gas) as the light water in conventional reactors does.

A pebble-bed reactor thus can have all of its supporting machinery fail, and the reactor will not crack, melt, explode or spew hazardous wastes. It simply goes up to a designed "idle" temperature, and stays there. In that state, the reactor vessel radiates heat, but the vessel and fuel spheres remain intact and undamaged. The machinery can be repaired or the fuel can be removed. These safety features were tested (and filmed) with the German AVR reactor. All the control rods were removed, and the coolant flow was halted. Afterward, the fuel balls were sampled and examined for damage - there was none.

PBRs are intentionally operated above the 250 °C annealing temperature of graphite, so that Wigner energy is not accumulated. This solves a problem discovered in an infamous accident, the Windscale fire. One of the reactors at the Windscale site in England (not a PBR) caught fire because of the release of energy stored as crystalline dislocations (Wigner energy) in the graphite. The dislocations are caused by neutron passage through the graphite. Windscale had a program of regular annealing in place to release accumulated Wigner energy, but since the effect was not anticipated during the construction of the reactor, and since the reactor was cooled by ordinary air in an open cycle, the process could not be reliably controlled, and led to a fire. The second generation of UK gas-cooled reactors, the AGRs, also operate above the annealing temperature of graphite.

Berkeley professor Richard A. Muller has called pebble-bed reactors "in every way safer than the present nuclear reactors".

Containment

Most pebble-bed reactor designs contain many reinforcing levels of containment to prevent contact between the radioactive materials and the biosphere:

- Most reactor systems are enclosed in a containment building designed to resist aircraft crashes and earthquakes.
- The reactor itself is usually in a two-meter-thick-walled room with doors that can be closed, and cooling plenums that can be filled from any water source.
- The reactor vessel is usually sealed.
- Each pebble, within the vessel, is a 60 millimetres (2.4 in) hollow sphere of pyrolytic graphite.
- A wrapping of fireproof silicon carbide.
- Low density porous pyrolytic carbon, high density nonporous pyrolytic carbon.
- The fission fuel is in the form of metal oxides or carbides.

Pyrolytic graphite is the main structural material in these pebbles. It sublimates at 4000 °C, more than twice the design temperature of most reactors. It slows neutrons very effectively, is strong, inexpensive, and has a long history of use in reactors and other very high temperature applications. For example, pyrolytic graphite is also used, unreinforced, to construct missile reentry nose-cones and large solid rocket nozzles. Its strength and hardness come from anisotropic crystals of carbon.

Pyrolytic carbon can burn in air when the reaction is catalyzed by a hydroxyl radical (e.g., from water). Infamous examples include the accidents at Windscale and Chernobyl—both graphite-moderated reactors. However, all pebble-bed reactors are cooled by inert gases to prevent fire. All pebble designs also have at least one layer of silicon carbide that serves as a fire break as well as a seal.

Production of Fuel

All kernels are precipitated from a sol-gel, then washed, dried and calcined. U.S. kernels use uranium carbide, while German (AVR) kernels use uranium dioxide. German-produced fuel-pebbles release about three orders of magnitude (1000 times) less radioactive gas than the U.S. equivalents, due to these different construction methods.

Magnox

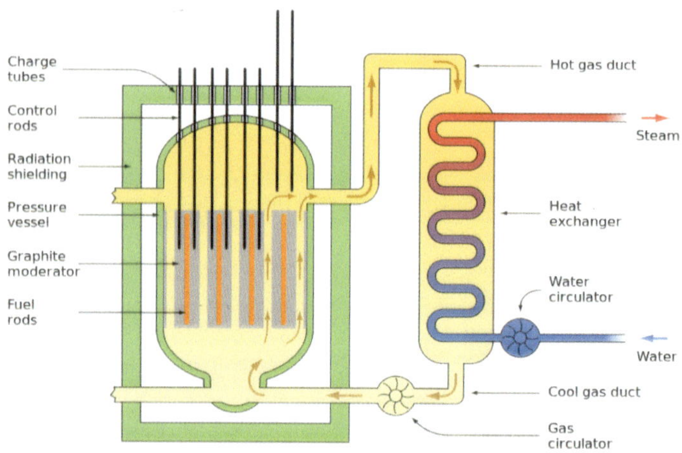

Schematic diagram of a Magnox nuclear reactor showing gas flow. The heat exchanger is outside the concrete radiation shielding. This represents an early Magnox design with a cylindrical, steel, pressure vessel.

Magnox is a type of nuclear power/production reactor that was designed to run on natural uranium with graphite as the moderator and carbon dioxide gas as the heat exchange coolant. It belongs to the wider class of gas cooled reactors. The name magnox comes from the magnesium-aluminium alloy used to clad the fuel rods inside the reactor. Like most other "Generation I nuclear reactors", the Magnox was designed with the dual purpose of producing electrical power and plutonium-239 for the nascent nuclear weapons program in Britain. The name refers specifically to the United Kingdom design but is sometimes used generically to refer to any similar reactor.

Like other plutonium-producing reactors, conserving neutrons is a key element of the design. In magnox, the neutrons are moderated in large blocks of graphite. The efficiency of graphite as a moderator allows the Magnox to run using natural uranium fuel, in contrast with the more common commercial light-water reactor which requires slightly enriched uranium. Graphite oxidizes readily in air, so the core is cooled with CO_2, which is then pumped into a heat exchanger to generate steam to drive conventional steam turbine equipment for power production. The core is open on one end, so fuel elements can be added or removed while the reactor is still running.

The "dual use" capability of the Magnox design led to the UK building up a large stockpile of fuel grade/"reactor grade" plutonium, with the aid of the B205 reprocessing facility. The low-to-interim burnup feature of the reactor design would become responsible for changes to US regulatory classifications after the US-UK "Reactor-grade" plutonium detonation test of the 1960s. Despite improving its electricity generating capabilities in later decades, marked by the transition to electric-power becoming the primary operational aim, magnox reactors were never capable of consistently generating high efficiency/high fuel "burn-ups" due to the handicap of its design and natural uranium heritage, compared with pressurized water reactors, the most widespread power-reactor design.

In total, only a few dozen reactors of this type were constructed, most of them in the UK from the 1950s to the 1970s, with very few exported to other countries. The first magnox reactor to come online was Calder Hall (at the Sellafield site) in 1956, frequently regarded as the "first commercial-scale electricity producing reactor in the world", while the last in Britain to shut down was Reactor 1 in Wylfa (on Ynys Môn) in 2015. North Korea remains the only operator to continue using Magnox style reactors as of 2016, at the Yongbyon Nuclear Scientific Research Center. The Magnox design was superseded by the Advanced Gas-cooled Reactor, which is similarly cooled but includes changes to improve its economic performance.

General Description

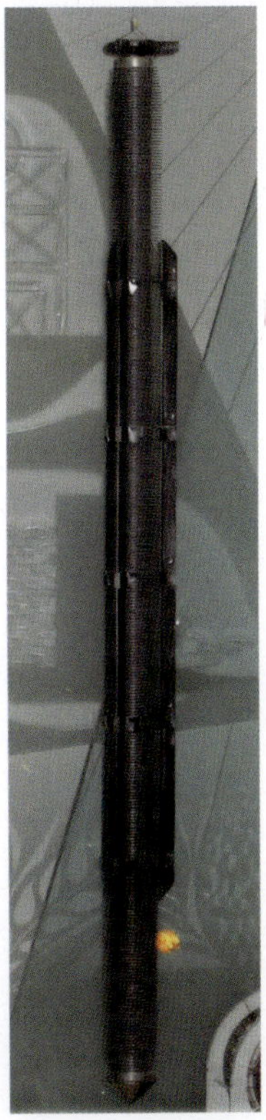

An early design Magnox fuel rod

Windscale

The UK's first full-scale nuclear reactor was the Windscale Pile in Sellafield. The pile was designed for the production of plutonium-239 which was bred in multi-week reactions taking place in natural uranium fuel. Under normal conditions, natural uranium is not sensitive enough to its own neutrons to maintain a chain reaction. To improve the fuel's sensitivity to neutrons, a neutron moderator is used, in this case highly purified graphite.

The reactors consisted of a huge cube of this material (the "pile") made up of many smaller blocks and drilled through horizontally to make a large number of *fuel channels*. Uranium fuel was placed in aluminium canisters and pushed into the channels in the front, pushing previous fuel canisters through the channel and out the back of the reactor where they fell into a pool of water. The system was designed to work at low temperatures and power levels and was air-cooled with the help of large fans.

Graphite is highly flammable and presents a serious safety risk. This was demonstrated on 10 October 1957 when Unit 1 of the now two-unit site caught fire. The reactor burned for three days, and massive contamination was only avoided due to the addition of filtering systems that had previously been derided as unnecessary "follies".

Magnox

As the UK nuclear establishment began to turn its attention to nuclear power, the need for more plutonium remained acute. This led to an effort to adapt the basic Windscale design to a power-producing version that would also produce plutonium. In order to be economically useful the plant would have to run at much higher power levels, and in order to efficiently convert that power to electricity, it would have to run at higher temperatures.

At these power levels, the fire risk is amplified and air cooling is no longer appropriate. In the case of the Magnox design, this led to the use of carbon dioxide (CO_2) as the coolant. There is no facility in the reactor to adjust the gas flow through the individual channels; this was set during construction in the diagrid, based on testing done on a mock up. Additional control over the reaction rate was provided by boron-steel control rods in vertical channels.

At higher temperatures, aluminium is no longer structurally sound, which led to the development of the magnox alloy fuel cladding. Unfortunately, magnox is increasingly reactive with increasing temperature, and the use of this material limited the operational temperatures to 360 °C (680 °F), much lower than desirable for efficient steam generation. This limit also meant that the reactors had to be very large in order to reach any given power level, which was further amplified by the use of gas for cooling, as the low thermal capacity of the fluid required very high flow rates.

The magnox fuel elements consisted of refined uranium enclosed in a loose-fitting magnox shell and then pressurized with helium. The outside of the shell was typically finned in order to improve heat exchange with the CO_2. Magnox alloy is reactive with water, which means it cannot be left in a cooling pond after extraction from the reactor for extended periods. In contrast to the Windscale layout, the Magnox design used vertical fuel channels. This required the fuel shells to lock together end-to-end to allow them to be pulled out of the channels from the top.

Like the Windscale designs, the Magnox reactors allowed access to the fuel channels and could be refuelled while operating. This was a key criterion for the design because its use of natural uranium leads to low burnup ratios and the requirement for frequent refuelling. For power use, the fuel canisters were left in the reactor as long as possible, while for plutonium production they were removed earlier. The complicated refuelling equipment proved to be less reliable than the reactor systems, and perhaps not advantageous overall.

The entire reactor assembly was placed in a large pressure vessel. Due to the size of the pile, only the reactor core itself was placed within the steel pressure assembly, which was then surrounded by a concrete confinement building (or "biological shield"). As there was no water in the core, and thus no possibility of a steam explosion, the building was able to tightly wrap the pressure vessel, which helped reduce construction costs. In order to keep the size of the confinement building down, the early Magnox designs placed the heat exchanger for the CO_2 gas outside the dome, connected through piping. Due to suspended particles of the reactor fuel and moderator in the gas, this causes the system to "shine" with gamma rays and neutrons.

The Magnox design was never truly finalised, and later units differ considerably from earlier ones. The cooling system was quickly moved inside an enlarged dome, while later units at Oldbury and Wylfa replaced the steel pressure vessels with prestressed concrete. Working pressure varies from 6.9 to 19.35 bar for the steel vessels, and 24.8 and 27 bar for the two concrete designs.

No British construction company at the time was large enough to build all the power stations, so various competing consortiums were involved, adding to the differences between the stations; for example, nearly every power station used a different design of Magnox fuel element.

For the initial start up of the reactor neutron sources were located within the core to provide sufficient neutrons to initiate the nuclear reaction. Other aspects of the design included the use of flux flattening bars to even out (to some extent) the neutron flux density across the core. If not used, the flux in the centre would be very high relative to the outer areas making control more difficult and leading to excessive central temperatures. Each fuel channel would have several elements stacked one upon another to form a *stringer*. This required the presence of a latching mechanism to allow the stack to be withdrawn and handled. This caused some problems as the Nimonic springs used contained cobalt, which became irradiated giving high gamma level when removed from the reactor. Additionally, thermocouples were attached to some elements and needed to be removed on fuel discharge from the reactor.

AGR

The "dual use" nature of the Magnox design leads to design compromises that limit its economic performance. As the Magnox design was being rolled out, work was already underway on the Advanced Gas-cooled Reactor (AGR) with the explicit intention of making the system more economical. Primary among the changes was the decision to run the reactor at much higher temperatures, about 650 °C (1,202 °F), which would greatly improve the efficiency when running the power-extracting steam turbines. This was too hot for the magnox alloy, and the AGR originally intended to use a new beryllium-based cladding, but this proved too brittle. This was replaced by a stainless steel cladding, but this absorbed enough neutrons to affect criticality, and in turn required the design to operate on slightly enriched uranium rather than the Magnox's natural uranium, driving up fuel costs. Ultimately the economics of the system proved little better than Magnox.

Technical Information

It differs from real:

Specification	Calder Hall	Wylfa	Oldbury
Thermal output (gross), MW	182	1875	835
Electrical output (gross), MW	46	590	280
Efficiency, %	23	33	34
Number of fuel channels	1696	6150	3320
Active core diameter	9,45 m	17,4 m	12,8 m
Active core height	6,4 m	9,2 m	8,5 m
Mean gas pressure	7 bar	26,2 bar	25,6 bar
Mean inlet gas temperature °C	140	247	245
Mean outlet gas temperature °C	336	414	410
Total gas flow	891 kg/s	10254 kg/s	4627 kg/s
Material	Natural uranium metal	Natural uranium metal	Natural uranium metal
Mass of uranium in tonnes	120	595	293
Pressure vessel internal diameter	11,28 m	29,3 m	23,5 m
Pressure vessel internal height	21,3 m	—	18,3 m
Gas circulators	4	4	4
Steam generators	4	1	4
Number of generators	2	2	1

Economics

The first Magnox reactors at Calder Hall were designed principally to produce plutonium for nuclear weapons. The production of plutonium from uranium by irradiation in a pile generates large quantities of heat which must be disposed of, and so generating steam from this heat, which could be used in a turbine to generate electricity, or as process heat in the nearby Windscale works, was seen as a kind of "free" by-product of an essential process.

The Calder Hall reactors had low efficiency by today's standards, only 18.8%.

The British government decided in 1957 that electricity generation by nuclear power would be promoted, and that there would be a building programme to achieve 5,000 to 6,000 MWe capacity by 1965, a quarter of UK's generating needs. Although Sir John Cockcroft had advised the government that electricity generated by nuclear power would be more expensive than that from coal, the government decided that nuclear power stations as alternatives to coal-fired power stations would be useful to reduce the bargaining power of the coal miners' unions, and so decided to go ahead. In 1960 a government white paper scaled back the building programme to 3,000 MWe, acknowledging that coal generation was 25% cheaper. A government statement to the House of Commons in 1963 stated that nuclear generation was more than twice as expensive as coal. The "plutonium credit" which assigned a value to the plutonium produced was used to improve the economic case, although the operators of the power stations were never paid this credit.

Once removed from the reactor the used fuel elements are stored in cooling ponds (with the exception of Wylfa which has dry stores in a carbon dioxide atmosphere) where the decay heat is transferred to the pond water, and then removed by the pond water circulation, cooling and filtration system. The fact that fuel elements can only be stored for a limited period in water before the Magnox cladding deteriorates, and must therefore inevitably be reprocessed, added to the costs of the Magnox programme.

A retrospective evaluation of costs, using a low 5% discount rate on capital, estimated Magnox electricity costs were nearly 50% higher than coal power stations would have provided.

Safety

The reactor buildings of Bradwell Magnox nuclear power station

The Magnox reactors were considered at the time to have a considerable degree of inherent safety because of their simple design, low power density, and gas coolant. Because of this they were not provided with secondary containment features. A safety design principle at the time was that of the "maximum credible accident", and the assumption was made that if the plant were designed to withstand that, then all other lesser but similar events would be encompassed. Loss of coolant accidents (at least those considered in the design) would not cause large-scale fuel failure as the Magnox cladding would retain the bulk of the radioactive material, assuming the reactor was rapidly shutdown (a SCRAM), because the decay heat could be removed by natural circulation of air. As the coolant is already a gas, explosive pressure buildup from boiling is not a risk, as happened in the catastrophic steam explosion at the Chernobyl accident. Failure of the reactor shutdown system to rapidly shut down the reactor, or failure of natural circulation, was not considered in the design. In 1967 Chapelcross experienced a fuel melt due to restricted gas flow in an individual channel and, although this was dealt with by the station crew without major incident, this event had not been designed or planned for, and the radioactivity released was greater than anticipated during the station design.

In the older steel pressure vessel design, boilers and gas ducting are outside the concrete biological shield. Consequently, this design emits a significant amount of direct gamma and neutron radiation, termed direct "shine", from the reactors. For example, the most exposed members of the public living near Dungeness Magnox reactor in 2002 received 0.56 mSv, over half the International Commission on Radiological Protection recommended maximum radiation dose limit for the public, from direct "shine" alone. The doses from the Oldbury and Wylfa reactors, which have concrete pressure vessels which encapsulate the complete gas circuit, are much lower.

Reactors Built

Sizewell A Magnox nuclear power station

In all, 11 power stations totalling 26 units were built in the United Kingdom where the design originated. In addition, one was exported to Tōkai in Japan and another to Latina in Italy. North Korea also developed their own Magnox reactors, based on the UK design which was made public at an Atoms for Peace conference.

The first Magnox power station, Calder Hall, was the world's first nuclear power station to generate electrical power on an industrial scale (a power station in Obninsk, Russia started supplying the grid in very small non-commercial quantities on 1 December 1954). First connection to the grid was on 27 August 1956, and the plant was officially opened by Queen Elizabeth II on 17 October 1956. When the station closed on 31 March 2003, the first reactor had been in use for nearly 47 years.

The first two stations (Calder Hall and Chapelcross) were originally owned by the UKAEA and primarily used in their early life to produce weapons-grade plutonium, with two fuel loads per year. From 1964 they were mainly used on commercial fuel cycles and in April 1995 the UK Government announced that all production of plutonium for weapons purposes had ceased.

The later and larger units were owned by CEGB and operated on commercial fuel cycles. However Hinkley Point A and two other stations were modified so that weapons-grade plutonium could be extracted for military purposes should the need arise.

Derating to Reduce Corrosion

In early operation it was found that there was significant oxidation of mild steel components by the high temperature carbon dioxide coolant, requiring a reduction in operating temperature and power output. For example, the Latina reactor was derated in 1969 by 24%, from 210 MWe to 160 MWe, by the reduction of operating temperature from 390 to 360 °C.

World's Last Operating Magnox Reactor

The Nuclear Decommissioning Authority (NDA) announced on December 30, 2015 that Wylfa Unit 1 - the world's last operating Magnox reactor - was closed. The unit had generated electricity for five years longer than originally planned. Two units at Wylfa were both scheduled to shut down at the end of 2012, but the NDA decided to shut down Unit 2 in April 2012 so that Unit 1 could continue operating in order to fully utilize existing stocks of fuel, which is no longer being manufactured.

The small 5 MWe experimental reactor, based on the Magnox design, at Yongbyon in North Korea, continues to operate.

Magnox Definitions

Magnox Alloy

Magnox is also the name of an alloy—mainly of magnesium with small amounts of aluminium and other metals—used in cladding unenriched uranium metal fuel with a non-oxidising covering to contain fission products. Magnox is short for Magnesium non-oxidising. This material has the advantage of a low neutron capture cross-section, but has two major disadvantages:

- It limits the maximum temperature, and hence the thermal efficiency, of the plant.
- It reacts with water, preventing long-term storage of spent fuel under water.

Magnox fuel incorporated cooling fins to provide maximum heat transfer despite low operating temperatures, making it expensive to produce. While the use of uranium metal rather than oxide made reprocessing more straightforward and therefore cheaper, the need to reprocess fuel a short time after removal from the reactor meant that the fission product hazard was severe. Expensive remote handling facilities were required to address this danger.

Magnox Plants

The term magnox may also loosely refer to:

- Three North Korean reactors, all based on the declassified blueprints of the Calder Hall Magnox reactors:
  - A small 5 MWe experimental reactor at Yongbyon, operated from 1986 to 1994, and restarted in 2003. Plutonium from this reactor's spent fuel has been used in the North Korea nuclear weapons program.
  - A 50 MWe reactor, also at Yongbyon, whose construction commenced in 1985 but was never finished in accord with the 1994 U.S.-North Korea Agreed Framework.
  - A 200 MWe reactor at Taechon, construction of which also halted in 1994.
- Nine UNGG power reactors built in France, all now shut down. These were carbon dioxide-cooled, graphite reactors with natural uranium metal fuel, very similar in design and purpose to the British Magnox reactors except that the fuel cladding was magnesium-zirconium alloy and that the bars were disposed horizontally (instead of vertically for Magnox).

Decommissioning

Chapelcross before the cooling towers were demolished in 2007

The Nuclear Decommissioning Authority (NDA) is responsible for the decommissioning of the UK Magnox power plants, at an estimated cost of £12.6 billion. There is currently debate about whether a 25 or 100 year decommissioning strategy should be adopted. After 80 years short-lifetime radioactive material in the defuelled core would have decayed to the point that human access to the reactor structure would be possible, easing dismantling work. A shorter decommissioning strategy would require a fully robotic core dismantling technique.

In addition the Sellafield site which, amongst other activities, reprocessed spent Magnox fuel, has an estimated decommissioning cost of £31.5 billion. Magnox fuel was produced at Springfields near Preston; estimated decommissioning cost is £371 million. The total cost of decommissioning Magnox activities is likely to exceed £20 billion, averaging about £2 billion per productive reactor site.

Calder Hall was opened in 1956 as the world's first commercial nuclear power station, and is a significant part of the UK's industrial heritage. The NDA is considering whether to preserve Calder Hall Reactor 1 as a museum site.

All the UK's Magnox Reactor Sites (apart from Calder Hall) are operated by Magnox Ltd, a Site Licence Company (SLC) of the NDA. Reactor Sites Management Company (RSMC) holds the contract to manage Magnox Ltd on behalf of the NDA. In 2007, RSMC was acquired by American nuclear fuel cycle service provider EnergySolutions from British Nuclear Fuels.

On 1 October 2008, Magnox Electric Ltd separated into two nuclear licensed companies, Magnox North Ltd and Magnox South Ltd.

Magnox North sites

- Chapelcross
- Hunterston A
- Oldbury
- Trawsfynydd
- Wylfa

Magnox South sites

- Berkeley
- Bradwell
- Dungeness A
- Hinkley Point A

In January 2011 Magnox North Ltd and Magnox South Ltd recombined as Magnox Ltd. Following procurement and management issues with the contract, Magnox Ltd will become a subsidiary of the NDA on September 2019.

Very-high-temperature Reactor

The Very High Temperature Nuclear Reactor (VHTR) is one of 6 technologies classified by the Generation IV International Forum as a promising reactor type likely to power our world in the coming decades. The fourth generation reactor designs are expected to be deployed between 2015-2025, and are identified as being economical, safe, efficient, and proliferation resistant. The fundamental physics governing the operation of the VHTR is identical to the current generation of fission reactors. The nuclei in a fissile fuel are split via neutron bombardment, converting nuclear binding energy into thermal energy (~200 million electron Volts per fission event) capable of being converted to useful work or process heat. The energy released by fission events is transferred out of the reactor core via a working fluid. The defining characteristic of a VHTR is the very high temperature of this working fluid, capable of running an efficient power cycle or being used as a high temperature input for a chemical transformation process (e.g. hydrogen production).

Benefits of the VHTR

The VHTR offers two advantages to modern day generation III reactor designs. The high temperature of the coolant exiting the reactor core enables high thermal efficiency for electricity generation, and can serve as process heat for hydrogen production.

Efficient Power Generation

Most nuclear power plants currently in operation harness the thermal energy released during fission events by boiling water to create high pressure, high temperature steam. This steam is used to drive a steam turbine in a Rankine cycle, whose efficiency is governed in part by the temperature of the steam at the turbine inlet. Typical outlet temperatures for pressurized light water (PWR) reactors are less than 400°C, leading to thermal efficiencies (ratio of electric power generated to thermal energy input) around 34 percent. Gas-cooled reactors have been built, in which gas (carbon dioxide or helium) is used as the coolant, enabling higher coolant outlet temperatures (up to about 850°C) for use in a Brayton cycle. The AVR reactor in Germany demonstrated Helium outlet temperatures above 900°C, but was plagued by contamination and only operated for four years. The VHTR is the next generation of high-temperature gas-cooled reactors (HTGRs). The VHTR uses helium as coolant, with design temperatures breaking 1000°C (eventually slated to reach

1500°C). Such high temperatures allow for higher thermal efficiencies (>50%), and therefore more electricity output per unit of nuclear fuel.

Efficient Hydrogen Production

The high coolant temperature also allows for efficient hydrogen production. In a carbon-constrained world, hydrogen presents an attractive option as an energy carrier. Hydrogen combustion does not produce CO_2 and Hydrogen is 3 times as energy dense as gasoline. However, current hydrogen production methods rely on fossil fuels. The majority of hydrogen in the U.S. is produced using the steam-methane reforming process, in which natural gas and steam react in the presence of a catalyst (usually Nickel) at high temperatures (800 to 1000°C) to form syngas (a mixture of carbon monoxide and hydrogen). Additional hydrogen is extracted from the syngas through a lower temperature water gas shift reaction. This process consumes 5% of the annual U.S. natural gas supply, and emits 74 million tons of CO_2. If we are to use hydrogen as an energy carrier in a carbon-constrained world, we must produce hydrogen in a clean and efficient manner.

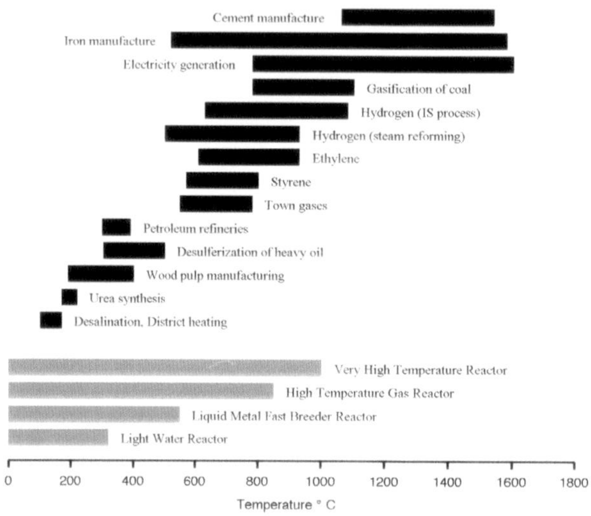

Table summarizing potential high-temperature applications of the VHTR.

Very High Temperature Reactors will supply process heat that can be used to generate the high temperatures required for clean hydrogen production. High temperature electrolysis and thermochemical water splitting offer two promising methods for nuclear powered hydrogen production. High temperature electrolysis works on the principal that the energy required to electrolyze water decreases with increasing temperature. The waste heat from a Brayton power cycle can be used as input to the electrolysis plant, enabling high overall hydrogen production efficiencies (45-50% thermal energy input to amount of hydrogen produced based on lower heating value of hydrogen). [10] Thermochemical water-splitting involves a series of thermally driven chemical reactions to separate water into hydrogen and oxygen. The Sulfur-Iodide cycle is a common thermochemical water-splitting cycle, in which heat is supplied as input to an endothermic reaction that dissociates sulfuric acid, eventually producing hydrogen and oxygen gas. Temperatures required for efficient S-I cycles are in excess of 950 °C and efficiencies have been reported above 50% (heat input to hydrogen output, again based on LHV). Thermochemical water-splitting efficiencies are very temperature dependent. As temperature increases from 900 to 1000 °C, hydrogen production efficiency increases from about 50 to 60 percent.

There are other hydrogen production methods that require high temperatures as input. Coal gasification presents an especially attractive option for hydrogen production in the U.S. given the abundance (484 billion tons of coal in reserves) of coal and current mining infrastructure. Coal gasification requires a coal temperatures between 800 and 1800 °C, which can in part be reached using the high coolant temperatures from a VHTR. [Clean coal gasification will require advanced carbon capture systems.

In addition to hydrogen production, high temperature coolant can be used for other useful applications including desalinating water, municipal heating, iron and glass manufacturing, urea synthesis, cement manufacturing, and much more.

Design Traits of the VHTR

The VHTR represents the evolution of traditional gas-cooled reactors. The design of the VHTR is therefore similar to current generation III+ gas-cooled reactors, which fall into two general categories: prismatic block reactors, and pebble bed reactors. The two designs refer primarily to the fuel arrangement. Prismatic block reactors involve a fuel core surrounded by a hexagonal graphite reflector, while pebble bed reactor cores use mobile fuel spheres about 6 cm in diameter that cycle throughout the core. The reference design for the prismatic block VHTR is the GT-MHR (Gas-Turbine Modular Helium Reactor) by General Atomics. Much of the following discussion is based on the GT-MHR reactor. The VHTR will use a once-through fuel cycle with a thermal neutron spectrum meaning U-235 nuclei will be split, emitting neutrons which will be moderated (or thermalized) to a lower energy level (nine orders of magnitude less than the initial energy released through fission), and then permanently stored in a nuclear repository of fuel rods, surrounded by a hexagonal graphite casing, which serves as a neutron reflector. The core will be approximately 26 feet tall and contain hundreds of fuel blocks, stacked on top of one another with a coolant jacket running around the outside of the core. The fuel in each of the blocks will consist of graphite coated fuel particles (totaling about 10 billion pebbles for a full load) suspended in a graphite matrix. These particles act as their own mini pressure vessels, isolating fission products both during and after operation, therefore minimizing radioactive release. The particles are also difficult to reprocess. A single fuel particle will contain a sphere of nuclear fuel only 650-850 microns in diameter. The 10 billion fuel pebbles will be difficult to separate from their graphite/SiC casings and fuel block assembly, thereby minimizing proliferation worries. The fuel column will be built inside of a ceramic core. The ceramic structure will be able to withstand higher temperatures than the casing around the control rod sheaths, supposedly providing inherent safety in an accident scenario.

The coolant will be helium, which is attractive due to its negligible neutron absorption cross-section (and therefore ability to remain non-radioactive), and its single phase during operation. The helium will directly power a Brayton cycle for power generation applications and will use an intermediate heat exchanger (IHX) for hydrogen production applications. The IHX simplifies the hydrogen production plant, since it will not have to be built to nuclear-standards, and provides a thermal buffer between the direct helium coolant line and the sensitive chemical processes in the hydrogen production process.

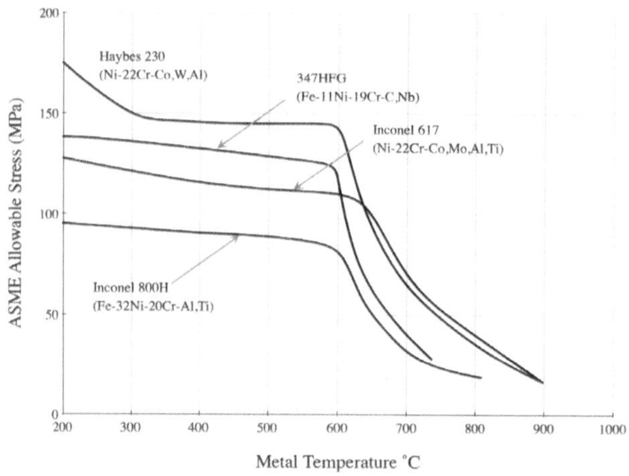

The American Society of Mechanical Engineers (ASME) provides maximum stress levels for different types of steels at different temperatures for use in pressure vessels. Note that the VHTR operates well beyond the 600°C point. After reference.

Technical Challenges

Many of the challenges of the VHTR have been addressed in prior plants. The Fort Saint Vrain helium cooled gas reactor operated in Colorado for 12 years, providing valuable insight into the challenges of a helium-cooled reactor. The Thorium High Temperature Reactor in Germany operated for 4 years, and Japan's High Temperature Engineering Test Reactor demonstrated 850°C outlet temperatures. The primary challenges of the VHTR are enumerated below:

- The high operating temperatures in a VHTR cause helium impurities to be aggravated. In order to avoid neutron poisoning, and accelerated core corrosion due to helium impurities, a helium monitoring and purification system will be required. The circulation system for the helium must include robust seals for the turbomachinery bearings and for the helium pathways. The Fort Saint Vrain reactor experienced problems with helium exfiltration. All turbomachinery will require magnetic bearings to prevent lubricant ingress. In addition, the turbine blades will have to be cooled, likely decreasing efficiency. Ceramic materials will likely need to be developed for the turbine. In short, the helium sealing system will be complex and expensive.

- Current fuel pebble design consists of a SiC coating over the graphite moderator that covers the fuel core. The maximum design temperatures for these SiC coatings are about 1600°C, which is comparable to the range of operating temperatures in the VHTR. ZrC is being researched as a possible alternative to SiC, which would enable design temperatures of 1800°C, but requires increased manufacturing and research costs.

- The operating conditions of the VHTR are beyond the temperature and irradiation conditions of current light water reactors. LWR pressure vessels are designed for 300°C. Higher temperatures will increase the rate of chemical reactions and atomic diffusion events on the surface of the metals in the VHTR, advancing corrosion and affecting material properties. The pressure vessel needs to be constructed from a metallic structure that has high creep strength and corrosion resistance. Figure depicts the effects of increasing temperature on maximum allowable stress as prescribed by the ASME pressure vessel code. Most steels experience plastic deformation at low stresses above about 600°C. The current reference design consists of 9% Cr steel pressure vessels. Tungsten, molybdenum and superplastic ceramics are being researched as structural members for the reactor. Carbon-fiber composites are being researched for control rod sheaths. In short, super alloy development will be necessary for the development of a VHTR.

As we look to the future of nuclear reactor design, we seek a technology that can both power our cities via electricity generation, but also provide an energy source for transportation fuels. The Very High Temperature Reactor marks the next evolutionary step in reactor design, capable of powering our cities with greater than 50% thermal efficiency, and producing clean-burning hydrogen via carbon-free processes. It is worth noting here that the majority of documentation on the VHTR is published by generation IV supporters, likely vying for funds for reactor development. The design of the VHTR is largely based on current gas-cooled reactor designs, but materials research and hydrogen production research need to be conducted before large-scale adoption of the VHTR.

Molten Salt Reactor

Molten Salt Reactors (MSRs) are nuclear reactors that use a fluid fuel in the form of very hot fluoride or chloride salt instead of the solid fuel used in most reactors. Since the fuel salt is liquid, it can be both the fuel (producing the heat) and the coolant (transporting the heat to the power plant). There are many different types of MSRs, but the most talked about one is definitely the Liquid Fluoride Thorium Reactor (LFTR). This MSR has Thorium and Uranium dissolved in a fluoride salt and can get planet-scale amounts of energy out of our natural resources of Thorium minerals, much like a fast breeder can get large amounts of energy out of our Uranium minerals. There are also fast breeder fluoride MSRs that don't use Th at all. And there are chloride salt based fast MSRs that are usually studied as nuclear waste-burners due to their extraordinary amount of very fast neutrons.

Benefits of Molten Salt Reactors

The benefits of MSRs are plentiful, hence their resilience as an interesting topic throughout reactor history. We break them down by topic here.

Sustainability

Sustainability is a measure of how efficiently a system can use natural resources. Most traditional reactors can only burn about 1% of the uranium on Earth. Many advanced reactors, including MSRs, can do much better. Here's why MSRs are good in this regard.

- Online fission product removal — Since the fuel is liquid, it can be processed during operation. This means that when atoms split into the smaller atoms (fission products), those small atoms can be collected and pulled out of the core very quickly. This prevents those atoms from absorbing neutrons that would otherwise continue the chain reaction. This allows very high fuel efficiency in MSRs.

- Good utilization of Thorium — As mentioned above, the MSR chemical plant can continuously remove fission products and other actinides during operation. This means that when Thorium absorbs a neutron and becomes Pa-233, the Protactinium can be removed from the core and allowed to decay to U-233 in peace, without any risk of causing parasitic neutron losses. While this is not the only way to burn Thorium, it is perhaps the most elegant.
- No neutron losses in structure — Since there is no structure like cladding, fuel ducts, grid spacers, etc., there are no neutron losses in these. This helps fuel efficiency and therefore sustainability.

Economics

Though economics are not truly known until a system is in commercial operation, there are reasons to think MSRs would have favorable low price.

- Online refueling — Where normal reactors have to shut down to move fuel around or put new fuel in, MSRs can do all this while at full power. You just dump in a new chunk of fuel into the vat (carefully, of course). This allows high capacity factors, improving economics. The reactors may still have to shut down to do maintenance, but they likely will have better uptimes.
- No fuel fabrication — Any commercial fuel fabricator will tell you that it is expensive to build fuel assemblies, fuel pellets, cladding tubes, core support structures, flow orifices, etc. MSRs are basically just vats of fuel, so they are much simpler and cheaper in this regard.
- High temperatures possible — Molten salts were first looked at for their ability to go to very high temperatures. At high temperatures, power cycles convert heat to electricity with much less loss, giving you more money for a given amount of heat. Additionally, many industrial processes require high-grade heat, and these reactors could be used to that while producing electricity. Best of all, MSRs can do high temperature without a pressurized coolant (as required in gas-cooled reactors).
- Smaller containment — Since the system pressure is low and the heat capacity is high, the containments can be smaller and thinner.

Safety

The most important aspect of a nuclear reactor is safety. Here's the good news for MSRs:

- Very low excess reactivity — Since they can be continually refueled, there is no need to load extra fissile material to allow the reactor to operate for a long time. This means that it is difficult to have something happen (like an earthquake) that could cause a shift in geometry that inserts reactivity and causes a power spike.
- Negative temperature coefficient of reactivity — In general, if the fuel heats up, it expands and becomes less reactive, keeping things stable. Note that this is not always true in graphite-moderated MSRs.
- Low pressure — Since the fuel and coolant are at atmospheric pressure, a leak in a tube

doesn't automatically result in the expulsion of a bunch of fuel and coolant. This is a major safety advantage that enables passive decay heat removal (preventing things like what happened at Fukushima). The salts generally have extremely high heat capacity as well, so they can absorb a lot of heat themselves. On the other hand, their thermal conductivity is about 60x worse than liquid metal sodium.

- No chemical reactivity with air or water — The fuel salt is generally not violently reactive with the environment. So where LWRs have hydrogen explosions and SFRs have sodium fires, MSRs do well. Of course, MSR leaks are still serious because it's not just coolant it's extremely radioactive fuel.
- Drain tank failure mechanism — If something goes wrong in a MSR and the temperature starts going up, a freeze plug can melt, pouring the entire core into subcritical drain tanks that are intimately linked to an ultimate heat sink, keeping them cool. This is an interesting accident mitigation feature that is possible only in fluid fuel reactors.

Problems with Molten Salt Reactors

All those wonderful benefits can't possibly come without a slew of problems. Lots of people promote these reactors without acknowledging the issues, but not us! A reactor concept has to stand on its two feet even in the face of disadvantages (and we think the MSR can do this). Let's go through them.

Mobile Fission Products

The primary concern with MSRs is that the radioactive fission products can get everywhere. They are not in fuel pins surrounded by cladding, but are just in a big, sealed vat. You can put a double-layer containment around it, sure, but it is still challenging to keep them all accounted for. Where some of these fission products and actinides are radioactive, others have chemical effects that can eat away at the containment. The implications of this are many.

- Material Degredation — with half the periodic table of the elements dissolved in salt and in contact with the containment vessel, there are lots of corrosion and related concerns. Noble metals will naturally plate out on cold metal surfaces. In a power reactor, a heat exchanger will be the coldest metal around, and so the heat transfer surfaces will need periodic replacement. At MSRE, Tellurium caused cracking of the Hastelloy-N material. This was mitigated with chemistry, but similar problems may show up in long-lived power reactors.
- Tritium production If lithium is used in the salt, tritium will be produced, which is radioactive and extremely mobile (since it is small, it can go through metal like a hot knife through butter). ORNL used a special sodium fluoroborate intermediate salt to capture most of it, but a large amount still escaped to the environment.
- Remote maintenance The chemical plants will need periodic maintenance, but all of the equipment will be highly radioactive. Expensive remote maintenance will be required. If graphite moderator is used, its replacement will also be remote and expensive.
- Complex chemical plant — Some of the fission product removal is simple, such as the gas

sparging to remove Xe and Kr, and noble metal plateout. But to do the more serious fission product (or actinide) separation, complex processes are required, such as the liquid Bismuth reductive process, volatilization, or electoplating. These have been studied in detail, but are complex enough to be a disadvantage. Don't make us post a process flow diagram. (You can find one from the MSBR on page 8 of ORNL-TM-6413.)

Proliferation

The main political barrier to MSRs is their perceived bomb-factory capabilities. If you talk to non-proliferation people, they will tell you that as soon as the (solid) fuel pins are cut open, a technology is considered proliferative. The problem with MSRs, then, is that the fuel is already completely cut open and melted. You're halfway to a bomb already, they think. Here's what they are worried about.

- Protactinium-233 decays to pure, weapons-grade U-233 — Many Thorium-cycle MSRs have to capture Pa as it is produced, removing it from the system while it decays to U-233 and then reinserting it into the reactor. They have to do this because otherwise the Pa-233 absorbs too many neutrons to maintain a breeding cycle. The problem here is that that ex-core U-233 is basically pure weapons-grade U-233 which could be used to make a bomb. It usually comes with Zr, but separating Pa from Zr is simple. Not many common reactors require such a proliferative step in their fuel cycle. Many MSR concepts do not do this, but LFTRs require it. Therefore, the owner of a LFTR could be producing bombs on the side. Many of the ideas for mitigating this problem (such as U-232 contamination and denaturing) only help against diversion by a nefarious third party. The owners of the plant could side-step these kinds of fixes easily, and that's really what matters.

- Inventory tracking is difficult -- Because a lot of materials plate out in the reactor and in the chemical plant, it is difficult to keep exact track of all of your actinides. The IAEA puts safeguards in reactors to make sure that all the actinides are accounted for (to verify that no one's making bombs on the side) but it will be difficult for the IAEA to distinguish plate-out losses from actual proliferative losses.

Other Minor Issues

There are a few other concerns, but these probably have practical solutions

- Unknown waste form — It's not clear what nuclear waste from MSRs will look like. The salt itself is not contained enough to be put in a repository so someone will have to come up with a stable waste form.

- Electrical heaters are required to stay liquid — in a prolonged power outage, the colder parts of the heat transfer loop might solidify. This could cause temperatures to rise over in the core (which will of course still be self-heated liquid).

Exotics in the Salt

For MSRs to breed in a thermal spectrum, the lithium in the salt must be enriched to very pure Li-7. Li-6 is a strong neutron poison and becomes tritium, the pesky mobile radiation source. Also,

Beryllium in Flibe is a controlled substance. It has some weapons applications and is a very dangerous material in terms of biological inhalation risks. In chloride salts, you must enrich to have pure Cl-37. Otherwise, Cl-35 has a strong (n,proton) threshold reaction that poisons the reactor. Also, the activation product Cl-36 is a long-lived, water soluble, hard beta emitter that complicates waste disposal. These enrichment needs increase the cost and complexity of MSR fuel cycles. Maybe we can find a more ideal salt someday.

Salt-cooled Reactors

The step-sister reactor of the MSR is the fluoride cooled high temperature reactor (FHR). These reactors have solid fuel in the form of TRISO particles (like the high temperature gas-cooled reactor), but rather than pumping high-pressure helium coolant through, they pump low-pressure liquid salt (usually Flibe) through. The idea here is to get the high-temperature and high-burnup benefits of the gas reactors without the risk of a depressurization accident. Salt coolant has incredible heat capacity (but low thermal conductivity), so it does a good job of handling accidents. These reactors are currently being studied by the DOE and many others (including the Chinese).

Ongoing Molten Salt Reactor Work

There is ongoing work in MSRs around the world. The Europeans are working on MOSART, the Japanese have FUJI, the Americans are focused on the FHR. In 2011, the Chinese Academy of Science started a project in Shanghai to put substantial resources into MSR technology. There are 300-500 people on the project, with a budget in the hundreds of millions. Most of the work is on FHRs, but there is definitely MSR work as well. They are manufacturing Hastelloy-N, separating lithium, building salt loops, and building advanced modeling capabilities. That project is most likely the most exciting in MSRs since 1969. They plan on having both an experimental 2MW FHR and an experimental 2MW MSR operational by 2020.

Several start-up companies have emerged to commercialize molten salt reactors. Hailing from MIT, Leslie Dewan and Mark Massie started Transatomic to build a MSR fueled largely by processed nuclear waste from traditional reactors. In Canada, Terrestrial Energy is working on small modular MSRs, featuring small reactor vessels that can be swapped out of a plant every 7 years or so. ThorCon is using its extensive experience in the ship-building industry to develop MSRs that need no new technology and can be built in the very near term. In Alabama, Kirk Sorensen himself runs a company called Flibe Energy targeting the development of small modular Thorium-MSRs. These companies represent the modern revival of interest in MSRs and we wish them all the best!

Liquid Fluoride Thorium Reactor

The modern concept of the Liquid-Fluoride Thorium Reactor (LFTR) uses uranium and thorium dissolved in fluoride salts of lithium and beryllium. These salts are chemically stable, impervious to radiation damage, and non-corrosive to the vessels that contain them. Because of their ability to tolerate heavy radiation, excellent temperature properties, minimal fuel loading requirements (i.e., easy of continual refueling) and other inherent factors, LFTR cores can be made much smaller than a typical light water reactor (LWR). In fact, liquid salt reactors, and LFTRs specifically, are listed as an unfunded part of the U.S. Department of Energy's Generation-4 Nuclear Solution Plan.

The Advantages

Some of the many advantages of the LFTR system over other nuclear reactor designs are outlined below. While LWRs can produce U233 from thorium, they will not provide the various advantages outlined below, because of their use of thorium in solid form. It is the unique combination of the thorium cycle and the liquid fluoride reactor that grants all of the following advantages only from the LFTR system.

Safety--LFTRs are designed to take advantage of the physics of the thorium cycle for optimum safety. The fluid in the core is not pressurized, thus eliminating the driving force of radiation release in conventional approaches. The LFTR reactor cannot melt down because of a runaway reaction or other nuclear reactivity accidents (such as at Chernobyl), because any increase in the reactor's operating temperature results in a reduction of reactor power, thus stabilizing the reactor without the need for human intervention. Further, the reactor is designed with a salt plug drain in the bottom of the core vessel. If the fluid gets too hot or for any other reason including power failures, the plug naturally melts, and the fluid dumps into a passively cooled containment vessel where decay heat is removed. This feature prevents any Three Mile Island-type accidents or radiation releases due to accident or sabotage and provides a convenient means to shut down and restart the system quickly and easily.

Proliferation Resistance--For all practical purposes, U233 is worthless as a nuclear weapons material, and indeed no nation has attempted to weaponize U233 because of the abundance of difficulties. U233 is considered an unsuitable choice for nuclear weapons material because whenever U233 is generated, uranium-232 (U232) contamination inevitably occurs. U232 rapidly decays into other elements, including thallium-208, a hard-gamma-ray emitter whose signature is easily detectable. The hard gamma rays from thallium-208 cause ionization of materials destroying the explosives and electronics of a nuclear weapon, and heavy lead shielding is required to protect personnel assembling the warhead. It is possible to generate U233 with little U232 contamination using specialized reactors (such as at the Hanford Site), but not with an LFTR. Any attempt to increase production of U233 in an LFTR reactor will generate U232 contamination and any attempt to steal quantities of U233 results in the reactor shutting down.

Energy Production--Because nearly all of the thorium is used up in an LFTR (versus only about 0.7% of uranium mined for an LWR), the reactor achieves high energy production per metric ton of fuel ore, on the order of 300 times the output of a typical uranium LWR. The LFTR allows much higher operating temperatures than does a typical LWR therefore a higher thermodynamic efficiency. The turbine system believed best suited for its operation is a triple-reheat closed-cycle helium turbine system, which should convert 50% of the reactor heat into electricity compared to today's steam cycle (~25% to 33%). This efficiency gain translates to about 4.11 million barrels of crude oil equivalent per year more than that generated by a steam system. Capital costs are lower due to smaller reactor & turbo-machinery size, low reactor pressures and minimal redundant safety systems. The greater energy production capability of LFTRs means we estimate the cost for electricity from a LFTR plant could be 25% to over 50% less than that from a LWR.

Waste--In theory, LFTRs would produce far less waste along their entire process chain, from ore extraction to nuclear waste storage, than LWRs. A LFTR power plant would generate 4,000 times less mining waste (solids and liquids of similar character to those in uranium mining) and would generate 1,000 to 10,000 times less nuclear waste than an LWR. Additionally, because LFTR burns all of its nuclear fuel, the majority of the waste products (83%) are safe within 10 years, and the remaining waste products (17%) need to be stored in geological isolation for only about 300 years (compared to 10,000 years or more for LWR waste). Additionally, the LFTR can be used to "burn down" waste from an LWR (nearly the entirety of the United States' nuclear waste stockpile) into the standard waste products of an LFTR, so long-term storage of nuclear waste would no longer be needed.

Supply--Thorium is abundant in the Earth's crust. It is the 36th most plentiful element in the crust--four times as common as uranium and 5,000 times as plentiful as gold. According to the U.S. Geological Survey's 2006 Mineral Yearbook, the United States is estimated to have 300,000 tons of thorium reserves (about 20% of the world's supply), more than half of which is easily extractable. Considering only the readily accessible portion, this national resource translates to nearly 1 trillion barrels of crude oil equivalent--five times the entire oil reserves of Saudi Arabia. In addition to the naturally occurring reserves, the United States currently has 3,200 metric tons of processed thorium nitrate buried in the Nevada desert. That supply is roughly equivalent to 21 billion barrels of crude oil equivalent when used in an LFTR with only minimal processing effort.

Secondary Products--Because an LFTR is so energy dense, the electricity and excess heat from the reactor can be used to fuel other industries beyond electricity production, including economical desalinization of water, cracking of hydrogen from water or hydrocarbons, generation of ammonia for fertilizer and fuel cells, and extraction of hydrocarbons from oil shale and tar sands. Additionally, the nuclear waste products from the LFTR include stable rhodium and ruthenium, rare elements needed in modern electronics; technetium-99, which offers great promise as a catalyst similar to platinum; iodine-131 and cesium-137 for medical applications; strontium-90 for radioisotope power; and xenon, used in commercial products and industrial processes.

The Risks

While LFTRs offer much promise, several economic and engineering issues need to be addressed before this technology can become a reality.

Thorium as a Fuel--Thorium has never actually been continually processed for fuel in a fully operational liquid fluoride reactor. The MSRE used U233 as a fuel, but the U233 was generated in another reactor. A follow-on reactor design was planned to do the full-system tests, which the MSRE was too cost-constrained to perform, but it was never funded. A prototype reactor based on the ORNL design work would need to be built and the continuous thorium cycle processing validated as the fuel source in an operational LFTR.

Turbine System--The gas turbo-machinery is similar engineering to the well-developed open-cycle turbine (e.g., jet aircraft engine). However, this kind of closed-cycle electric generation system has never been built. A new triple-reheat closed-cycle Brayton system would need to be built and tested along with the LFTR. However, this is a minimal engineering risk in obtaining the overall efficiency of the electricity generation system. If the close cycle turbine system proves not to be economically viable, a steam system can be used.

Cost of Thorium--The price of thorium ore is difficult to quantify. On one hand, some will argue that it is expensive, citing the lack of demand and the consequently limited market supply. On the other hand, the case can be made that thorium is nearly worthless in light of the U.S. government's decision to essentially "throw away" 3,200 metric tons of processed thorium by burying it in the Nevada desert. We cannot predict how the price of thorium would be affected if the world's thorium reserves were exploited for use in LFTRs. However, thorium does not incur a cost of enrichment as uranium does, mostly due to the fact that natural thorium occurs only in one isotope. We believe that if the world's thorium supplies were exploited for energy, its price would drop to be comparable to--or even lower than--current uranium ore prices.

Cost of Thorium Reactors--Even though a full-scale LFTR has never been built, we expect the lifecycle cost of thorium reactors could be at least 30% to 50% less than equivalent-power uranium-based LWRs. Nevertheless, the engineering, fabrication and licensing of any energy-dense endeavourer is never certain and subject to many outside factors. Because of the various advantages afforded by the LFTR technology, we expect there will be a reduced regulatory burden, which would lessen costs and accelerate startups. For full-scale construction of LFTRs, factory-built modular construction can be used to provide scalable reactors from 100-kilowatt to multi-gigawatt production. This flexibility in site location eliminates the largest risk facing new U.S. commercial power plants today. Further, LFTRs have operational cost advantages over both types of reactors currently licensed. Unlike pressurized water reactors, LFTRs will not have to be shut down for extensive periods for refueling. Unlike boiling water reactors, LFTRs do not radioactively contaminate the turbines used for electrical generation, which should translate into significantly reduced operational and maintenance costs for this portion of the power plant and reduced amounts of low-level waste for end-of-life disposal.

Molten-salt Reactor Experiment

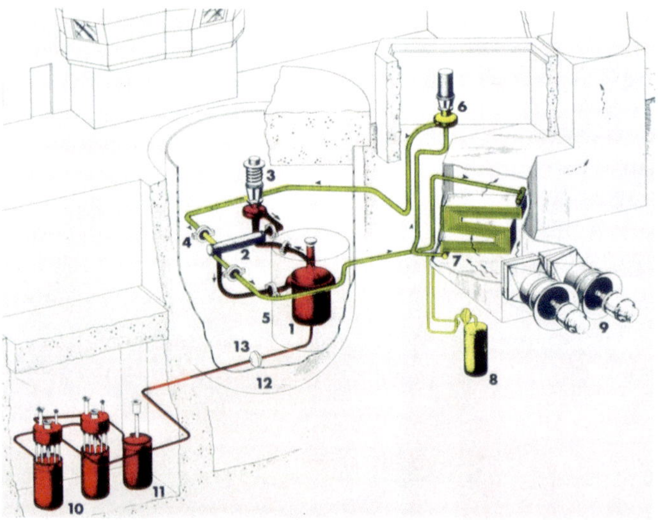

MSRE plant diagram: (1) Reactor vessel, (2) Heat exchanger, (3) Fuel pump, (4) Freeze flange, (5) Thermal shield, (6) Coolant pump, (7) Radiator, (8) Coolant drain tank, (9) Fans, (10) Fuel drain tanks, (11) Flush tank, (12) Containment vessel, (13) Freeze valve. Also note Control area in upper left and Chimney upper right.

The Molten-Salt Reactor Experiment (MSRE) was an experimental molten salt reactor at the Oak Ridge National Laboratory (ORNL) researching this technology through the 1960s; constructed by 1964, it went critical in 1965 and was operated until 1969.

The MSRE was a 7.4 MW_{th} test reactor simulating the neutronic "kernel" of a type of inherently safer epithermal thorium breeder reactor called the liquid fluoride thorium reactor. It primarily used two fuels: first uranium-235 and later uranium-233. The latter $^{233}UF_4$ was the result of breeding from thorium in other reactors. Since this was an engineering test, the large, expensive breeding blanket of thorium salt was omitted in favor of neutron measurements.

In the MSRE, the heat from the reactor core was shed via a cooling system using air blown over radiators. It is thought similar reactors could power high-efficiency heat engines such as closed-cycle gas turbines.

The MSRE's piping, core vat and structural components were made from Hastelloy-N and its moderator was a pyrolytic graphite core. The fuel for the MSRE was $LiF\text{-}BeF_2\text{-}ZrF_4\text{-}UF_4$ (65-29-5-1), the graphite core moderated it, and its secondary coolant was FLiBe ($2LiF\text{-}BeF_2$), it operated as hot as 650 °C and operated for the equivalent of about 1.5 years of full power operation.

The result promised to be a simple, reliable reactor. The purpose of the Molten-Salt Reactor Experiment was to demonstrate that some key features of the proposed molten-salt power reactors could be embodied in a practical reactor that could be operated safely and reliably and be maintained without excessive difficulty. For simplicity, it was to be a fairly small, one-fluid (i.e. non-breeding) reactor operating at 10 MW_{th} or less, with heat rejection to the air via a secondary (fuel-free) salt.

Reactor Description

Molten salt reactor

Core

Graphite MSRE core

The pyrolytic graphite core, grade CGB, also served as the moderator.

Before the MSRE development began, tests had shown that salt would not permeate graphite in which the pores were on the order of a micrometer. Graphite with the desired pore structure was available only in small, experimentally prepared pieces, however, and when a manufacturer set out to produce a new grade (CGB) to meet the MSRE requirements, difficulties were encountered.

Fuel/Primary Coolant

The fuel was ^{7}LiF-BeF_2-ZrF_4-UF_4 (65-29.1-5-0.9 mole %).

The first fuel was 33% ^{235}U, later a smaller amount of $^{233}UF_4$ was used.

By 1960 a better understanding of fluoride salt based molten-salt reactors had emerged due to earlier molten salt reactor research for the Aircraft Reactor Experiment.

Fluoride salts are strongly ionic, and when melted, are stable at high temperatures, low pressures, and high radiation fluxes. Low pressure stability permits less robust reactor vessels and increases reliability. The high reactivity of fluorine traps most fission reaction by products.

It appeared that the fluid salt would permit on-site chemical separation of the fuel and wastes.

The fuel system was located in sealed cells, laid out for maintenance with long-handled tools through openings in the top shielding. A tank of LiF-BeF_2 salt was used to flush the fuel circulating system before and after maintenance. In a cell adjacent to the reactor was a simple facility for bubbling gas through the fuel or flush salt: H_2-hydrogen fluoride to remove oxide, fluorine to remove uranium as uranium hexafluoride. Haubenreich and Engel, Robertson, and Lindauer provide more detailed descriptions of the reactor and processing plant.

Secondary Coolant

Molten FLiBe

The secondary salt was LiF-BeF_2 (66–34 mole %).

Pump

The bowl of the fuel pump was the surge space for the circulating loop, and here about 50 gal/min of fuel was sprayed into the gas space to allow xenon and krypton to escape from the salt.

Removing the most significant neutron poison xenon-135 made the reactor safer and easier to restart. In solid-fuel reactors, on restart the ^{135}Xe in the fuel absorbs neutrons, followed by a sudden jump in reactivity as the ^{135}Xe is burned out. Conventional reactors may have to wait hours until xenon-135 decays after shutting down and not immediately restarting (so-called *iodine pit*).

Also in the pump bowl was a port through which salt samples could be taken or capsules of concentrated fuel-enriching salt (UF_4-LiF or PuF_3) could be introduced.

Air-cooled Heat Exchangers

MSRE air-cooled heat exchanger glowing a dull red due to high temperature.

At the time, the high temperatures were seen almost as a disadvantage, because they hampered use of conventional steam turbines. Now, such temperatures are seen as an opportunity to use high-efficiency closed-cycle gas turbines. After two months of high-power operation, the reactor was down for 3 months because of the failure of one of the main cooling blowers.

Neutronics and Thermal-hydraulics

The reactor experienced stable neutronic operation. If either temperatures increased, or bubbles formed, the volume of the fluid fuel salts would increase and some fluid fuel salts would be forced out of the core, thereby reducing the reactivity.

The MSRE development program did not include reactor physics experiments or heat transfer measurements. There was enough latitude in the MSRE that deviations from predictions would not compromise safety or accomplishment of the objectives of the experimental reactor.

Building Grounds

Aircraft Reactor Experiment building at ORNL that was retrofitted to house the MSRE.

Construction of the primary system components and alterations of the old Aircraft Reactor Experiment building (which had been partly remodeled for a proposed 60-MW(t) aircraft reactor) were started in 1962. Installation of the salt systems was completed in mid-1964. ORNL was responsible for quality assurance, planning, and management of construction. The primary systems were installed by ORNL personnel; subcontractors modified the building and installed ancillary systems.

Structural Alloy Hastelloy-N

A low chromium, nickel–molybdenum alloy, Hastelloy-N, was used in the MSRE and proved compatible with the fluoride salts FLiBe and FLiNaK. All metal parts contacting salt were made of Hastelloy-N.

The choice of Hastelloy-N for the MSRE was on the basis of the promising results of tests at aircraft nuclear propulsion conditions and the availability of much of the required metallurgical data. Development for the MSRE generated the further data required for ASME code approval. It also included preparation of standards for Hastelloy-N procurement and for component fabrication. Almost 200,000 lb (90,000 kg) in a variety of shapes of material for the MSRE was produced commercially. Requests for bids on component fabrication went to several companies in the nuclear fabrication industry, but all declined to submit lump-sum bids because of lack of experience with the new alloy. Consequently, all major components were fabricated in U.S. Atomic Energy Commission-owned shops at Oak Ridge and Paducah.

At the time that design stresses were set for the MSRE, the data that was available indicated that the strength and creep rate of Hastelloy-N were hardly affected by irradiation. After the construction was well along, the stress-rupture life and fracture strain were found to be drastically reduced by thermal neutron irradiation. The MSRE stresses were reanalyzed, and it was concluded that the reactor would have adequate life to reach its goals. At the same time a program was launched to improve the resistance of Hastelloy-N to the embrittlement.

An out-of-pile corrosion test program was carried out for Hastelloy-N which indicated extremely low corrosion rates at MSRE conditions. Capsules exposed in the Materials Testing Reactor showed that salt fission power densities of more than 200 W/cm^3 had no adverse effects on compatibility of fuel salt, Hastelloy-N, and graphite. Fluorine gas was found to be produced by radiolysis of frozen salts, but only at temperatures below about 100 °C.

Components that were developed especially for the MSRE included flanges for 5-inch (130 mm) lines carrying molten salt, freeze valves (an air-cooled section where salt could be frozen and thawed), flexible control rods to operate in thimbles at 1200 °F (650 °C), and the fuel sampler-enricher. Centrifugal pumps were developed similar to those used successfully in the aircraft reactor program, but with provisions for remote maintenance, and including a spray system for xenon removal. Remote maintenance considerations pervaded the MSRE design, and developments included devices for remotely cutting and brazing together 1½-inch pipe, removable heater-insulation units, and equipment for removing specimens of metal and graphite from the core.

Development and Construction Timeline

Most of the MSRE effort from 1960 through 1964 was devoted to design, development, and construction of the MSRE. Production and further testing of graphite and Hastelloy-N, both in-pile and out, were major development activities. Others included work on reactor chemistry, development of fabrication techniques for Hastelloy-N, development of reactor components, and remote-maintenance planning and preparations.

Operation

Alvin M. Weinberg noting "6000 full-power hours!" of MSRE operation, in 1967.

The MSRE operated for 5 years. The salt was loaded in 1964 and nuclear operation ended in December 1969, and all the objectives of the experiment were achieved during this period.

Checkout and prenuclear tests included 1,000 hours of circulation of flush salt and fuel carrier salt. Nuclear testing of the MSRE began in June 1965, with the addition of enriched ^{235}U as UF_4-LiF eutectic to the carrier salt to make the reactor critical. After zero-power experiments to measure rod worth and reactivity coefficients, the reactor was shut down and final preparations made for power operation. Power ascension was delayed when vapors from oil that had leaked into the fuel pump were polymerized by the radioactive offgas and plugged gas filters and valves. Maximum power, which was limited to 7.4 MW(t) by the capability of the heat-rejection system, was reached in May 1966.

After two months of high-power operation, the reactor was down for three months because of the failure of one of the main cooling blowers. Some further delays were encountered because of offgas line plugging, but by the end of 1966 most of the startup problems were behind. During the next 15 months, the reactor was critical 80% of the time, with runs of 1, 3, and 6 months that were uninterrupted by a fuel drain. By March 1968, the original objectives of the MSRE had been accomplished, and nuclear operation with ^{235}U was concluded.

By this time, ample ^{233}U had become available, so the MSRE program was extended to include substitution of ^{233}U for the uranium in the fuel salt, and operation to observe the new nuclear characteristics. Using the on-site processing equipment the flush salt and fuel salt were fluorinated to recover the uranium in them as UF_6. $^{233}UF_4$-LiF eutectic was then added to the carrier salt, and in October 1968, the MSRE became the world's first reactor to operate on ^{233}U.

The ^{233}U zero-power experiments and dynamics tests confirmed the predicted neutronic characteristics. An unexpected consequence of processing the salt was that its physical properties were altered slightly so that more than the usual amount of gas was entrained from the fuel pump into the circulating loop. The circulating gas and the power fluctuations that accompanied it were eliminated by operating the fuel pump at slightly lower speed. Operation at high power for several months permitted accurate measurement of the capture-to-fission ratio, for ^{233}U in this reactor, completing the objectives of the ^{233}U operation.

AEC Chairman Seaborg at the MSRE controls in 1968 for startup with U-233.

In the concluding months of operation, xenon stripping, deposition of fission products, and tritium behavior were investigated. The feasibility of using plutonium in molten-salt reactors was emphasized by adding PuF_3 as makeup fuel during this period.

After the final shutdown in December 1969, the reactor was left in standby for nearly a year. A limited examination program was then carried out, including a moderator bar from the core, a control rod thimble, heat exchanger tubes, parts from the fuel pump bowl, and a freeze valve that had developed a leak during the final reactor shutdown. The radioactive systems were then closed to await ultimate disposal.

Statistics

Other operational statistics:

- Hours critical: 17,655
- Circulating fuel loop hours: 21,788
- Core volume: less than 2 m^3

U-235 fuel operation

- Critical June 1, 1965
- Full power May 23, 1966
- End operation March 26, 1968
- Equivalent full power hours: 9,005

U-233 fuel operation

- Critical October 2, 1968
- Full power January 28, 1969
- Reactor shutdown December 12, 1969
- Equivalent full power hours: 4,167

Results

The broadest and perhaps most important conclusion from the MSRE experience was that a molten salt fueled reactor concept was viable. It ran for considerable periods of time, yielding valuable information, and maintenance was accomplished safely and without excessive delay.

The MSRE confirmed expectations and predictions. For example, it was demonstrated that: the fuel salt was immune to radiation damage, the graphite was not attacked by the fuel salt, and the corrosion of Hastelloy-N was negligible. Noble gases were stripped from the fuel salt by a spray system, reducing the ^{135}Xe poisoning by a factor of about 6. The bulk of the fission product elements remained stable in the salt. Additions of uranium and plutonium to the salt during operation were quick and uneventful, and recovery of uranium by fluorination was efficient. The neutronics, including critical loading, reactivity coefficients, dynamics, and long-term reactivity changes, agreed with prior calculations.

In other areas, the operation resulted in improved data or reduced uncertainties. The ^{233}U capture-to-fission ratio in a typical MSR neutron spectrum is an example of basic data that was improved. The effect of fissioning on the redox potential of the fuel salt was resolved. The deposition of some elements ("noble metals") was expected, but the MSRE provided quantitative data on relative deposition on graphite, metal, and liquid-gas interfaces. Heat transfer coefficients measured in the MSRE agreed with conventional design calculations and did not change over the life of the reactor. Limiting oxygen in the salt proved effective, and the tendency of fission products to be dispersed from contaminated equipment during maintenance was low.

Operation of the MSRE provided insights into the problem of tritium in a molten-salt reactor. It was observed that about 6–10% of the calculated 54 Ci/day (2.0 TBq) production diffused out of the fuel system into the containment cell atmosphere and another 6–10% reached the air through the heat removal system. The fact that these fractions were not higher, indicated that something partially negated the transfer of tritium through hot metals.

One unexpected finding was shallow, inter-granular cracking in all metal surfaces exposed to the fuel salt. The cause of the embrittlement was tellurium - a fission product generated in the fuel. This was first noted in the specimens that were removed from the core at intervals during the reactor operation. Post-operation examination of pieces of a control-rod thimble, heat-exchanger tubes and pump bowl parts revealed the ubiquity of the cracking and emphasized its importance to the MSR concept. The crack growth was rapid enough to become a problem over the planned thirty-year life of a follow-on thorium breeder reactor. This cracking could be reduced by adding small amounts of niobium to the Hastelloy-N.

Decommissioning

After shutdown, the salt was believed to be in long-term safe storage. At low temperatures, radiolysis can free fluorine from the salt. As a countermeasure, the salt was annually reheated to about 150 °C until 1989. But beginning in the mid-1980s, there was concern that radioactivity was migrating through the system. Sampling in 1994 revealed concentrations of uranium that created a potential for a nuclear criticality accident, as well as a potentially dangerous build-up of fluorine gas — the environment above the solidified salt was approximately one atmosphere of fluorine. The ensuing decontamination and decommissioning project was called "the most technically challenging" activity assigned to Bechtel Jacobs under its environmental management contract with the U.S. Department of Energy's Oak Ridge Operations organization. In 2003, the MSRE cleanup project was estimated at about $130 million, with decommissioning expected to be completed in 2009. Removal of uranium from the salt was finally complete in March 2008, however still leaving the salt with the fission products in the tanks.

Much of the high cost was caused by the unpleasant surprise of fluorine and uranium hexafluoride evolution from cold fuel salt in storage that ORNL did not defuel and store correctly, but this has now been taken into consideration in MSR design.

A potential decommissioning process has been described; uranium is to be removed from the fuel as the hexafluoride by adding excess fluorine, and plutonium as the plutonium dioxide by adding sodium carbonate.

IMSR

The IMSR uses molten fluoride salt, a highly stable, inert liquid with robust coolant properties and high intrinsic radionuclide retention properties, for its primary fuel salt. A secondary, coolant salt loop, also using a fluoride salt (but without fuel), transfers heat away from the primary heat exchangers integrated inside the core-unit. The coolant salt loop, in turn, transfers its heat load to a solar salt loop, which is pumped out of the nuclear island to a separate building where it either heats steam generators that generate superheated steam for power generation or is used for process heat applications. The safety philosophy behind the IMSR is to produce a nuclear power plant with generation IV reactor levels of safety. For ultimate safety, there is no dependence on operator intervention, powered mechanical components, coolant injection or their support systems, such as electricity supply or instrument air in dealing with upset conditions. This is achieved through a combination of design features: the inert, stable properties of the salt; an inherently stable nuclear core; fully passive back-up core and containment cooling systems; and an integral reactor architecture.

How the Integral Molten Salt Reactor Works

- The term 'Integral' refers to the company's proprietary design in which all the primary components (pumps, moderator and primary HX) of the reactor core are sealed in a compact and replaceable component, the IMSR Core-unit.
- A new Core-unit is exchanged every 7 years with the old Core-unit stored on site.
- The IMSR is a "pool-type" reactor with no penetrations into the reactor vessel.
- IMSR fuel, the Fuel Salt, is a liquid, high-temperature fluoride salt that operates at ~700C.

- These salts have high thermal stability and are excellent heat transfer liquids.
- IMSR Fuel salt can be produced today with current methods and within current regulations.
- The Fuel Salt, which contains the nuclear fuel, never leaves the reactor core vessel during operation.
- Fuel Salt is circulated in a closed loop up through the graphite core, where the fuel fissions in a thermal neutron spectrum creating heat within the fuel, which then circulates back down through heat exchangers giving up heat to a secondary salt in a primary heat exchanger and isolated loop. The Fuel Salt circulates back into the core.
- The reactor core contains a graphite moderator - outside of the moderated area, the salt is no longer active.
- A secondary heat exchanger exchanges heat via secondary salt to a third loop containing a 600C industrial salt that can be transported up to 5 kilometers.
- The IMSR has a low level of tritium production. Furthermore, the IMSR's three successive isolation loops further ensures no tritium moves beyond the nuclear island.
- Uranium enriched to less than 5% is currently intended to fuel the IMSR; however, with future iterations of IMSR technology, the IMSR is very capable of using a diverse array of fuel forms, including thorium-based fuels and spent nuclear fuels from existing nuclear fleets.
- An IMSR power plant can rapidly load follow grid power demand.
- An IMSR power plant is anticipated to deliver power at less than $50 per MWh, which is highly competitive with fossil fuel combustion.

References

- Gilbert, Richard J.; Kahn, Edward P. (18 January 2007). International Comparisons of Electricity Regulation. Cambridge University Press. p. 47. ISBN 9780521030779. Retrieved 6 October 2017
- Owen, Geoffrey (7 March 2016). "Book review: 'The Fall and Rise of Nuclear Power in Britain'". Financial Times. Retrieved 16 March 2016
- Boiling-water-reactor: energyeducation.ca, Retrieved 29 June 2018
- Murray, P. (1981). "Developments in oxide fuels at Harwell". Journal of Nuclear Materials. 100 (1–3): 67–71. Bibcode:1981JNuM..100...67M. doi:10.1016/0022-3115(81)90521-3
- Association of German Engineers (VDI), the Society for Energy Technologies (publ.) (1990). AVR - Experimental High-Temperature Reactor, 21 Years of Successful Operation for A Future Energy Technology. Association of German Engineers (VDI), The Society for Energy Technologies. pp. 9–23. ISBN 3-18-401015-5
- Pressurized-water-reactor: energyeducation.ca, Retrieved 21 March 2018
- Rainer Moormann (1 April 2009). "PBR safety revisited". Nuclear Engineering International. Archived from the original on 30 May 2012. Retrieved 2009-04-02
- Steven B Krivit; Jay H Lehr; Thomas B Kingery, eds. (2011). Nuclear Energy Encyclopedia: Science, Technology, and Applications. Wiley. p. 28. ISBN 978-1-118-04347-9.
- Reactor-types, lwr-light-water-reactor, nuclear-power-plant: nuclear-power.net, Retrieved 31 March 2018

3

Components of Nuclear Reactors

Nuclear power reactor has a number of components, which are essential for its functioning, such as nuclear reactor core, neutron moderator, nuclear fuel, neutron reflector, etc. This chapter discusses in detail about such components for an understanding of the working of a nuclear power reactor.

Nuclear Reactor Core

A nuclear reactor core is a key component of a nuclear reactor. In reactor physics, the nuclear core is a bounded region, where a neutron multiplication occurs and where chain reactions take place. The reactor core contains especially the nuclear fuel (fuel assemblies), the moderator and the control rods. The core of the reactor contains all the nuclear fuel assemblies and generates most of the heat (fraction of the heat is genarated outside the reactor – e.g. gamma rays energy). The assemblies are exactly placed in the reactor according to a fuel loading pattern.

The core is located inside a reactor pressure vessel and is carried by a core support barrel. A bottom plate of the support barrel is called lower core plate. This structure consist of N holes for fixing fuel assemblies (N is number of fuel assemblies in specific reactor – e.g. 157 for AP1000).

There is a neutron reflector (core baffle or heavy reflector) around the core (around peripheral fuel assemblies), which is usually formed of stainless rings accurately surrounding the nuclear fuel assemblies.

The top of the nuclear core is defined by an upper guide structure assembly. This assembly is made of stainless steel and has many purposes. The upper guide structure assembly exerts an axial force on fuel assemblies (through springs in the fuel assembly), thus defines the exact position of the fuel assembly in the core. The upper guide structure assembly also guides and protects control rod assemblies and in-core instrumentation.

Components of a Nuclear Reactor Core

The key components common to most PWR types of nuclear reactor cores are:

- Nuclear fuel. The nuclear fuel is composed of nuclear fuel assemblies located in exact positions in the reactor.
- Neutron moderator. The moderator slows down neutrons from fission to thermal energies.
- Control rods. Control rods are rods, or tubes containing a neutron absorbing material such as boron, hafnium, cadmium, etc., used to control the power of a nuclear reactor.

- Neutron Reflector. The neutron reflector (core baffle) surrounds the core. The reflector reduces the non-uniformity of the power distribution in the peripheral fuel assemblies and reduces a coolant flow bypass of the core.
- Burnable neutron poison. Burnable poisons / absorbers are primarily used to control large amounts of excess fuel reactivity, are used to reduce a concentration of a boric acid at the beginning of the fuel cycle.
- Neutron source. Neutron sources increase a neutron flux during subcritical state.
- In-core instrumentation. The in-core instrumentation system measures neutron flux distribution and temperatures in the reactor core. This helps determine safety margins of the reactor.

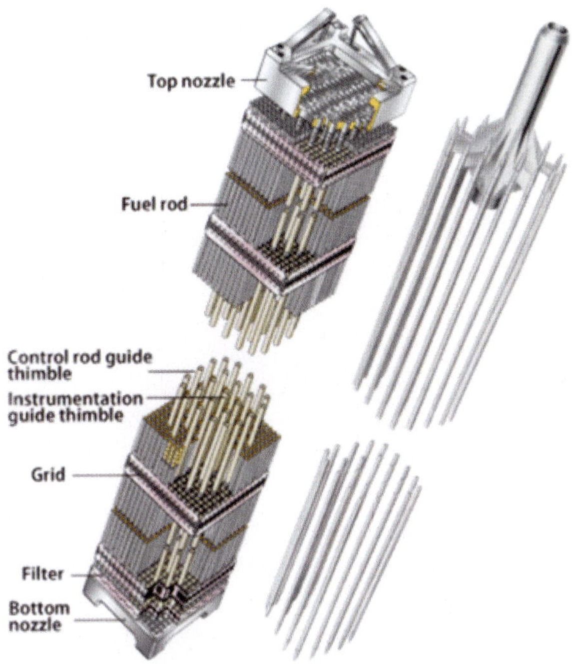

The nuclear fuel assembly with the rod cluster control assembly.

Top view of the WWER-1000 nuclear core. The lower support structure, the neutron reflector and 11 fuel assemblies.

Nuclear Fuel

Nuclear fuel is the fuel that is used in a nuclear reactor to sustain a nuclear chain reaction. These fuels are fissile, and the most common nuclear fuels are the radioactive metals uranium-235 and plutonium-239. All processes involved in obtaining, refining, and using this fuel make up a cycle known as the nuclear fuel cycle.

Uranium-235 is used as a fuel in different concentrations. Some reactors, such as the CANDU reactor, can use natural uranium with uranium-235 concentrations of only 0.7%, while other reactors require the uranium to be slightly enriched to levels of 3% to 5%. Plutonium-239 is produced and used in reactors (specifically fast breeder reactors) that contain significant amounts of uranium-238. It can also be recycled and used as a fuel in thermal reactors. Current research is being done to investigate how thorium-232 can be used as a fuel.

Production

Fuel fabrication plants are facilities that convert enriched uranium into fuel for nuclear reactors. For light water reactors, uranium is received from an enrichment plant in solid form. It is then converted into a gas and chemically converted into a uranium dioxide powder. This powder is then pressed into pellets and packed into fuel assemblies. A mixed oxide fuel can also be created when the uranium powder is packed along with plutonium oxide. The hazards present at fuel fabrication facilities—mainly chemical and radiological—are similar to the hazards at enrichment plants. These facilities generally pose a low risk to the public.

Use

When used in a reactor, the fuels used can have a variety of different forms a metal, an alloy, or some sort of oxide. Most nuclear reactors are fueled with a compound known as uranium dioxide. This uranium dioxide is put together in a fuel assembly and inserted into the nuclear reactor—where it can stay for several months or up to a few years. While in the reactor the fuel undergoes nuclear fission and releases energy. This released energy is used to generate electricity. Neutrons released during the fission process allow for a fission chain reaction to occur, allowing energy to be generated continually. The fuel is removed from the reactor after large amounts of the fuel—whether it is uranium-235 or plutonium-239—have undergone fission. The "used" nuclear fuel is known as *spent* or *irradiated* fuel. After use, the fuel must be cooled for a few years as it is extremely hot.

The spent fuel is placed in large, deep pools of water that act as a coolant and a radiation shield. The coolant property allows the water to remove the decay heat and the shielding abilities protect workers from the radioactivity of the fuel. After cooling, the fuel can be re-purposed or sent to storage depending on regulations.

Fuel Assembly

Nuclear reactors are powered by powdered uranium dioxide that has been compressed into small pellets, shown in Figure. However, a power plant requires *many* of these pellets to run. Thus large numbers of these pellets are bundled into a fuel rod. A single uranium fuel pellet, only as big as a fingertip, contains as much energy as 481 cubic meters of natural gas, 807 kilograms of coal or 564 liters of oil. These rods are composed of numerous pellets of fissionable uranium fuel and can be several meters in length and about a centimeter in diameter. Then several of these rods, generally a dozen or more, are held together by strong metallic brackets in a fuel assembly. These rods are not bunched tightly together, rather there are several millimeters between each rod to allow coolant to flow between them. The tubes containing the pellets of uranium are generally composed of zirconium.

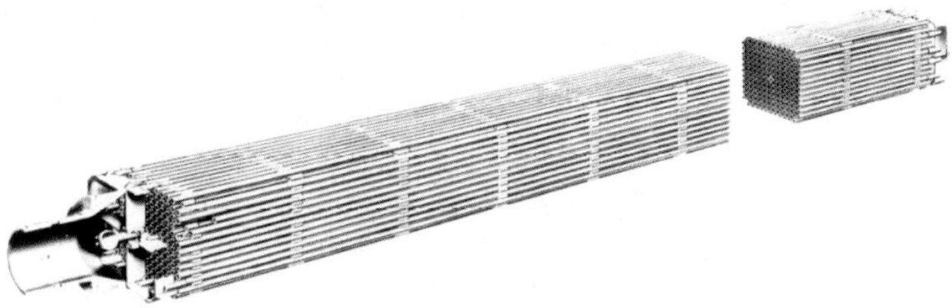

A nuclear fuel bundle.

Oxide Fuel

For fission reactors, the fuel (typically based on uranium) is usually based on the metal oxide; the oxides are used rather than the metals themselves because the oxide melting point is much higher than that of the metal and because it cannot burn, being already in the oxidized state.

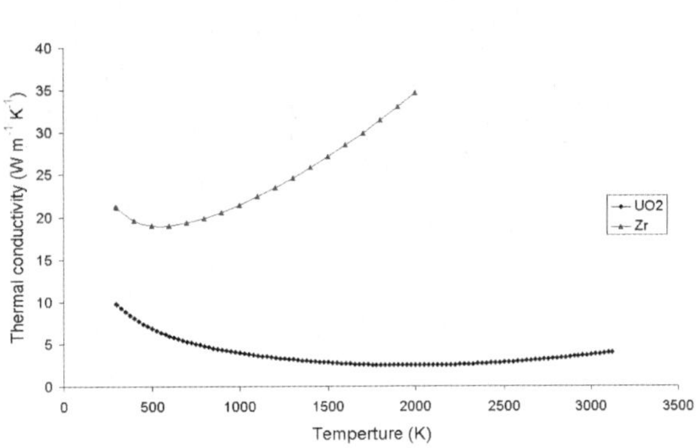

The thermal conductivity of zirconium metal and uranium dioxide as a function of temperature

UOX

Uranium dioxide is a black semiconducting solid. It can be made by reacting uranyl nitrate with a base (ammonia) to form a solid (ammonium uranate). It is heated (calcined) to form U_3O_8 that can then be converted by heating in an argon / hydrogen mixture (700 °C) to form UO_2. The UO_2 is then mixed with an organic binder and pressed into pellets, these pellets are then fired at a much higher temperature (in H_2/Ar) to sinter the solid. The aim is to form a dense solid which has few pores.

The thermal conductivity of uranium dioxide is very low compared with that of zirconium metal, and it goes down as the temperature goes up.

Corrosion of uranium dioxide in water is controlled by similar electrochemical processes to the galvanic corrosion of a metal surface.

MOX

Mixed oxide, or MOX fuel, is a blend of plutonium and natural or depleted uranium which behaves similarly (though not identically) to the enriched uranium feed for which most nuclear reactors were designed. MOX fuel is an alternative to low enriched uranium (LEU) fuel used in the light water reactors which predominate nuclear power generation.

Some concern has been expressed that used MOX cores will introduce new disposal challenges, though MOX is itself a means to dispose of surplus plutonium by transmutation.

Reprocessing of commercial nuclear fuel to make MOX was done in the Sellafield MOX Plant (England). As of 2015, MOX fuel is made in France, and to a lesser extent in Russia, India and Japan. China plans to develop fast breeder reactors and reprocessing.

The Global Nuclear Energy Partnership, was a U.S. proposal in the George W. Bush Administration to form an international partnership to see spent nuclear fuel reprocessed in a way that renders the plutonium in it usable for nuclear fuel but not for nuclear weapons. Reprocessing of spent commercial-reactor nuclear fuel has not been permitted in the United States due to nonproliferation considerations. All of the other reprocessing nations have long had nuclear weapons from military-focused "research"-reactor fuels except for Japan. Normally, with the fuel being changed every three years or so, about half of the Pu-239 is 'burned' in the reactor, providing about one third of the total energy. It behaves like U-235 and its fission releases a similar amount of energy. The higher the burn-up, the more plutonium in the spent fuel, but the lower the fraction of fissile plutonium. Typically about one percent of the used fuel discharged from a reactor is plutonium, and some two thirds of this is fissile (c. 50% Pu-239, 15% Pu-241). Worldwide, some 70 tonnes of plutonium contained in used fuel is removed when refueling reactors each year.

Metal Fuel

Metal fuels have the advantage of a much higher heat conductivity than oxide fuels but cannot survive equally high temperatures. Metal fuels have a long history of use, stretching from the Clementine reactor in 1946 to many test and research reactors. Metal fuels have the potential for the highest fissile atom density. Metal fuels are normally alloyed, but some metal fuels have been made with pure uranium metal. Uranium alloys that have been used include uranium aluminum, uranium zirconium, uranium silicon, uranium molybdenum, and uranium zirconium hydride. Any of the aforementioned fuels can be made with plutonium and other actinides as part of a closed nuclear fuel cycle. Metal fuels have been used in water reactors and liquid metal fast breeder reactors, such as EBR-II.

TRIGA Fuel

TRIGA fuel is used in TRIGA (Training, Research, Isotopes, General Atomics) reactors. The TRIGA reactor uses UZrH fuel, which has a prompt negative fuel temperature coefficient of reactivity, meaning that as the temperature of the core increases, the reactivity decreases—so it is highly unlikely for a meltdown to occur. Most cores that use this fuel are "high leakage" cores where the excess leaked neutrons can be utilized for research. TRIGA fuel was originally designed to use highly enriched uranium, however in 1978 the U.S. Department of Energy launched its Reduced

Enrichment for Research Test Reactors program, which promoted reactor conversion to low-enriched uranium fuel. A total of 35 TRIGA reactors have been installed at locations across the USA. A further 35 reactors have been installed in other countries.

Actinide Fuel

In a fast neutron reactor, the minor actinides produced by neutron capture of uranium and plutonium can be used as fuel. Metal actinide fuel is typically an alloy of zirconium, uranium, plutonium, and minor actinides. It can be made inherently safe as thermal expansion of the metal alloy will increase neutron leakage.

Molten Plutonium

Molten plutonium, alloyed with other metals to lower its melting point and encapsulated in tantalum, was tested in two experimental reactors, LAMPRE I and LAMPRE II, at LANL in the 1960s. "LAMPRE experienced three separate fuel failures during operation."

Non-oxide Ceramic Fuels

Ceramic fuels other than oxides have the advantage of high heat conductivities and melting points, but they are more prone to swelling than oxide fuels and are not understood as well.

Uranium Nitride

This is often the fuel of choice for reactor designs that NASA produces, one advantage is that UN has a better thermal conductivity than UO_2. Uranium nitride has a very high melting point. This fuel has the disadvantage that unless ^{15}N was used (in place of the more common ^{14}N) that a large amount of ^{14}C would be generated from the nitrogen by the (n,p) reaction. As the nitrogen required for such a fuel would be so expensive it is likely that the fuel would have to be reprocessed by pyroprocessing to enable the ^{15}N to be recovered. It is likely that if the fuel was processed and dissolved in nitric acid that the nitrogen enriched with ^{15}N would be diluted with the common ^{14}N.

Uranium Carbide

Much of what is known about uranium carbide is in the form of pin-type fuel elements for liquid metal fast reactors during their intense study during the 1960s and 1970s. However, recently there has been a revived interest in uranium carbide in the form of plate fuel and most notably, micro fuel particles (such as TRISO particles).

The high thermal conductivity and high melting point makes uranium carbide an attractive fuel. In addition, because of the absence of oxygen in this fuel (during the course of irradiation, excess gas pressure can build from the formation of O_2 or other gases) as well as the ability to complement a ceramic coating (a ceramic-ceramic interface has structural and chemical advantages), uranium carbide could be the ideal fuel candidate for certain Generation IV reactors such as the gas-cooled fast reactor.

Liquid Fuels

Liquid fuels are liquids containing dissolved nuclear fuel and have been shown to offer numerous operational advantages compared to traditional solid fuel approaches.

Liquid-fuel reactors offer significant safety advantages due to their inherently stable "self-adjusting" reactor dynamics. This provides two major benefits: - virtually eliminating the possibility of a run-away reactor meltdown, - providing an automatic load-following capability which is well suited to electricity generation and high-temperature industrial heat applications.

Another major advantage of the liquid core is its ability to be drained rapidly into a passively safe dump-tank. This advantage was conclusively demonstrated repeatedly as part of a weekly shutdown procedure during the highly successful 4 year Molten Salt Reactor Experiment.

Another huge advantage of the liquid core is its ability to release xenon gas which normally acts as a neutron absorber and causes structural occlusions in solid fuel elements (leading to the early replacement of solid fuel rods with over 98% of the nuclear fuel unburned, including many long-lived actinides). In contrast, Molten Salt Reactors (MSR) are capable of retaining the fuel mixture for significantly extended periods, which not only increases fuel efficiency dramatically but also incinerates the vast majority of its own waste as part of the normal operational characteristics.

Molten Salts

Molten salt fuels have nuclear fuel dissolved directly in the molten salt coolant. Molten salt-fueled reactors, such as the liquid fluoride thorium reactor (LFTR), are different from molten salt-cooled reactors that do not dissolve nuclear fuel in the coolant.

Molten salt fuels were used in the LFTR known as the Molten Salt Reactor Experiment, as well as other liquid core reactor experiments. The liquid fuel for the molten salt reactor was a mixture of lithium, beryllium, thorium and uranium fluorides: LiF-BeF_2-ThF_4-UF_4 (72-16-12-0.4 mol%). It had a peak operating temperature of 705 °C in the experiment, but could have operated at much higher temperatures since the boiling point of the molten salt was in excess of 1400 °C.

Aqueous Solutions of Uranyl Salts

The aqueous homogeneous reactors (AHRs) use a solution of uranyl sulfate or other uranium salt in water. Historically, AHRs have all been small research reactors, not large power reactors. An AHR known as the Medical Isotope Production System is being considered for production of medical isotopes.

Common Physical Forms of Nuclear Fuel

Uranium dioxide (UO_2) powder is compacted to cylindrical pellets and sintered at high temperatures to produce ceramic nuclear fuel pellets with a high density and well defined physical properties and chemical composition. A grinding process is used to achieve a uniform cylindrical geometry with narrow tolerances. Such fuel pellets are then stacked and filled into the metallic tubes. The metal used for the tubes depends on the design of the reactor. Stainless steel was used in the past, but most reactors now use a zirconium alloy which, in addition to being highly corrosion-resistant, has low neutron absorption. The tubes containing the fuel pellets are sealed: these tubes are called fuel rods. The finished fuel rods are grouped into fuel assemblies that are used to build up the core of a power reactor.

Cladding is the outer layer of the fuel rods, standing between the coolant and the nuclear fuel. It is made of a corrosion-resistant material with low absorption cross section for thermal neutrons, usually Zircaloy or steel in modern constructions, or magnesium with small amount of aluminium and other metals for the now-obsolete Magnox reactors. Cladding prevents radioactive fission fragments from escaping the fuel into the coolant and contaminating it.

Nuclear Regulatory Commission (NRC) photo of unirradiated (fresh) fuel pellets.

NRC photo of fresh fuel pellets ready for assembly.

NRC photo of fresh fuel being inspected.

PWR Fuel

Pressurized water reactor (PWR) fuel consists of cylindrical rods put into bundles. A uranium oxide ceramic is formed into pellets and inserted into Zircaloy tubes that are bundled together. The Zircaloy tubes are about 1 cm in diameter, and the fuel cladding gap is filled with helium gas to improve the conduction of heat from the fuel to the cladding. There are about 179-264 fuel rods per fuel bundle and about 121 to 193 fuel bundles are loaded into a reactor core. Generally, the fuel bundles consist of fuel rods bundled 14×14 to 17×17. PWR fuel bundles are about 4 meters long. In PWR fuel bundles, control rods are inserted through the top directly into the fuel bundle. The fuel bundles usually are enriched several percent in ^{235}U. The uranium oxide is dried before inserting into the tubes to try to eliminate moisture in the ceramic fuel that can lead to corrosion and hydrogen embrittlement. The Zircaloy tubes are pressurized with helium to try to minimize pellet-cladding interaction which can lead to fuel rod failure over long periods.

BWR Fuel

In boiling water reactors (BWR), the fuel is similar to PWR fuel except that the bundles are "canned". That is, there is a thin tube surrounding each bundle. This is primarily done to prevent local density variations from affecting neutronics and thermal hydraulics of the reactor core. In modern BWR fuel bundles, there are either 91, 92, or 96 fuel rods per assembly depending on the manufacturer. A range between 368 assemblies for the smallest and 800 assemblies for the largest U.S. BWR forms the reactor core. Each BWR fuel rod is backfilled with helium to a pressure of about three atmospheres (300 kPa).

CANDU Fuel

CANDU fuel bundles Two CANDU ("CANada Deuterium Uranium") fuel bundles, each about 50 cm long, 10 cm in diameter.

CANDU fuel bundles are about a half meter long and 10 cm in diameter. They consist of sintered (UO_2) pellets in zirconium alloy tubes, welded to zirconium alloy end plates. Each bundle is roughly 20 kg, and a typical core loading is on the order of 4500-6500 bundles, depending on the design. Modern types typically have 37 identical fuel pins radially arranged about the long axis of the bundle, but in the past several different configurations and numbers of pins have been used. The CANFLEX bundle has 43 fuel elements, with two element sizes. It is also about 10 cm (4 inches) in diameter, 0.5 m (20 in) long and weighs about 20 kg (44 lb) and replaces the 37-pin standard bundle. It has been designed specifically to increase fuel performance by utilizing two different pin diameters. Current CANDU designs do not need enriched uranium to achieve criticality (due to their more efficient heavy water moderator), however, some newer concepts call for low enrichment to help reduce the size of the reactors.

Less-common Fuel Forms

Various other nuclear fuel forms find use in specific applications, but lack the widespread use of those found in BWRs, PWRs, and CANDU power plants. Many of these fuel forms are only found in research reactors, or have military applications.

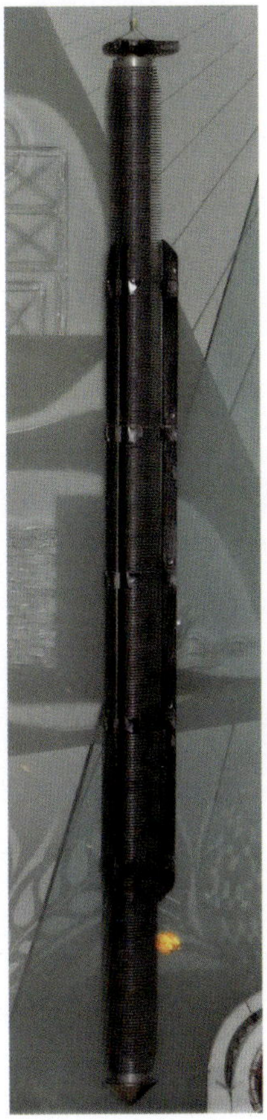

A magnox fuel rod

Magnox Fuel

Magnox reactors are pressurised, carbon dioxide–cooled, graphite-moderated reactors using natural uranium (i.e. unenriched) as fuel and Magnox alloy as fuel cladding. Working pressure varies from 6.9 to 19.35 bar for the steel pressure vessels, and the two reinforced concrete designs operated at 24.8 and 27 bar. Magnox alloy consists mainly of magnesium with small amounts of aluminium and other metals—used in cladding unenriched uranium metal fuel with a non-oxidising covering to contain fission products. Magnox is short for Magnesium non-oxidising. This material has the advantage of a low neutron capture cross-section, but has two major disadvantages:

- It limits the maximum temperature, and hence the thermal efficiency, of the plant.
- It reacts with water, preventing long-term storage of spent fuel under water.

Magnox fuel incorporated cooling fins to provide maximum heat transfer despite low operating temperatures, making it expensive to produce. While the use of uranium metal rather than oxide made reprocessing more straightforward and therefore cheaper, the need to reprocess fuel a short time after removal from the reactor meant that the fission product hazard was severe. Expensive remote handling facilities were required to address this danger.

TRISO Fuel

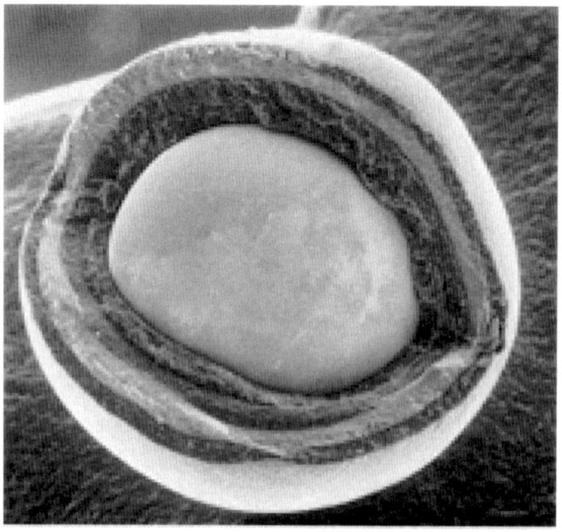

TRISO fuel particle which has been cracked, showing the multiple coating layers

Tristructural-isotropic (TRISO) fuel is a type of micro fuel particle. It consists of a fuel kernel composed of UO_X (sometimes UC or UCO) in the center, coated with four layers of three isotropic materials. The four layers are a porous buffer layer made of carbon, followed by a dense inner layer of pyrolytic carbon (PyC), followed by a ceramic layer of SiC to retain fission products at elevated temperatures and to give the TRISO particle more structural integrity, followed by a dense outer layer of PyC. TRISO fuel particles are designed not to crack due to the stresses from processes (such as differential thermal expansion or fission gas pressure) at temperatures up to and beyond 1600 °C, and therefore can contain the fuel in the worst of accident scenarios in a properly designed reactor. Two such reactor designs are the pebble-bed reactor (PBR), in which thousands of TRISO fuel particles are dispersed into graphite pebbles, and the prismatic-block gas-cooled reactor (such as the GT-MHR), in which the TRISO fuel particles are fabricated into compacts and placed in a graphite block matrix. Both of these reactor designs are high temperature gas reactors (HTGRs). These are also the basic reactor designs of very-high-temperature reactors (VHTRs), one of the six classes of reactor designs in the Generation IV initiative that is attempting to reach even higher HTGR outlet temperatures.

TRISO fuel particles were originally developed in the United Kingdom as part of the Dragon reactor project. The inclusion of the SiC as diffusion barrier was first suggested by D. T. Livey. The first nuclear reactor to use TRISO fuels was the Dragon reactor and the first powerplant was the THTR-300. Currently, TRISO fuel compacts are being used in the experimental reactors, the HTR-10 in China, and the High-temperature engineering test reactor in Japan. Spherical fuel elements utilizing a TRISO particle with a UO_2 and UC solid solution kernel are being used in the Xe-100 in the United States.

QUADRISO Fuel

In QUADRISO particles a burnable neutron poison (europium oxide or erbium oxide or carbide) layer surrounds the fuel kernel of ordinary TRISO particles to better manage the excess of reactivity. If the core is equipped both with TRISO and QUADRISO fuels, at beginning of life neutrons do not reach the fuel of the QUADRISO particles because they are stopped by the burnable poison. During irradiation, the poison depletes and more neutrons are able to stream into the fuel kernel of QUADRISO particles inducing fission reactions. This mechanism compensates for the normal depletion of fissile material during fuel burnup. In the generalized QUADRISO fuel concept the poison can eventually be mixed with the fuel kernel or the outer pyrocarbon. The QUADRISO concept has been conceived at Argonne National Laboratory.

QUADRISO Particle

RBMK Fuel

RBMK reactor fuel was used in Soviet-designed and built RBMK-type reactors. This is a low-enriched uranium oxide fuel. The fuel elements in an RBMK are 3 m long each, and two of these sit back-to-back on each fuel channel, pressure tube. Reprocessed uranium from Russian VVER reactor spent fuel is used to fabricate RBMK fuel. Following the Chernobyl accident, the enrichment of fuel was changed from 2.0% to 2.4%, to compensate for control rod modifications and the introduction of additional absorbers.

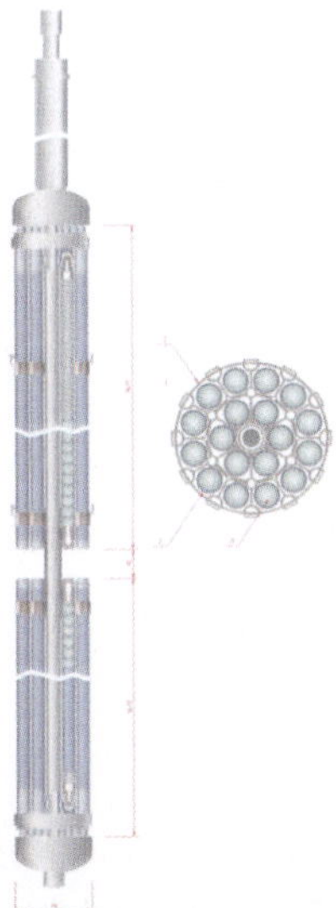

RBMK reactor fuel rod holder 1 – distancing armature; 2 – fuel rods shell; 3 – fuel tablets.

CerMet Fuel

CerMet fuel consists of ceramic fuel particles (usually uranium oxide) embedded in a metal matrix. It is hypothesized that this type of fuel is what is used in United States Navy reactors. This fuel has high heat transport characteristics and can withstand a large amount of expansion.

ATR Core The Advanced Test Reactor at Idaho National Laboratory uses plate-type fuel in a clover leaf arrangement. The blue glow around the core is known as Cherenkov radiation.

Plate-type Fuel

Plate-type fuel has fallen out of favor over the years. Plate-type fuel is commonly composed of enriched uranium sandwiched between metal cladding. Plate-type fuel is used in several research reactors where a high neutron flux is desired, for uses such as material irradiation studies or isotope production, without the high temperatures seen in ceramic, cylindrical fuel. It is currently used in the Advanced Test Reactor (ATR) at Idaho National Laboratory, and the nuclear research reactor at the University of Massachusetts Lowell Radiation Laboratory.

Sodium-bonded Fuel

Sodium-bonded fuel consists of fuel that has liquid sodium in the gap between the fuel slug (or pellet) and the cladding. This fuel type is often used for sodium-cooled liquid metal fast reactors. It has been used in EBR-I, EBR-II, and the FFTF. The fuel slug may be metallic or ceramic. The sodium bonding is used to reduce the temperature of the fuel.

Spent Nuclear Fuel

Used nuclear fuel is a complex mixture of the fission products, uranium, plutonium, and the transplutonium metals. In fuel which has been used at high temperature in power reactors it is common for the fuel to be *heterogeneous*; often the fuel will contain nanoparticles of platinum group metals such as palladium. Also the fuel may well have cracked, swollen, and been heated close to its melting point. Despite the fact that the used fuel can be cracked, it is very insoluble in water, and is able to retain the vast majority of the actinides and fission products within the uranium dioxide crystal lattice.

Oxide Fuel Under Accident Conditions

Two main modes of release exist, the fission products can be vaporised or small particles of the fuel can be dispersed.

Fuel Behavior and Post-irradiation Examination

Post-Irradiation Examination (PIE) is the study of used nuclear materials such as nuclear fuel. It has several purposes. It is known that by examination of used fuel that the failure modes which occur during normal use (and the manner in which the fuel will behave during an accident) can be studied. In addition information is gained which enables the users of fuel to assure themselves of its quality and it also assists in the development of new fuels. After major accidents the core (or what is left of it) is normally subject to PIE to find out what happened. One site where PIE is done is the ITU which is the EU centre for the study of highly radioactive materials.

Materials in a high-radiation environment (such as a reactor) can undergo unique behaviors such as swelling and non-thermal creep. If there are nuclear reactions within the material (such as what happens in the fuel), the stoichiometry will also change slowly over time. These behaviors can lead to new material properties, cracking, and fission gas release.

The thermal conductivity of uranium dioxide is low; it is affected by porosity and burn-up. The burn-up results in fission products being dissolved in the lattice (such as lanthanides), the precipitation of fission products such as palladium, the formation of fission gas bubbles due to fission products such as xenon and krypton and radiation damage of the lattice. The low thermal conductivity can lead to overheating of the center part of the pellets during use. The porosity results in a decrease in both the thermal conductivity of the fuel and the swelling which occurs during use.

According to the International Nuclear Safety Center the thermal conductivity of uranium dioxide can be predicted under different conditions by a series of equations.

The bulk density of the fuel can be related to the thermal conductivity

Where ρ is the bulk density of the fuel and ρ_{td} is the theoretical density of the uranium dioxide.

Then the thermal conductivity of the porous phase (K_f) is related to the conductivity of the perfect phase (K_o, no porosity) by the following equation. Note that s is a term for the shape factor of the holes.

$$K_f = K_o(1 - p/1 + (s - 1)p)$$

Rather than measuring the thermal conductivity using the traditional methods such as Lees' disk, the Forbes' method, or Searle's bar, it is common to use Laser Flash Analysis where a small disc of fuel is placed in a furnace. After being heated to the required temperature one side of the disc is illuminated with a laser pulse, the time required for the heat wave to flow through the disc, the density of the disc, and the thickness of the disk can then be used to calculate and determine the thermal conductivity.

$$\lambda = \rho C_p \alpha$$

- λ thermal conductivity
- ρ density
- C_p heat capacity
- α thermal diffusivity

If $t_{1/2}$ is defined as the time required for the non illuminated surface to experience half its final temperature rise then.

$$\alpha = 0.1388\, L^2/t_{1/2}$$

- L is the thickness of the disc

Radioisotope Decay Fuels

Radioisotope Battery

The terms atomic battery, nuclear battery and radioisotope battery are used interchangeably to describe a device which uses the radioactive decay to generate electricity. These systems use radioisotopes that produce low energy beta particles or sometimes alpha particles of varying energies. Low energy beta particles are needed to prevent the production of high energy penetrating bremsstrahlung radiation that would require heavy shielding. Radioisotopes such as plutonium-238, curium-242, curium-244 and strontium-90 have been used. Tritium, nickel-63, promethium-147, and technetium-99 have been tested.

There are two main categories of atomic batteries: thermal and non-thermal. The non-thermal atomic batteries, which have many different designs, exploit charged alpha and beta particles. These designs include the direct charging generators, betavoltaics, the optoelectric nuclear battery, and the radioisotope piezoelectric generator. The thermal atomic batteries on the other hand, convert the heat from the radioactive decay to electricity. These designs include thermionic converter, thermophotovoltaic cells, alkali-metal thermal to electric converter, and the most common design, the radioisotope thermoelectric generator.

Radioisotope Thermoelectric Generators

Inspection of Cassini spacecraft RTGs before launch

A *radioisotope thermoelectric generator* (RTG) is a simple electrical generator which converts heat into electricity from a radioisotope using an array of thermocouples.

^{238}Pu has become the most widely used fuel for RTGs, in the form of plutonium dioxide. It has a half-life of 87.7 years, reasonable energy density, and exceptionally low gamma and neutron radiation levels. Some Russian terrestrial RTGs have used ^{90}Sr; this isotope has a shorter half-life and a much lower energy density, but is cheaper. Early RTGs, first built in 1958 by the U.S. Atomic Energy Commission, have used ^{210}Po. This fuel provides phenomenally huge energy density, (a single gram of polonium-210 generates 140 watts thermal) but has limited use because of its very short half-life and gamma production, and has been phased out of use for this application.

Photo of a disassembled RHU

Radioisotope Heater Units (RHU)

Radioisotope heater units normally provide about 1 watt of heat each, derived from the decay of a few grams of plutonium-238. This heat is given off continuously for several decades.

Their function is to provide highly localised heating of sensitive equipment (such as electronics in outer space). The Cassini–Huygens orbiter to Saturn contains 82 of these units (in addition to its 3 main RTGs for power generation). The Huygens probe to Titan contains 35 devices.

Fusion Fuels

Fusion fuels include deuterium (^{2}H) and tritium (^{3}H) as well as helium-3 (^{3}He). Many other elements can be fused together, but the larger electrical charge of their nuclei means that much higher temperatures are required. Only the fusion of the lightest elements is seriously considered as a future energy source. Fusion of the lightest atom, ^{1}H hydrogen, as is done in the Sun and stars, has also not been considered practical on Earth. Although the energy density of fusion fuel is even higher than fission fuel, and fusion reactions sustained for a few minutes have been achieved, utilizing fusion fuel as a net energy source remains only a theoretical possibility.

First-generation Fusion Fuel

Deuterium and tritium are both considered first-generation fusion fuels; they are the easiest to fuse, because the electrical charge on their nuclei is the lowest of all elements. The three most commonly cited nuclear reactions that could be used to generate energy are:

$$^{2}H + ^{3}H \rightarrow n\ (14.07\ MeV) + ^{4}He\ (3.52\ MeV)$$

$$^{2}H + ^{2}H \rightarrow n\ (2.45\ MeV) + ^{3}He\ (0.82\ MeV)$$

$$^{2}H + ^{2}H \rightarrow p\ (3.02\ MeV) + ^{3}H\ (1.01\ MeV)$$

Second-generation Fusion Fuel

Second-generation fuels require either higher confinement temperatures or longer confinement time than those required of first-generation fusion fuels, but generate fewer neutrons. Neutrons are an unwanted byproduct of fusion reactions in an energy generation context, because they are absorbed by the walls of a fusion chamber, making them radioactive. They cannot be confined by magnetic fields, because they are not electrically charged. This group consists of deuterium and helium-3. The products are all charged particles, but there may be significant side reactions leading to the production of neutrons.

$$^{2}H + {}^{3}He \rightarrow p\ (14.68\ MeV) + {}^{4}He\ (3.67\ MeV)$$

Third-generation Fusion Fuel

Third-generation fusion fuels produce only charged particles in the primary reactions, and side reactions are relatively unimportant. Since a very small amount of neutrons is produced, there would be little induced radioactivity in the walls of the fusion chamber. This is often seen as the end goal of fusion research. ^{3}He has the highest Maxwellian reactivity of any 3rd generation fusion fuel. However, there are no significant natural sources of this substance on Earth.

$$^{3}He + {}^{3}He \rightarrow 2\ p + {}^{4}He\ (12.86\ MeV)$$

Another potential aneutronic fusion reaction is the proton-boron reaction:

$$p + {}^{11}B \rightarrow 3\ {}^{4}He\ (8.7\ MeV)$$

Under reasonable assumptions, side reactions will result in about 0.1% of the fusion power being carried by neutrons. With 123 keV, the optimum temperature for this reaction is nearly ten times higher than that for the pure hydrogen reactions, the energy confinement must be 500 times better than that required for the D-T reaction, and the power density will be 2500 times lower than for D-T.

Advantages and Disadvantages of Nuclear Fuels

Worldwide, there are extensive reserves of uranium left to mine. While nuclear fuel is not renewable, it is sustainable since there is so much of it. It will run out eventually, but not for centuries. Unlike fossil fuels, using nuclear fuels to produce energy does not directly produce carbon dioxide or sulfur dioxide. It should be mentioned that the processes of mining, transporting, and refining the fuel have carbon emissions associated with them, comparable to those of wind and solar power. Although the carbon footprint of using nuclear fuels is smaller, there are still disadvantages of using nuclear fuel. The waste, while a much lower volume must be handled very carefully because of its radioactivity. Nuclear fuels require far more complicated systems to extract their energy, which calls for greater regulation. These complex systems and regulation make for very long build times. In addition, public opinions on nuclear energy tend to be more negative than with other energy sources. The over-estimation of the dangers associated with releases of radioactive material is a significant issue, as large-scale nuclear incidents are rare.

Neutron Moderator

Neutron moderators are a type of material in a nuclear reactor that work to slow down the fast neutrons (produced by splitting atoms in fissile compounds like uranium-235), to make them more effective in the fission chain reaction. This slowing or moderation of the neutrons allows them to be more easily absorbed by fissile nuclei, creating more fission events.

Materials used for moderation need to a very specific set of properties. First, a moderator cannot absorb neutrons itself. This means that the moderator should have a *low* neutron absorption cross-section. However, the moderator should be able to slow down neutrons to an acceptable speed. Thus, in an ideal moderator the neutron scattering cross-section is high. This neutron scattering is a measure of how likely a neutron will interact with an atom of the moderator. If the collisions between neutrons and nuclei are elastic collisions, it implies that the closer in size the nucleus of an atom is to a neutron, the more the neutron will be slowed. For this reason, lighter elements tend to be more efficient moderators. The table below shows that common moderators have a low neutron absorption cross-section but a comparatively large neutron scattering cross-section.

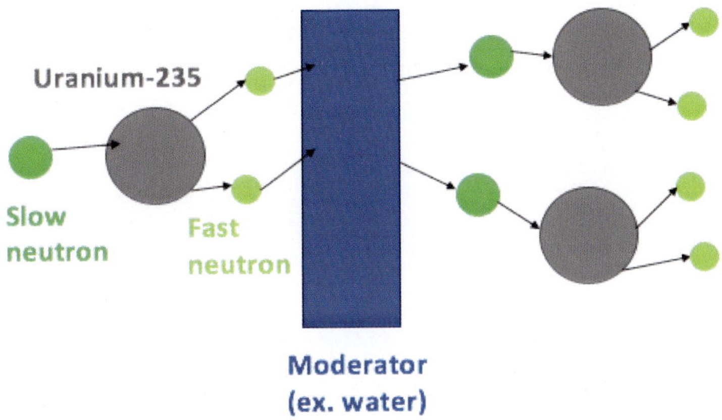

When a slow neutron collides with a fissile material like Uranium-235, it produces fast neutrons. The moderator will then slow these fast neutrons, and produce more slow neutrons to continue the nuclear chain reaction. When this process is repeated the fissile events are doubled each time.

Properties of Common Neutron Moderators		
	Neutron scattering cross section (σ_s) in barns	**Neutron absorption cross section (σ_s) in barns**
Light water (H_2O)	49	0.66
Heavy water (D_2O)	10.6	0.0013
Graphite (C)	4.7	0.0035

Types of Moderating Materials

There are several different types of moderating materials, and each have places where they are used more effectively. Typically-used moderator materials include heavy water, light water, and graphite. The relative properties of these materials are compared below. The moderators vary in terms of their moderating abilities, as well as in their costs.

Light Water

Light water (no different than regular water) is used in many reactors because it contains large amounts of hydrogen. Hydrogen works well as a neutron moderator because its mass is almost identical to that of a neutron. This means that one collision will significantly reduce the speed of the neutron because of the laws of conservation of energy and momentum. In addition, light water is abundant and fairly inexpensive. One drawback is that hydrogen has a relatively high neutron absorption cross-section because of its ability to form deuterium. Thus light water can only be used as a moderator along with enriched fuels. Reactors that use light water are known as light water reactors and include the pressurized water reactor (PWR), the boiling water reactor (BWR), and the supercritical water cooled reactor (SCWR).

Heavy Water

Heavy water is used in reactors because its benefits are similar to light water, but since it contains deuterium atoms, its neutron absorption cross section is much lower. The main disadvantage to the use of heavy water is its high cost of production, as it is made using the Girder-Sulfide process. Reactors that use heavy water include the CANDU designs and the pressurized heavy water reactor.

Graphite

Graphite has been a popular moderator in the past, however, one drawback is that it needs to be extremely pure to be effective. Graphite can be made artificially using boron electrodes, however, since boron is a very good neutron absorber— a small amount of contamination will make the graphite an ineffective moderator. One benefit of graphite is that even at the high purity that is necessary for graphite to perform well, it is available at a fairly low price. In addition, graphite is a good moderator as it is thermally stable and conducts heat well. However, at high temperatures the graphite can react with oxygen and carbon dioxide in the reactor and this decreases its effectiveness. Another potential issue with using graphite as a moderator is its ability to oxidize in the presence of air, and its low strength and density which could cause it to change dimensions in the reactor.

Reactors that use graphite moderator include the RBMK, pebble bed reactors, and the magnox reactor.

Alternatives

Nuclear reactors can be either thermal or fast. Currently, almost all operating reactors are thermal and thus require a moderator to slow down fast neutrons to the thermal level so that nuclear fission can continue. However, in fast reactors a moderator is not needed, and the neutrons within it move much more quickly. Fast reactors are beneficial as they enhance the sustainability of nuclear power. This is because they have the ability to get more neutrons out of their fuel, can transform nuclear waste into products that decay more quickly, and they respond better to potentially catastrophic equipment failures. However, they are more expensive, and they can overheat fairly easily.

Moderation

Neutrons are normally bound into an atomic nucleus, and do not exist free for long in nature. The unbound neutron has a half-life of 10 minutes and 11 seconds. The release of neutrons from the nucleus requires exceeding the binding energy of the neutron, which is typically 7-9 MeV for most isotopes. Neutron sources generate free neutrons by a variety of nuclear reactions, including nuclear fission and nuclear fusion. Whatever the source of neutrons, they are released with energies of several MeV.

According to the equipartition theorem, the average kinetic energy, $\bar{E}$, can be related to temperature, T ,via:

$$\bar{E} = \frac{1}{2} m_n \langle v^2 \rangle = \frac{3}{2} k_B T,$$

where m_n is the neutron mass, v is the average neutron speed, and k_B is the Boltzmann constant. The characteristic neutron temperature of several-MeV neutrons is several tens of billions kelvin.

Moderation is the process of the reduction of the initial high speed (high kinetic energy) of the free neutron. Since energy is conserved, this reduction of the neutron speed takes place by transfer of energy to a material called a *moderator*.

The probability of scattering of a neutron from a nucleus is given by the scattering cross section. The first couple of collisions with the moderator may be of sufficiently high energy to excite the nucleus of the moderator. Such a collision is inelastic, since some of the kinetic energy is transformed to potential energy by exciting some of the internal degrees of freedom of the nucleus to form an excited state. As the energy of the neutron is lowered, the collisions become predominantly elastic, i.e., the total kinetic energy and momentum of the system (that of the neutron and the nucleus) is conserved.

Given the mathematics of elastic collisions, as neutrons are very light compared to most nuclei, the most efficient way of removing kinetic energy from the neutron is by choosing a moderating nucleus that has near identical mass.

Reactor Moderators

In a thermal-neutron reactor, the nucleus of a heavy fuel element such as uranium absorbs a slow-moving free neutron, becomes unstable, and then splits ("fissions") into two smaller atoms ("fission products"). The fission process for ^{235}U nuclei yields two fission products, two to three fast-moving free neutrons, plus an amount of energy primarily manifested in the kinetic energy of the recoiling fission products. The free neutrons are emitted with a kinetic energy of ~2 MeV each. Because more free neutrons are released from a uranium fission event than thermal neutrons are required to initiate the event, the reaction can become self-sustaining — a chain reaction — under controlled conditions, thus liberating a tremendous amount of energy.

The probability of further fission events is determined by the fission cross section, which is dependent upon the speed (energy) of the incident neutrons. For thermal reactors, high-energy neutrons in the MeV-range are much less likely to cause further fission. (Note: It is not *impossible* for fast neutrons to cause fission, just much less likely.) The newly released fast neutrons, moving at roughly 10% of the speed of light, must be slowed down or "moderated," typically to speeds of a few kilometres per second, if they are to be likely to cause further fission in neighbouring ^{235}U nuclei and hence continue the chain reaction. This speed happens to be equivalent to temperatures in the few hundred Celsius range.

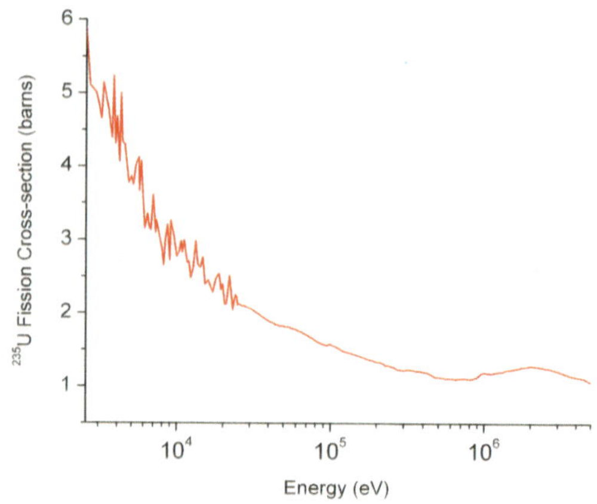

Fission cross section, measured in barns (a unit equal to 10^{-28} m^2), is a function of the energy (so-called excitation function) of the neutron colliding with a ^{235}U nucleus.

Fission probability decreases as neutron energy (and speed) increases. This explains why most reactors fueled with ^{235}U need a moderator to sustain a chain reaction and why removing a moderator can shut down a reactor.

In all moderated reactors, some neutrons of all energy levels will produce fission, including fast neutrons. Some reactors are more fully *thermalised* than others; for example, in a CANDU reactor nearly all fission reactions are produced by thermal neutrons, while in a pressurized water reactor (PWR) a considerable portion of the fissions are produced by higher-energy neutrons. In the proposed water-cooled supercritical water reactor (SCWR), the proportion of fast fissions may exceed 50%, making it technically a fast neutron reactor.

A fast reactor uses no moderator, but relies on fission produced by unmoderated fast neutrons to sustain the chain reaction. In some fast reactor designs, up to 20% of fissions can come from direct fast neutron fission of uranium-238, an isotope which is not fissile at all with thermal neutrons.

Moderators are also used in non-reactor neutron sources, such as plutonium-beryllium and spallation sources.

Moderator Impurities

Good moderators are also free of neutron-absorbing impurities such as boron. In commercial nuclear power plants the moderator typically contains dissolved boron. The boron concentration of the reactor coolant can be changed by the operators by adding boric acid or by diluting with water to manipulate reactor power. The Nazi Nuclear Program suffered a substantial setback when its inexpensive graphite moderators failed to work. At that time, most graphites were deposited on boron electrodes, and the German commercial graphite contained too much boron. Since the war-time German program never discovered this problem, they were forced to use far more expensive heavy water moderators. In the U.S., Leó Szilárd, a former chemical engineer, discovered the problem.

Non-graphite Moderators

Some moderators are quite expensive, for example beryllium, and reactor-grade heavy water. Reactor-grade heavy water must be 99.75% pure to enable reactions with unenriched uranium. This is difficult to prepare because heavy water and regular water form the same chemical bonds in almost the same ways, at only slightly different speeds.

The much cheaper light water moderator (essentially very pure regular water) absorbs too many neutrons to be used with unenriched natural uranium, and therefore uranium enrichment or nuclear reprocessing becomes necessary to operate such reactors, increasing overall costs. Both enrichment and reprocessing are expensive and technologically challenging processes, and additionally both enrichment and several types of reprocessing can be used to create weapons-usable material, causing proliferation concerns. Reprocessing schemes that are more resistant to proliferation are currently under development.

The CANDU reactor's moderator doubles as a safety feature. A large tank of low-temperature, low-pressure heavy water moderates the neutrons and also acts as a heat sink in extreme loss-of-coolant accident conditions. It is separated from the fuel rods that actually generate the heat. Heavy water is very effective at slowing down (moderating) neutrons, giving CANDU reactors their important and defining characteristic of high "neutron economy."

Neutron Poison

Neutron poisons are substances used in nuclear reactors to exert control of reactivity. The chemistry of these poisons can be a determining factor in whether or not a poison is used for a particular reactor. Boron carbide (B_4C), boric acid (H_3BO_3), and gadolinia (Gd_2O_3) are three of the most prominent neutron poisons and each has their advantages and flaws. Boric acid has the most potential for hazardous damage but is one of the few useful poisons that are dissolved in the reactor coolant. Boron carbide and gadolinia are inserted into fuel assemblies via separated control rods or directly mixed into the fissionable fuel, respectively. Boron carbide, though chemically resilient, produces helium, the buildup of which can cause pressure issues in control rods. Gadolinia and rare earth poisons can produce undesirable decay products due to neutron irradiation.

Boron as a Poison

Boron can be used in many different forms in order to fulfill its role as a neutron poison in reactors. Elemental boron is 19.9% ^{10}B and 80.1% ^{11}B, so it should be noted that ^{10}B is the isotope of interest for manufacturing neutron poisons.[1] Boron-10 is of interest to the study of neutron poisons as a result of its relatively high thermal (room temperature) absorption cross section, with a value of 3835 barns (3.835×10^{-21} cm^2).[1]

Boron is used in various compounds in reactors to control excess reactivity. Boric acid is a common boron compound used in reactor control. Boric acid can be dissolved in reactor coolant in order to decrease probability of fission of reactor fuel by increasing absorption probability of neutrons traveling through coolant. While boric acid serves its purpose of absorbing neutrons well, its use can have potentially hazardous side effects. In March 2002 the Davis-Besse nuclear power station was subject to a Nuclear Regulatory Commission (NRC) inspection and boric acid crystal deposits were found at several openings on the pressure vessel.

Exposure to boric acid poses a huge threat to reactor materials due to the speed at which it can corrode metals if left unattended. While boric acid concentrations *inside* the reactor vessel are trivial due to carefully controlled amounts of acid; if there is a leak and acid accumulates on the outside of the vessel, the acid can become concentrated as water is boiled off. "These factors can increase the corrosion rate of exposed carbon steel from <0.001 in/yr (<0.024 mm/yr) to as much as 10 in/yr (254 mm/yr)." Cracks as large as 7 in. (177.8 mm) were found on the Davis Besse pressure vessel, a large portion of which can be attributed to corrosion via boric acid.

Boric acid corrodes metals by increasing the amount of positive hydrogen ions (H^+) available in water. The increase in H^+ ions by definition lowers the pH of the water, thus increasing the ability of the water to absorb negatively charged electrons. This leads to an increase in electrons removed from the iron used in vessel materials, causing an increase in the amount of iron ions created which react with dissolved oxygen in water to produce rust. The inspection of the Davis-Besse reactor shows that temperature can drastically affect concentrations of boric acid and in turn corrosion rates. Figure shows a plot of pH versus concentration of boric acid as well as a plot of pH versus temperature for a solution of boric acid from room temperature to 100 °C.

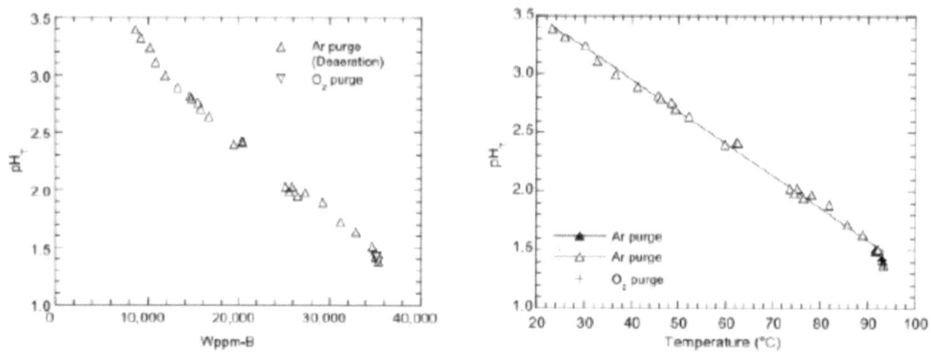

Figure: Plots of pH versus boron concentration and pH versus Temperature in a solution of either aerated or deaerated solutions of boric acid.

From the plots it can be concluded that as temperature in the reactor vessel increases, pH decreases. A decrease in pH leads to an increase in the concentration of boric acid which increases the corrosion effects experienced by exposed reactor materials. The combination of these effects could lead to a corrosion rate of up to 125.2 mm/yr (on A533 Gr.-B low-alloy steel), although this will generally decrease with time as corrosion rates are often higher in the short-term due to higher reactivity of a fresh solution.

Boron carbide is another boron based neutron poison used to control reactivity. Boron carbide differs from boric acid in terms of the way it is used as well as its chemistry. Boron carbide is an incredibly hard solid material that is inserted into rods in fuel assemblies to reduce excess reactivity. Due to its incredible chemical stability boron carbide poses less of a risk when used in reactors. Boron carbide is not affected by cold chemical reagents and oxidation begins at 600 °C, as well as attack by halides such as chlorine resulting in boron trihalide formation.

Although safer chemically, boron carbide still has drawbacks. The most notable disadvantage of boron carbide would be helium production. When irradiated by neutrons produced by fission, boron produces helium atoms via the following reaction:

${}^{10}_{5}B + {}^{1}_{0}n \rightarrow {}^{7}_{3}Li + {}^{4}_{2}He$, resulting in a pressure buildup in the control rods if the helium is contained. If left unchecked, boron carbide can also pose problems for operating reactors.

Poisons Added to Fuel

Neutron poisons can also be implemented in reactors by combining the poison and the fuel pellets inside the fuel assembly. This method simplifies reactivity control due to a predetermined amount of poison being preloaded into the fuel, but poison-fuel mixing adds another step to fuel manufacturing. Oxide compounds of elements such as erbium, dysprosium, and hafnium are all candidates for mixed fuel poisons, but the most common poison mixed with fuel is gadolinium. Gadolinia is made with natural gadolinium which as has two naturally occurring isotopes,Gd and Gd, with thermal absorption cross sections of 58,000 b and 240,000 b respectively.

Benefits to using gadolinia include a high melting point of 2850 °C, which is ideal due to resilience under operating reactor temperatures, as well as gadolinium being the most abundant of the rare earth metals.[6] Gadolinium is ideal as a neutron poison considering its incredibly high absorption cross section. These high absorption cross sections for gadolinium isotopes *should* result in a smaller amount of gadolinium needed to mix with fuel to ensure reactivity control. Counterintuitively, even though gadolinium has a much larger thermal absorption cross section than boron, it's relative efficiency in absorbing neutrons is roughly the same when mixed with UO_2(0.16 neutrons absorbed per fission neutron produced) due to boron's absorption cross section surpassing gadolinium's for neutron energies above 0.2 electron volts.[6] These differences can be attributed to the atomic density of boron carbide being "15 times larger than the atomic densities of the two absorbing gadolinium isotopes in gadolinia."

Other rare earths such as erbium and dysprosium are also candidates that can be considered for neutron poisons. Erbium-167 and ^{164}Dy have thermal absorption cross sections (800 barns and 1000 barns respectively) that are not as high as gadolinium's so their benefits are limited.5 Erbium in particular is considered over gadolinium more because its "efficiency as a function of the content and number of poisoned rods is nearly linear, and that whereas cycle length effect(s) and initial efficiency are clearly separated when using gadolinium, with erbium, the poison may be selected according to both parameters and the lowest possible content can be selected or, in any case, the content that will not downgrade the thermal conductivity."[7] Essentially this means that erbium could be a more efficient poison due to a lower concentration of poison needed to be mixed in each fuel rod, as well as minimal effects on the heat conduction of the fuel.

Limitations of Current Methods

Neutron poisons can be difficult to study due to the purpose they are meant to serve. Compounds such as boric acid are relatively easy to study because its chemistry and effects are more easily studied without the aid of a nuclear reactor. Compounds such as gadolinia are more difficult to study without the aid of a reactor or computer simulation because it is typically mixed directly with reactor fuel to form Gd_2O_3-UO_2. This limitation also applies to other rare earth oxide poisons such as erbium and dysprosium, meaning a reactor loaded with different rare earth oxide poisons is necessary or preferable to study behavior of these poisons, or to verify if they behave in accordance with simulations. Rare earth poisons are also limited by their natural abundances; for a neutron poison to be commercially viable it must have an abundance within a practical range for consumer purposes, as well as a chemistry that allows for practical separation of the element or compound.

Control Rod

A control rod is a device that is used to absorb neutrons so that the nuclear chain reaction taking place within the reactor core can be slowed down or stopped completely by inserting the rods further, or accelerated by removing them slightly. Essentially, control rods provide real-time control of the fission process, ensuring that it remains active while preventing it from accelerating out of control.

Uranium-235 fission releases 2.5 neutrons on average, but only one neutron is needed to sustain the nuclear chain reaction at a steady rate. The control rods absorb these extra neutrons and can be used to adjust the power output of the reactor. When inserted the standard amount, their position is at criticality and the power output remains the same. If the rod is pushed in the number of neutrons decreases along with the power output and the reactor is below criticality. The opposite is true if the rods are pulled out slightly as fission goes beyond criticality. This is shown in Figure.

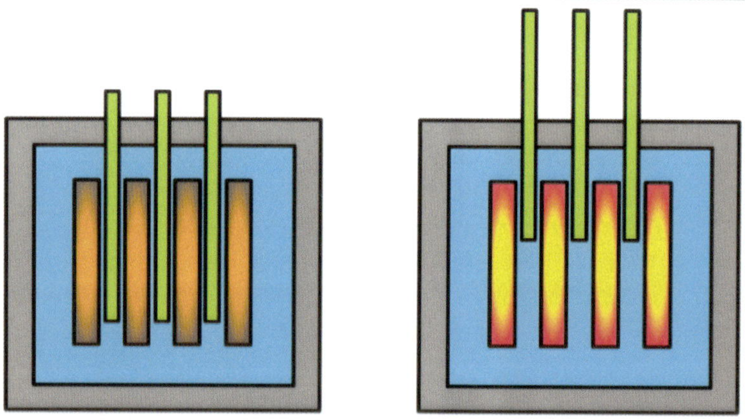

A schematic showing how reactor power output changes with how much the control rods (shown in green) are inserted. On the left, the control rods are inserted more than usual, reducing the power output of the reactor. On the right, the control rods are inserted less than usual, increasing the power output.

Design

Control rods are rods, plates, or tubes containing a material (like boron or hafnium) that absorbs neutrons. Several of these control rods, about the size of fuel rods, are spaced evenly and connected on one end by a metal bracket known as a spider. Typical reactors can contain around 50 of these clusters with 20 individual control rods in each cluster.

The ability of a control rod to absorb neutrons to control the fission chain reaction requires a choice of material that has high neutron-absorbing abilities. The measure of how well a material absorbs neutrons is known as its neutron absorption cross section, or σa, measured in barns (equal to 10^{-28} square meters). Generally speaking, control rods are made using cadmium, hafnium, or enriched boron.

Along with the choice in material, the mechanical properties and cost are important when designing a control rod. For example, boron-10 is one of the best neutron absorbers, however it is not ideal for building control rods as it is extremely brittle. Additionally, natural boron must be enriched to have reasonable absorption levels, which becomes very expensive.

Why the Need to Control?

It does not require much reflection to imagine why proper control is necessary with in nuclear reactor. Some of the basic reasons are as follows:

- A nuclear chain reaction should be started when a reactor fires from the cold condition. In the absence of such a reaction the process would soon die out.
- It is not only necessary and sufficient to start the chain reaction but it is equally necessary to ensure that the reaction is sustained in the long run as long as the power requirements are present.
- In case of emergency situations such as a sudden mechanical or structural damage, the reactor needs to be shut down quickly in order to prevent any major disaster like say Chernobyl which could be very costly in terms of loss to life and environment.

Fuel rods inside the reactor should be prevented from melting or getting disintegrated and therefore a control mechanism is absolutely necessary.

We have seen the reasons for controlling and taming the wild nuclear power and the best method to achieve this is through the use of control rods which can be inserted or withdrawn from the core and help to control the nuclear reactions taking place inside the reactor.

Safety Precautions

Control rods are important safety features of reactors as they allow the user to control the output of the reactor. In some types of reactors, most notably CANDU reactors, the control rods are held up by electromagnets. This means that if there is some sort of power failure or loss of signal the control rods are immediately released and fall into the reactor core because of gravity. This prevents the fission reaction from continuing and acts as the primary shutdown system in case of emergency. This dropping motion can also be induced manually if the machinery holding the rods up fails in some way. When the control rods are dropped into the reactor, it is a process known as scramming. In boiling water reactors, the design is different as the control rods must be pushed up into the reactor since they are located on the bottom of the reactor in that design.

Neutron Reflector

It is well known that each reactor core is surrounded by a neutron reflector or reactor core baffle. The reflector reduces the non-uniformity of the power distribution in the peripheral fuel assemblies, reduces neutron leakage and reduces a coolant flow bypass of the core. The neutron reflector is a non-multiplying medium, whereas the reactor core is a multiplying medium.

Neutron reflector inside a reactor core of LWR.

The neutron reflector scatters back (or reflects) into the core many neutrons that would otherwise escape (i.e. reduces the neutron leakage). By reducing neutron leakage, the reflector increases k_{eff} and reduces the amount of fuel necessary to maintain the reactor critical for a long period. In LWRs the neutron reflector is installed for following purposes:

- The neutron flux distribution is "flattened", i.e., the ratio of the average flux to the maximum flux is increased. Therefore reflectors reduce the non-uniformity of the power distribution.
- Because of the higher flux at the edge of the core, there is much better utilization in the peripheral fuel assemblies. This fuel, in the outer regions of the core, now contributes much more to the total power production.
- The neutron reflector scatters back (or reflects) into the core many neutrons that would otherwise escape. The neutrons reflected back into the core are available for chain reaction. This means that the minimum critical size of the reactor is reduced. Alternatively, if the

core size is maintained, the reflector makes additional reactivity available for higher fuel burnup. The decrease in the critical size of core required is known as the reflector savings.

- Neutron reflectors reduce neutron leakage i.e. to reduce the neutron fluence on a reactor pressure vessel.
- Neutron reflectors reduce a coolant flow bypass of a core.
- Neutron reflectors serve as a thermal and radiation shield of a reactor core.

The efficiency of a reflector is measured by the ratio of the number of neutrons reflected back into the reactor to the number entering the reflector. This ratio is called the albedo, or the reflection coefficient. The albedo can be expressed it in terms of neutron currents as:

$$\alpha = \frac{J_{exit}}{J_{enter}} = \frac{J_x^-(0)}{J_x^+(0)}$$

The value of the albedo will depend on the composition and thickness of the reflector. An infinite reflector will have the maximum albedo, but in all practical cases the geometry of a reflector must conform to the geometry of a reactor core and a reactor pressure vessel. For sufficiently thick reflectors, it can be derived, that albedo becomes:

$$\alpha = \frac{1 - 2D_{refl} / L_{refl}}{1 + 2D_{refl} / L_{refl}}$$

where D_{refl} is the diffusion coefficient in the reflector and the L_{refl} is the diffusion length in the reflector.

Materials and Types of Neutron Reflectors

Top view of the WWER-1000 nuclear core. The lower support structure, the neutron reflector and 11 fuel assemblies."

There are a variety of materials which are used as a reflecting medium for neutrons in nuclear reactors. In general, reflector materials are normally not fissionable, have a high scattering cross section, and have a low absorption cross section. Since main differences among reactor types arise from differences in their neutron energy spectra, we have to specify whether we mean reflectors in fast reactors or in thermal reactors. Since the moderation of neutrons is undesirable in fast reactors, they use only high Z materials.

To be an effective reflector, its material must possess the following properties.

- Low absorption cross-section
- High reflection coefficient
- Radiation stability – A radiation stability is very important property of reflectors, because the reflector material will be exposed to high levels of radiation (especially gamma radiation and fast neutrons)
- Resistance to Oxidation – The reflector is often situated in a chemically aggressive environment (especially in fast reactors), therefore the material should not get oxidized.

Elastic Scattering Cross-sections for Light and Heavy Elements

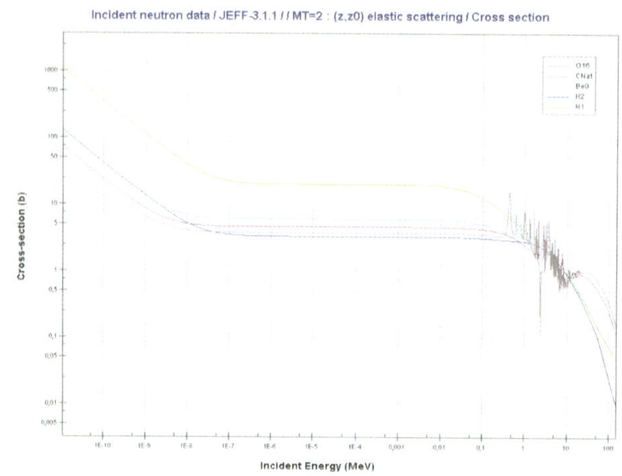

Elastic scattering cross-sections for light elements are more or less independent of neutron energy up to 1 MeV.

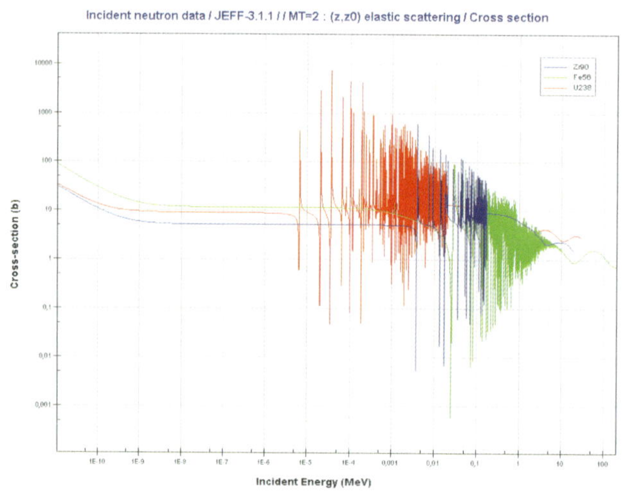

For intermediate and heavy elements, the elastic cross-section is constant at low energy with some specifics at higher energy.

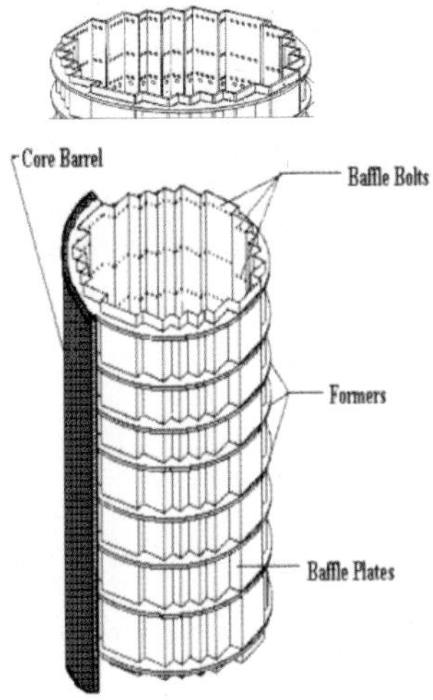

Core Baffle

Essentially, for thermal reactors a good moderator is also a good reflector, because most of the moderators also possess the above mentioned properties of a good reflector as well. Water, heavy water, beryllium, or graphite are commonly used as reflectors. In pressurized water reactors water serves as an axial reflector. This axial reflector does not form any special device, the neutrons are simply reflected by core inlet and outlet coolant.

On the other hand common water volume cannot be used as a reflector in radial direction, because it is of the highest importance to maintain high flow rates in the core and not to bypass fuel assemblies. Therefore the neutron reflectors are installed in PWR and BWR reactor cores. Design of neutron reflectors may vary, but we can distinguish between two basic types:

- Core Baffle: Core baffle consists of baffle and former assembly that consist of vertical plates called baffles and horizontal support plates called formers.This assembly forms the interface between the core and the core barrel and there is water between the baffle and the core barrel. Since the coolant flow in the former region is significantly reduced by the former, the high flow rate through fuel assemblies is maintained. A secondary benefit is that water in the former region reduces the neutron flux on the pressure vessel, which causes an irradiation embrittlement of pressure vessel material.

Visualisation of a heavy reflector. It is only an illustrative example.

Heavy Reflector: Heavy reflector is a structure that is installed inside a core barrel (similarly as a core baffle). But the heavy reflector is a wall made of stainless steel slabs stacked vertically surrounding the reactor core. Due to higher atomic number density heavy reflectors reduce neutron leakage (especially of fast neutrons) from the core more efficiently than core baffle. This provides additional protection of the reactor vessel from irradiation embrittlement, caused especially by fast neutrons. While acting as a neutron shield, the heavy reflector is heated due to absorption of the gamma radiation. In order to avoid overheating, the heat in the reflector is removed by water flowing through cooling channels drilled through the reflector.

In fast reactors moderation of neutrons is undesirable, therefore reflectors are not composed of moderating materials. In fast breeder reactors the core is surrounded by a radial core reflector. The radial core reflector is usually composed of fuel assemblies with natural uranium (so called the radial blanket zone), for which the diffusion coefficient is about the same as that of the core. This reflector improves the radial flux distribution, reduces the fissile inventory, and increases the internal breeding ratio.

Neutron Sources

A neutron source is any device that emits neutrons. Neutron sources have many applications, they can be used in research, engineering, medicine, petroleum exploration, biology, chemistry and nuclear power. A neutron source is characterized by a number of factors:

- Significance of the source.
- Intensity. The rate of neutrons emitted by the source.
- Energy distribution of emitted neutrons.
- Angular distribution of emitted neutrons.
- Mode of emission. Continuous or pulsed operation.

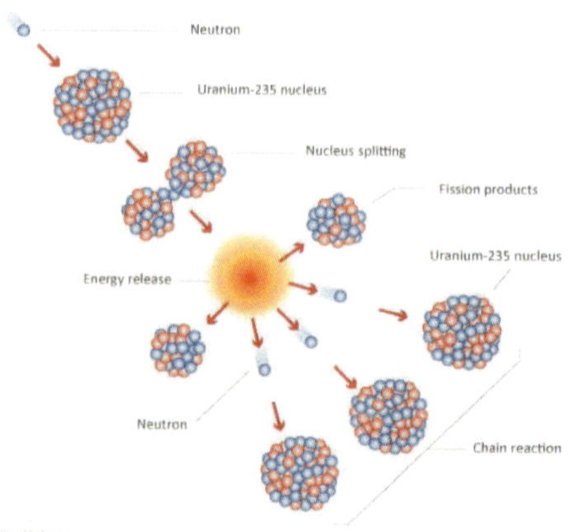

A nuclear chain reaction occurs when one single nuclear reaction causes an average of one or more subsequent nuclear reactions.

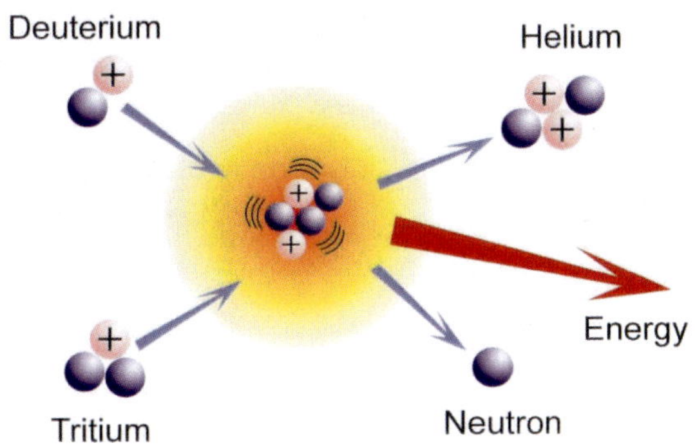

Classification by Significance of the Source

- Large (Significant) neutron sources:
    - Nuclear Reactors: There are nuclei that can undergo fission on their own spontaneously, but only certain nuclei, like uranium-235, uranium-233 and plutonium-239, can sustain a fission chain reaction. This is because these nuclei release neutrons when they break apart, and these neutrons can induce fission of other nuclei. Uranium-235 which exists as 0.7% of naturally occurring uranium undergoes nuclear fission with thermal neutrons with the production of, on average, 2.4 fast neutrons and the release of ~ 180 MeV of energy per fission. Free neutrons released by each fission play very important role as a trigger of the reaction, but they can be also used fo another purpose. For example: One neutron is required to trigger a further fission. Part of free neutrons (let say 0.5 neutrons/fission) is absorbed in other material, but an excess of neutrons (0.9 neutrons/fission) is able to leave the surface of the reactor core and can be used as a neutron source.
- Fusion Systems: Nuclear fusion is a nuclear reaction in which two or more atomic nuclei (e.g. D+T) collide at a very high energy and fuse together. Thy byproduct of DT fusion is a free neutron (see picture), therefore also nuclear fusion reaction has the potential to produces large quantities of neutrons.
- Spallation Sources: A spallation source is a high-flux neutron source in which protons that have been accelerated to high energies hit a heavy target material, causing the emission of neutrons. The reaction occurs above a certain energy threshold for the incident particle, which is typically 5 – 15 MeV.
- Medium neutron sources:
    - Bremssstrahlung from Electron Accelerators / Photofission: Energetic electrons when slowed down rapidly in a heavy target emit intense gamma radiation during the deceleration process. This is known as Bremsstrahlung or braking radiation. The interaction of the gamma radiation with the target produces neutrons via the (γ,n) reaction, or the (γ,fission) reaction when a fissile target is used. e-→Pb → γ→ Pb →(γ,n) and (γ,fission). The Bremsstrahlung γ energy exceeds the binding energy of the "last" neutron in the target. A source strength of 10^{13} neutrons/second produced in short (i.e. < 5 μs) pulses can be readily realised.

- Dense plasme focus: The dense plasma focus (DPF) is a device that is known as an efficient source of neutrons from fusion reactions. Mechanism of dense plasma focus (DPF) is based on nuclear fusion of short-lived plasma of deuterium and/or tritium. This device produces a short-lived plasma by electromagnetic compression and acceleration that is called a pinch. This plasma is during the pinch hot and dense enough to cause nuclear fusion and the emission of neutrons.

- Light ion accelerators: Neutrons can be also produced by particle accelerators using targets of deuterium, tritium, lithium, beryllium, and other low-Z materials. In this case the target must be bombarded with accelerated hydrogen (H), deuterium (D), or tritium (T) nuclei.

• Small neutron sources:

- Neutron Generators: Neutrons are produced in the fusion of deuterium and tritium in the following exothermic reaction. $^{2}D + {}^{3}T \rightarrow {}^{4}He + n + 17.6$ MeV. The neutron is produced with a kinetic energy of 14.1 MeV. This can be achieved on a small scale in the laboratory with a modest 100 kV accelerator for deuterium atoms bombarding a tritium target. Continuous neutron sources of ~10^{11} neutrons/second can be achieved relatively simply.

- Radioisotope source – (α,n) reactions: In certain light isotopes the 'last' neutron in the nucleus is weakly bound and is released when the compound nucleus formed following α-particle bombardment decays. The bombardment of beryllium by α-particles leads to the production of neutrons by the following exothermic reaction: $^{4}He + {}^{9}Be \rightarrow {}^{12}C + n + 5.7$ MeV. This reaction yields a weak source of neutrons with an energy spectrum resembling that from a fission source and is used nowadays in portable neutron sources. Radium, plutonium or americium can be used as an α-emitter.

- Radioisotope source – (γ,n) reactions: (γ,n) reactions can also be used for the same purpose. In this type of source, because of the greater range of the γ-ray, the two physical components of the source can be separated making it possible to 'switch off' the reaction if so required by removing the radioactive source from the beryllium. (γ,n) sources produce a monoenergetic neutrons unlike (α,n) sources. The (γ,n) source uses antimony-124 as the gamma emitter in the following endothermic reaction.

$$^{124}Sb \rightarrow {}^{124}Te + \beta - + \gamma$$

$$\gamma + {}^{9}Be \rightarrow {}^{8}Be + n - 1.66 \text{ MeV}$$

- Radioisotope source – spontaneous fission: Certain isotopes undergo spontaneous fission with emission of neutrons. The most commonly used spontaneous fission source is the radioactive isotope californium-252. Cf-252 and all other spontaneous fission neutron sources are produced by irradiating uranium or another transuranic element in a nuclear reactor, where neutrons are absorbed in the starting material and its subsequent reaction products, transmuting the starting material into the SF isotope.

Neutron Source	Example	Neutron Yield	Released Energy MeV / neutron
(α,n) reactions	Radium-Beryllium source	8 x 10-5 / α	6,600,000
(D,T) fusion	400 keV deuterons in neutron generator	4 x 10-5 / deuteron	10,000
Electron Bremsstrahlung / Photofission	100 MeV electrons on uranium	5 x 10-2 / electron	2,000
Fission	Induced fission of U-235 in thermal reactor	1 / fission	180
Spallation	800 MeV protons on uranium	30 / proton	55
(D,T) fusion	Inertial confinement fusion	1 / fusion	18

References

- Hargraves, Robert. "Liquid Fuel Nuclear Reactors". Forum on Physics and Society. APS Physics. Retrieved 14 July 2018
- Power-plants, components-of-nuclear-power-plant-control-rods-2724: brighthubengineering.com, Retrieved 11July 2018
- Price, M. S. T. (2012). "The Dragon Project origins, achievements and legacies". Nucl. Eng. Design. 251: 60–68. doi:10.1016/j.nucengdes.2011.12.024
- Chemistry-of-Neutron-Poisons-220302405: scribd.com, Retrieved 30 March 2018
- DOE Fundamentals Handbook: Nuclear Physics and Reactor Theory. Vol. 2 (DOE-HDBK-1019/2-93) (PDF). U.S. Department of Energy. January 1993. Retrieved November 29, 2013
- Nuclear-reactor-core, nuclear-power-plant: nuclear-power.net, Retrieved 28 May 2018

4

Concepts in Nuclear Reactor Physics

The field of nuclear reactor physics is concerned with the study of controlled chain reactions for generating a desired rate of fission for the production of energy in a nuclear reactor. The aim of this chapter is to give an overview of the principles and fundamental concepts of nuclear reactor physics, such as nuclear reactions and nuclear chain reaction.

Nuclear Reactor Physics

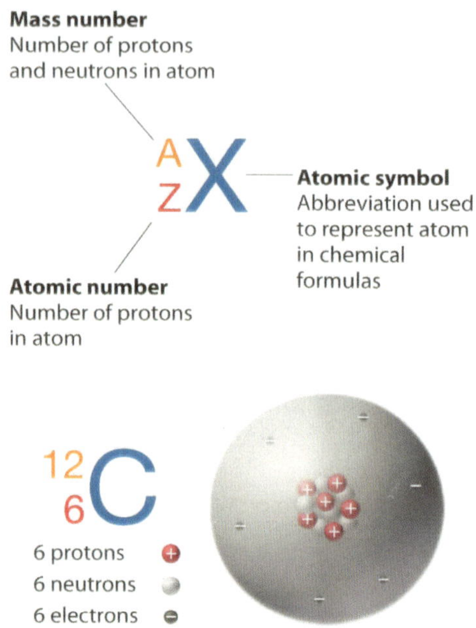

Notation of nuclei

A knowledge of atomic and nuclear physics is essential to nuclear engineers, who deal with nuclear reactors. It should be noted that atomic and nuclear physics is very extensive branch of science. Nuclear reactor physics belongs to an applied physics as a particle physics or nuclear chemistry. These branches have common fundamentals. Atomic and nuclear physics describes fundamental particles (i.e. electrons, protons, neutrons), their structure, properties and behavior. These physical fundamentals consist of:

- Fundamental Particles
- Atomic and Nuclear Structure
- Mass and Energy

- Radiation
- Nuclear Stability
- Radioactive Decay
- Nuclear Reactions
- Binding Energy

Interaction of Radiation with Matter

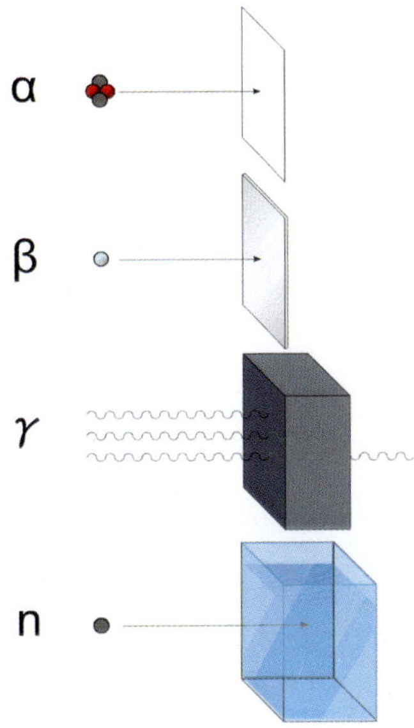

Interaction of Radiation with Matter

The design of all nuclear reactors and other nuclear systems depends fundamentally on the way in which nuclear radiation interacts with matter. In nuclear reactor we can typically encounter with one of these types of radiation:

- Interactions of heavy charged particles
- Interactions of beta radiation
- Interactions of gamma rays
- Interactions of neutrons
- Reactor antineutrinos

Neutron Diffusion Theory

Neutron Diffusion Theory is the branch of science that deals with the study and application of neutrons and its behavior within the nuclear core or in various environments. The Diffusion Theory provides theoretical basis for a neutron-physical computing of nuclear cores. The term theoretical has to be emphasized.

In order to design a nuclear reactor properly, it is necessary to be able to predict how the neutrons will be distributed throughout the system. This, in general, is a difficult problem, for the neutrons move in a reactor in complicated paths as a result of repeated collisions. To a first approximation the overall effect of these collisions is that the neutrons undergo a kind of diffusion in the reactor medium, much like the diffusion of one gas in another. This approximation allows solving such problem using the diffusion equation.

Nuclear Reactor Theory

The nuclear reactor theory is based on diffusion theory. The key term of the reactor theory is "criticality" of the reactor. Using the term "criticality" may seem counter-intuitive as a way to describe normalcy. The word often describes situations with potential for disaster. Nevertheless, in the context of nuclear power, "criticality" indicates that a reactor is operating safely and the neutron flux is stable at a given power level. In a critical reactor, there is a balance between the number of neutrons produced in fission and the number lost, either by absorbtion in the reactor or by leaking from its surface. The nuclear reactor theory solves the central problem in the design of a reactor, what is the calculation of the size and composition (nuclear fuel) of the system required to maintain this balance.

Reactor Kinetics – Zero Power Physics

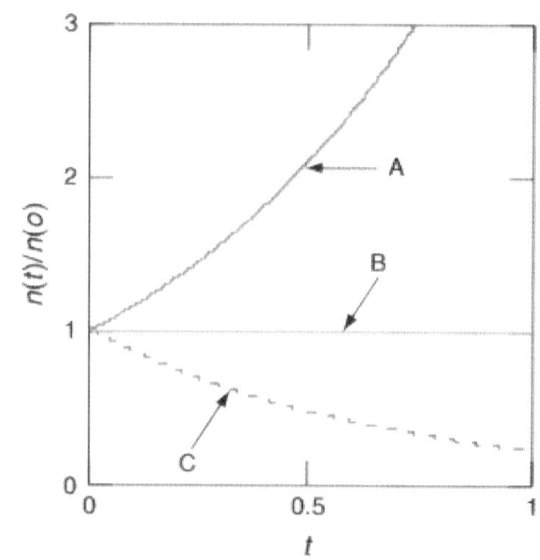

Reactor criticality. A – a supercritical state; B – a critical state; C – a subcritical state

The nuclear reactor theory is concerned with critical nuclear reactors operating at constant power. But the reactor must not be always in the critical state. The reactor must be noncritical (supercritical or subcritical) in order to change the reactor power. The study of the behavior of neutron populations in a noncritical ractor is called reactor kinetics. This theory introduces a new unit – reactivity – of a nuclear reactor. The reactivity characterizes the deflection of reactor from the critical state.

It should be noted this theory can be used for a reactor at low power levels, hence "zero power criticality" (i.e. a reactor without reactivity feedbacks, 10E-8% – 1% of rated power). First periods of a reactor operation (reactor startup and power maneuvering at low power levels) are well described by this theory. Once the reactor core heats up (thermal power outweighs a decay power), reactivity feedbacks change the dynamics of the reactor.

Reactor operation – Full Power Physics

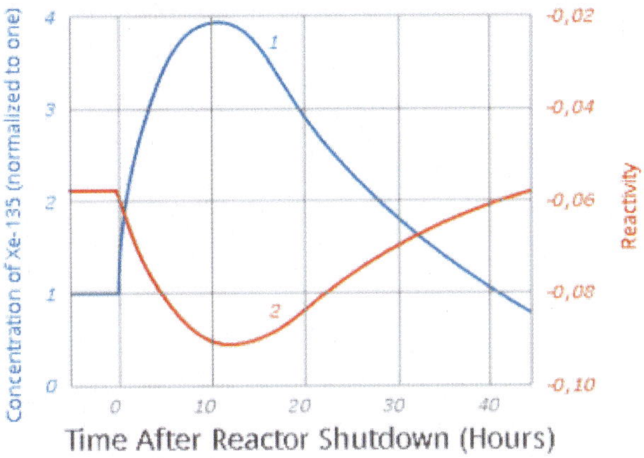

Iodine pit – Xenon poisoning

This branch of nuclear engineering deals is very specific and dependent on a reactor type. PWRs have much in common with Boiling Water Reactors, but the operation of the reactors is not the same. On the other hand for example fast neutron reactors have completely different physics than PWRs and BWRs.

The degree of criticality at high power levels (~1% – 100% of rated power) is influenced by many variables. The most important are summarized below:

- Nuclear Fuel and Fuel Loading Pattern.
- Fuel Depletion.
- Neutron Leakage.
- Control Rods and Chemical Shim.
- Reactivity Feedbacks.
- Burnable Absorbers.
- Xenon and Samarium Poisoning.

Basic reactor operation physics includes these operation states:

- Reactor Startup ("going critical").
- Control of the Reactor and Power Maneuvering.
- Axial Offset Control.
- Xenon Oscillations.

- Reactor Shutdown.
- Emergency Shutdown – SCRAM.

Thermohydraulics of Nuclear Core

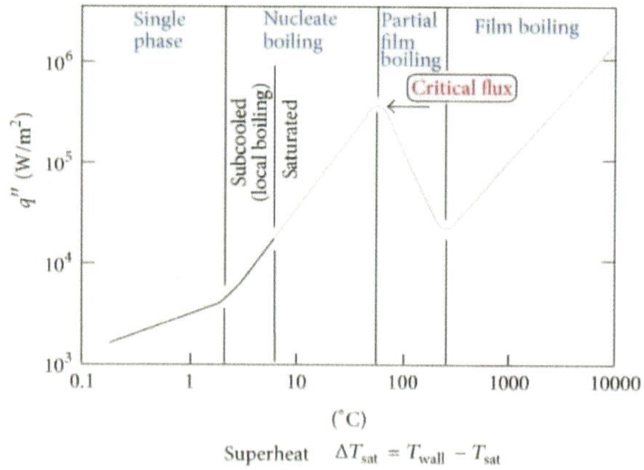

If the heat flux of a boiling system is higher than the critical heat flux then DNB (Departure from Nucleate Boiling) may occur.

Heat removal from nuclear reactors is also key issue of reactor physics. For a reactor to operate in a steady state, with an internal power distribution that is independent of time, all of the heat released in the system must be removed as fast as it is produced. This is accomplished by passing a coolant (water in case of PWRs or BWRs) through the core and through other regions where heat is generated.

The thermohydraulics deals with thermal behavior (heat generation, heat removal) and hydraulic behavior (Delta P, cross-flow issue, flow pattern) of the nuclear core. The nature and operation of this coolant system is one of the most important considerations on the design of a nuclear reactor. In the PWR reactors, there are phenomena which must be avoided. Namely DNB (Departure from Nucleate Boiling) and coolant saturation are very dangerous phenomena.

Behavior of Fuel Rod

Thermal and mechanical behavior of nuclear fuel constitute one of three key core design disciplines. Nuclear fuel is operated under very inhospitable conditions (thermal, radiation, mechanical) and must withstand more than normal conditions operation. For example temperatures in the centre of fuel pellets reach more than 1000°C (1832°F) accompanied by fission-gas releases. Behavior of nuclear fuel is associated with many phenomena. The most important are summarized below:

- Fuel swelling
- Fuel densification
- Thermal expansion
- Fuel relocation

- High Burnup Structure
- Pellet-Cladding Interaction
- Pellet-Coolant Interaction
- Fission-Gas Releases

Applied Core Physics

Core physics is the branch of science that deals with the study of behavior of specific nuclear core. The core physics provides methodological basis for a core design. Each nuclear core is due to a fuel reloading.

Full-scale physical computing of reactor core includes these types of computations:

- Core Design computing
    - Fuel cross-section computing. Neutron cross-sections must be calculated for for fresh fuel assemblies usually in two-Dimensional models of the fuel lattice.
    - Loading Pattern search. The best arrangement of new fuel assemblies and used fuel assemblies is searched for in the core design from the viewpoints of safety and economy.
    - Fuel Cycle depletion. Behavior and key safety parameters of each proposed fuel cycle (power distribution, reactor kinetics parameters, control rods efficiency, etc.) must be analyzed for entire duration of the proposed cycle.
- Reload Safety Assessment
    - Neutron-physical computing.
    - Thermal-hydraulic computing.
    - Fuel rod behaviour computing.

Criticality

In a nuclear reactor, the neutron population at any instant is a function of the rate of neutron production (due to fission processes) and the rate of neutron losses (due to non-fission absorption mechanisms and leakage from the system). When a reactor's neutron population remains steady from one generation to the next (creating as many new neutrons as are lost), the fission chain reaction is self-sustaining and the reactor's condition is referred to as "critical". When the reactor's neutron production exceeds losses, characterized by increasing power level, it is considered "supercritical", and when losses dominate, it is considered "subcritical" and exhibits decreasing power.

The "Six-factor formula" is the neutron life-cycle balance equation, which includes six separate factors, the product of which is equal to the ratio of the number of neutrons in any generation to that of the previous one; this parameter is called the effective multiplication factor k, also denoted by K_{eff}, where k = ϵ L_f ρ L_{th} f η, where ϵ = "fast-fission factor", L_f = "fast non-leakage factor", ρ = "resonance escape probability", L_{th} = "thermal non-leakage factor", f = "thermal fuel utilization factor", and η = "reproduction factor". This equation's factors are roughly in order of potential occurrence for a fission born neutron during critical operation. As already mentioned before, k = (Neutrons produced in one generation)/(Neutrons produced in the previous generation). In other words, when the reactor is critical, k = 1; when the reactor is subcritical, k < 1; and when the reactor is supercritical, k > 1.

Reactivity is an expression of the departure from criticality. δk = (k – 1)/k. When the reactor is critical, δk = 0. When the reactor is subcritical, δk < 0. When the reactor is supercritical, δk > 0. Reactivity is also represented by the lowercase Greek letter rho (ρ). Reactivity is commonly expressed in decimals or percentages or pcm (per cent mille) of Δk/k. When reactivity ρ is expressed in units of delayed neutron fraction β, the unit is called the *dollar*.

If we write 'N' for the number of free neutrons in a reactor core and τ for the average lifetime of each neutron (before it either escapes from the core or is absorbed by a nucleus), then the reactor will follow the differential equation (*evolution equation*)

$$\frac{dN}{dt} = \frac{\alpha N}{\tau}$$

where α is a constant of proportionality, and dN / dt is the rate of change of the neutron count in the core. This type of differential equation describes exponential growth or exponential decay, depending on the sign of the constant α , which is just the expected number of neutrons after one average neutron lifetime has elapsed:

$$\alpha = P_{impact} P_{fission} n_{avg} - P_{absorb} - P_{escape}$$

Here, P_{impact} is the probability that a particular neutron will strike a fuel nucleus, $P_{fission}$ is the probability that the neutron, having struck the fuel, will cause that nucleus to undergo fission, P_{absorb} is the probability that it will be absorbed by something other than fuel, and P_{escape} is the probability that it will "escape" by leaving the core altogether. n_{avg} is the number of neutrons produced, on average, by a fission event—it is between 2 and 3 for both ^{235}U and ^{239}Pu.

If α is positive, then the core is *supercritical* and the rate of neutron production will grow exponentially until some other effect stops the growth. If α is negative, then the core is "subcritical" and the number of free neutrons in the core will shrink exponentially until it reaches an equilibrium at zero (or the background level from spontaneous fission). If α is exactly zero, then the reactor is *critical* and its output does not vary in time ($dN / dt = 0$, from above).

Nuclear reactors are engineered to reduce $_{escape}$ and P_{absorb} . Small, compact structures reduce the probability of direct escape by minimizing the surface area of the core, and some materials (such as graphite) can reflect some neutrons back into the core, further reducing P_{escape} .

The probability of fission, $P_{fission}$, depends on the nuclear physics of the fuel, and is often expressed as a cross section. Reactors are usually controlled by adjusting P_{absorb} . Control rods made of a strongly neutron-absorbent material such as cadmium or boron can be inserted into the core: any neutron that happens to impact the control rod is lost from the chain reaction, reducing α . P_{absorb} is also controlled by the recent history of the reactor core itself (see below).

Starter Sources

The mere fact that an assembly is supercritical does not guarantee that it contains any free neutrons at all. At least one neutron is required to "strike" a chain reaction, and if the spontaneous fission rate is sufficiently low it may take a long time (in ^{235}U reactors, as long as many minutes) before a chance neutron encounter starts a chain reaction even if the reactor is supercritical. Most nuclear reactors include a "starter" neutron source that ensures there are always a few free neutrons in the reactor core, so that a chain reaction will begin immediately when the core is made critical. A common type of startup neutron source is a mixture of an alpha particle emitter such as ^{241}Am (americium-241) with a lightweight isotope such as ^{9}Be (beryllium-9).

The primary sources described above have to be used with fresh reactor cores. For operational reactors, secondary sources are used; most often a combination of antimony with beryllium. Antimony becomes activated in the reactor and produces high-energy gamma photons, which produce photoneutrons from beryllium.

Uranium-235 undergoes a small rate of natural spontaneous fission, so there are always some neutrons being produced even in a fully shutdown reactor. When the control rods are withdrawn and criticality is approached the number increases because the absorption of neutrons is being progressively reduced, until at criticality the chain reaction becomes self-sustaining. Note that while a neutron source is provided in the reactor, this is not essential to start the chain reaction, its main purpose is to give a shutdown neutron population which is detectable by instruments and so make the approach to critical more observable. The reactor will go critical at the same control rod position whether a source is loaded or not.

Once the chain reaction is begun, the primary starter source may be removed from the core to prevent damage from the high neutron flux in the operating reactor core; the secondary sources usually remains in situ to provide a background reference level for control of criticality.

Subcritical Multiplication

Even in a subcritical assembly such as a shut-down reactor core, any stray neutron that happens to be present in the core (for example from spontaneous fission of the fuel, from radioactive decay of fission products, or from a neutron source) will trigger an exponentially decaying chain reaction. Although the chain reaction is not self-sustaining, it acts as a multiplier that increases the equilibrium number of neutrons in the core. This *subcritical multiplication* effect can be used in two ways: as a probe of how close a core is to criticality, and as a way to generate fission power without the risks associated with a critical mass.

As a measurement technique, subcritical multiplication was used during the Manhattan Project in early experiments to determine the minimum critical masses of ^{235}U and of ^{239}Pu. It is still used today to calibrate the controls for nuclear reactors during startup, as many effects can change the required control settings to achieve criticality in a reactor. As a power-generating technique, subcritical multiplication allows generation of nuclear power for fission where a critical assembly is undesirable for safety or other reasons. A subcritical assembly together with a neutron source can serve as a steady source of heat to generate power from fission.

Including the effect of an external neutron source ("external" to the fission process, not physically external to the core), one can write a modified evolution equation:

$$\frac{dN}{dt} = \frac{\alpha N}{\tau} + R_{ext}$$

where R_{ext} is the rate at which the external source injects neutrons into the core. In equilibrium, the core is not changing and dN/dt is zero, so the equilibrium number of neutrons is given by:

$$N = -\frac{\tau R_{ext}}{\alpha}$$

If the core is subcritical, then α is negative so there is an equilibrium with a positive number of neutrons. If the core is close to criticality, then α is very small and thus the final number of neutrons can be made arbitrarily large.

Neutron Moderators

To improve $P_{fission}$ and enable a chain reaction, uranium-fueled reactors must include a neutron moderator that interacts with newly produced fast neutrons from fission events to reduce their kinetic energy from several MeV to thermal energies of less than one eV, making them more likely to induce fission. This is because ^{235}U is much more likely to undergo fission when struck by one of these thermal neutrons than by a freshly produced neutron from fission.

Neutron moderators are thus materials that slow down neutrons. Neutrons are most effectively slowed by colliding with the nucleus of a light atom, hydrogen being the lightest of all. To be effective, moderator materials must thus contain light elements with atomic nuclei that tend to scatter neutrons on impact rather than absorb them. In addition to hydrogen, beryllium and carbon atoms are also suited to the job of moderating or slowing down neutrons.

Hydrogen moderators include water (H_2O), heavy water(D_2O), and zirconium hydride (ZrH_2), all of which work because a hydrogen nucleus has nearly the same mass as a free neutron: neutron-H_2O or neutron-ZrH_2 impacts excite rotational modes of the molecules (spinning them around). Deuterium nuclei (in heavy water) absorb kinetic energy less well than do light hydrogen nuclei, but they are much less likely to absorb the impacting neutron. Water or heavy water have the advantage of being transparent liquids, so that, in addition to shielding and moderating a reactor core, they permit direct viewing of the core in operation and can also serve as a working fluid for heat transfer.

Carbon in the form of graphite has been widely used as a moderator. It was used in Chicago Pile-1, the world's first man-made critical assembly, and was commonplace in early reactor designs including the Soviet RBMK nuclear power plants such as the Chernobyl plant.

Moderators and Reactor Design

The amount and nature of neutron moderation affects reactor controllability and hence safety. Because moderators both slow and absorb neutrons, there is an optimum amount of moderator to include in a given geometry of reactor core. Less moderation reduces the effectiveness by reducing the $P_{fission}$ term in the evolution equation, and more moderation reduces the effectiveness by increasing the P_{escape} term.

Most moderators become less effective with increasing temperature, so *under-moderated* reactors are stable against changes in temperature in the reactor core: if the core overheats, then the quality of the moderator is reduced and the reaction tends to slow down (there is a "negative temperature coefficient" in the reactivity of the core). Water is an extreme case: in extreme heat, it can boil, producing effective voids in the reactor core without destroying the physical structure of the core; this tends to shut down the reaction and reduce the possibility of a fuel meltdown. *Over-moderated* reactors are unstable against changes in temperature (there is a "positive temperature coefficient" in the reactivity of the core), and so are less inherently safe than under-moderated cores.

Some reactors use a combination of moderator materials. For example, TRIGA type research reactors use ZrH_2 moderator mixed with the ^{235}U fuel, an H_2O-filled core, and C (graphite) moderator and reflector blocks around the periphery of the core.

Delayed neutrons and controllability

Fission reactions and subsequent neutron escape happen very quickly; this is important for nuclear weapons, where the objective is to make a nuclear core release as much energy as possible before it physically explodes. Most neutrons emitted by fission events are prompt: they are emitted effectively instantaneously. Once emitted, the average neutron lifetime (τ) in a typical core is on the order of a millisecond, so if the exponential factor α is as small as 0.01, then in one second the reactor power will vary by a factor of $(1 + 0.01)^{1000}$, or more than ten thousand. Nuclear weapons are engineered to maximize the power growth rate, with lifetimes well under a millisecond and exponential factors close to 2; but such rapid variation would render it practically impossible to control the reaction rates in a nuclear reactor.

Fortunately, the *effective* neutron lifetime is much longer than the average lifetime of a single neutron in the core. About 0.65% of the neutrons produced by ^{235}U fission, and about 0.20% of the neutrons produced by ^{239}Pu fission, are not produced immediately, but rather are emitted from an excited nucleus after a further decay step. In this step, further radioactive decay of some of the fission products (almost always negative beta decay), is followed by immediate neutron emission from the excited daughter product, with an average life time of the beta decay (and thus the neutron emission) of about 15 seconds. These so-called delayed neutrons increase the effective average lifetime of neutrons in the core, to nearly 0.1 seconds, so that a core with α of 0.01 would increase in one second by only a factor of $(1 + 0.01)^{10}$, or about 1.1: a 10% increase. This is a controllable rate of change.

Most nuclear reactors are hence operated in a *prompt subcritical, delayed critical* condition: the prompt neutrons alone are not sufficient to sustain a chain reaction, but the delayed neutrons make up the small difference required to keep the reaction going. This has effects on how reactors are controlled: when a small amount of control rod is slid into or out of the reactor core, the power level changes at first very rapidly due to *prompt subcritical multiplication* and then more gradually, following the exponential growth or decay curve of the delayed critical reaction. Furthermore, *increases* in reactor power can be performed at any desired rate simply by pulling out a sufficient length of control rod. However, without addition of a neutron poison or active neutron-absorber, *decreases* in fission rate are limited in speed, because even if the reactor is taken deeply subcritical to stop prompt fission neutron production, delayed neutrons are produced after ordinary beta decay of fission products already in place, and this decay-production of neutrons cannot be changed.

Kinetics

The kinetics of the reactor is described by the balance equations of neutrons and nuclei (fissile, fission products).

Reactor Poisons

Any element that strongly absorbs neutrons is called a reactor poison, because it tends to shut down (poison) an ongoing fission chain reaction. Some reactor poisons are deliberately inserted into fission reactor cores to control the reaction; boron or cadmium control rods are the best example. Many reactor poisons are produced by the fission process itself, and buildup of neutron-absorbing fission products affects both the fuel economics and the controllability of nuclear reactors.

Long-lived Poisons and Fuel Reprocessing

In practice, buildup of reactor poisons in nuclear fuel is what determines the lifetime of nuclear fuel in a reactor: long before all possible fissions have taken place, buildup of long-lived neutron absorbing fission products damps out the chain reaction. This is the reason that nuclear reprocessing is a useful activity: spent nuclear fuel contains about 96% of the original fissionable material present in newly manufactured nuclear fuel. Chemical separation of the fission products restores the nuclear fuel so that it can be used again.

Nuclear reprocessing is useful economically because chemical separation is much simpler to accomplish than the difficult isotope separation required to prepare nuclear fuel from natural uranium ore, so that in principle chemical separation yields more generated energy for less effort than mining, purifying, and isotopically separating new uranium ore. In practice, both the difficulty of handling the highly radioactive fission products and other political concerns make fuel reprocessing a contentious subject. One such concern is the fact that spent uranium nuclear fuel contains significant quantities of ^{239}Pu, a prime ingredient in nuclear weapons.

Short-lived Poisons and Controllability

Short-lived reactor poisons in fission products strongly affect how nuclear reactors can operate. Unstable fission product nuclei transmute into many different elements (*secondary fission products*) as they undergo a decay chain to a stable isotope. The most important such element is xenon, because the isotope ^{135}Xe, a secondary fission product with a half-life of about 9 hours, is an extremely strong neutron absorber. In an operating reactor, each nucleus of ^{135}Xe becomes ^{136}Xe (which may later sustain beta decay) by neutron capture almost as soon as it is created, so that there is no buildup in the core. However, when a reactor shuts down, the level of ^{135}Xe builds up in the core for about 9 hours before beginning to decay. The result is that, about 6–8 hours after a reactor is shut down, it can become physically impossible to restart the chain reaction until the ^{135}Xe has had a chance to decay over the next several hours. This temporary state, which may last several days and prevent restart, is called the iodine pit or xenon-poisoning. It is one reason why nuclear power reactors are usually operated at an even power level around the clock.

^{135}Xe buildup in a reactor core makes it extremely dangerous to operate the reactor a few hours after it has been shut down. Because the ^{135}Xe absorbs neutrons strongly, starting a reactor in a high-Xe condition requires pulling the control rods out of the core much farther than normal. However, if the reactor does achieve criticality, then the neutron flux in the core becomes high and ^{135}Xe is destroyed rapidly—this has the same effect as very rapidly removing a great length of control rod from the core, and can cause the reaction to grow too rapidly or even become prompt critical.

^{135}Xe played a large part in the Chernobyl accident: about eight hours after a scheduled maintenance shutdown, workers tried to bring the reactor to a zero power critical condition to test a control circuit. Since the core was loaded with ^{135}Xe from the previous day's power generation, it was necessary to withdraw more control rods to achieve this. As a result, the overdriven reaction grew rapidly and uncontrollably, leading to steam explosion in the core, and violent destruction of the facility.

Uranium Enrichment

While many fissionable isotopes exist in nature, the only usefully fissile isotope found in any quantity is ^{235}U. About 0.7% of the uranium in most ores is the 235 isotope, and about 99.3% is the non-fissile 238 isotope. For most uses as a nuclear fuel, uranium must be *enriched* - purified so that it contains a higher percentage of ^{235}U. Because ^{238}U absorbs fast neutrons, the critical mass needed to sustain a chain reaction increases as the ^{238}U content increases, reaching infinity at 94% ^{238}U (6% ^{235}U). Concentrations lower than 6% ^{235}U cannot go fast critical, though they are usable in a nuclear reactor with a neutron moderator. A nuclear weapon primary stage using uranium uses HEU enriched to ~90% ^{235}U, though the secondary stage often uses lower enrichments. Nuclear reactors with water moderator require at least some enrichment of ^{235}U. Nuclear reactors with heavy water moderation can operate with natural uranium, eliminating altogether the need for enrichment and preventing the fuel from being useful for nuclear weapons; the CANDU power reactors used in Canadian power plants are an example of this type.

Uranium enrichment is difficult because the chemical properties of ^{235}U and ^{238}U are identical, so physical processes such as gaseous diffusion, gas centrifuge or mass spectrometry must be used for isotopic separation based on small differences in mass. Because enrichment is the main technical hurdle to production of nuclear fuel and simple nuclear weapons, enrichment technology is politically sensitive.

Oklo: A Natural Nuclear Reactor

Modern deposits of uranium contain only up to ~0.7% ^{235}U (and ~99.3% ^{238}U), which is not enough to sustain a chain reaction moderated by ordinary water. But ^{235}U has a much shorter half-life (700 million years) than ^{238}U (4.5 billion years), so in the distant past the percentage of ^{235}U was much higher. About two billion years ago, a water-saturated uranium deposit (in what is now the Oklo mine in Gabon, West Africa) underwent a naturally occurring chain reaction that was moderated by groundwater and, presumably, controlled by the negative void coefficient as the water boiled from the heat of the reaction. Uranium from the Oklo mine is about 50% depleted compared to other locations: it is only about 0.3% to 0.7% ^{235}U; and the ore contains traces of stable daughters of long-decayed fission products.

Nuclear Reactions

A nuclear reaction is considered to be the process in which two nuclear particles (two nuclei or a nucleus and a nucleon) interact to produce two or more nuclear particles or γ-rays (gamma rays). Thus, a nuclear reaction must cause a transformation of at least one nuclide to another. Sometimes if a nucleus interacts with another nucleus or particle without changing the nature of any nuclide, the process is referred to a nuclear scattering, rather than a nuclear reaction. Perhaps the most notable nuclear reactions are the nuclear fusion reactions of light elements that power the energy production of stars and the Sun. Natural nuclear reactions occur also in the interaction between cosmic rays and matter.

The most notable man-controlled nuclear reaction is the fission reaction which occurs in nuclear reactors. Nuclear reactors are devices to initiate and control a nuclear chain reaction, but there are not only manmade devices. The world's first nuclear reactor operated about two billion years ago. The natural nuclear reactor formed at Oklo in Gabon, Africa, when a uranium-rich mineral deposit became flooded with groundwater that acted as a neutron moderator, and a nuclear chain reaction started. These fission reactions were sustained for hundreds of thousands of years, until a chain reaction could no longer be supported. This was confirmed by existence of isotopes of the fission-product gas xenon and by different ratio of U-235/U-238 (enrichment of natural uranium).

Notation of Nuclear Reactions

Standard nuclear notation shows the chemical symbol, the mass number and the atomic number of the isotope.

If the initial nuclei are denoted by *a* and *b*, and the product nuclei are denoted by c and *d*, the reaction can be represented by the equation:

$$a + b \rightarrow c + d$$

$$^{10}_{5}B + ^{1}_{0}n \rightarrow ^{7}_{3} Li + ^{4}_{2} He + 2.8\ Mev$$

This equation describes neutron capture in the boron, which is diluted in the coolant. Boric acid is used in nuclear power plants as a long-term compensator of nuclear fuel reactivity.

Instead of using the full equations in the style above, in many situations a compact notation is used to describe nuclear reactions. This style of the form a(b,c)d is equivalent to a + b producing c + d. Light particles are often abbreviated in this shorthand, typically p means proton, n means neutron, d means deuteron, α means an alpha particle or helium-4, β means beta particle or electron, γ means gamma photon, etc. The reaction above would be written as 10B(n,α)7Li.

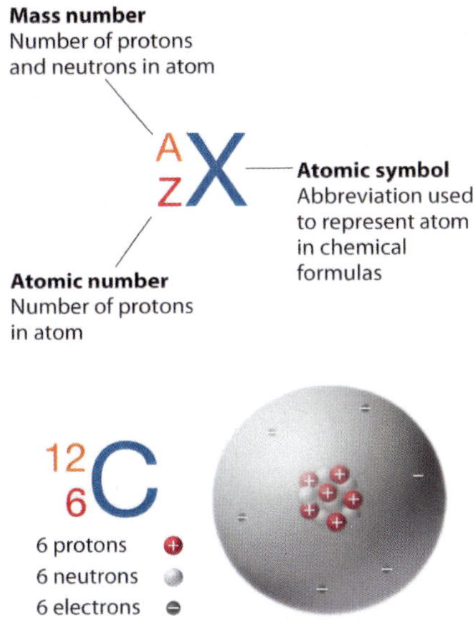

Notation of nuclei

Basic Classification of Nuclear Reactions

In order to understand the nature of neutron nuclear reactions, the classification according to the time scale of of these reactions has to be introduced. Interaction time is critical for defining the reaction mechanism.

There are two extreme scenarios for nuclear reactions (not only neutron reactions):

- A projectile and a target nucleus are within the range of nuclear forces for the very short time allowing for an interaction of a single nucleon only. These type of reactions are called the direct reactions.
- A projectile and a target nucleus are within the range of nuclear forces for the time allowing for a large number of interactions between nucleons. These type of reactions are called the compound nucleus reactions.

In fact, there is always some non-direct (multiple internuclear interaction) component in all reactions, but the direct reactions have this component limited.

Energy Conservation

Kinetic energy may be released during the course of a reaction (exothermic reaction) or kinetic energy may have to be supplied for the reaction to take place (endothermic reaction). This can be calculated by reference to a table of very accurate particle rest masses, as follows: according to the reference tables, the $^{6}_{3}Li$ nucleus has a standard atomic weight of 6.015 atomic mass units (abbreviated u), the deuterium has 2.014 u, and the helium-4 nucleus has 4.0026 u. Thus:

- the sum of the rest mass of the individual nuclei = 6.015 + 2.014 = 8.029 u;

- the total rest mass on the two helium-nuclei = 2 × 4.0026 = 8.0052 u;
- missing rest mass = 8.029 – 8.0052 = 0.0238 atomic mass units.

In a nuclear reaction, the total (relativistic) energy is conserved. The "missing" rest mass must therefore reappear as kinetic energy released in the reaction; its source is the nuclear binding energy. Using Einstein's mass-energy equivalence formula $E = mc^2$, the amount of energy released can be determined. We first need the energy equivalent of one atomic mass unit:

$$1\ \mathrm{u}\ c^2 = (1.66054 \times 10^{-27}\ \mathrm{kg}) \times (2.99792 \times 10^8\ \mathrm{m/s})^2$$

$$= 1.49242 \times 10^{-10}\ \mathrm{kg\ (m/s)^2} = 1.49242 \times 10^{-10}\ \mathrm{J\ (joule)} \times (1\ \mathrm{MeV} / 1.60218 \times 10^{-13}\ \mathrm{J})$$

$$= 931.49\ \mathrm{MeV},$$

so $1\ \mathrm{u}\ c^2 = 931.49$ MeV.

Hence, the energy released is 0.0238 × 931 MeV = 22.2 MeV.

Expressed differently: the mass is reduced by 0.3%, corresponding to 0.3% of 90 PJ/kg is 270 TJ/kg.

This is a large amount of energy for a nuclear reaction; the amount is so high because the binding energy per nucleon of the helium-4 nucleus is unusually high, because the He-4 nucleus is "doubly magic". (The He-4 nucleus is unusually stable and tightly bound for the same reason that the helium atom is inert: each pair of protons and neutrons in He-4 occupies a filled 1s nuclear orbital in the same way that the pair of electrons in the helium atom occupy a filled 1s electron orbital). Consequently, alpha particles appear frequently on the right hand side of nuclear reactions.

The energy released in a nuclear reaction can appear mainly in one of three ways:

- kinetic energy of the product particles (fraction of the kinetic energy of the charged nuclear reaction products can be directly converted into electrostatic energy);
- emission of very high energy photons, called gamma rays;
- some energy may remain in the nucleus, as a metastable energy level.

When the product nucleus is metastable, this is indicated by placing an asterisk ("*") next to its atomic number. This energy is eventually released through nuclear decay.

A small amount of energy may also emerge in the form of X-rays. Generally, the product nucleus has a different atomic number, and thus the configuration of its electron shells is wrong. As the electrons rearrange themselves and drop to lower energy levels, internal transition X-rays (X-rays with precisely defined emission lines) may be emitted.

Q-value and Energy Balance

In writing down the reaction equation, in a way analogous to a chemical equation, one may in addition give the reaction energy on the right side:

Target nucleus + projectile → Final nucleus + ejectile + Q.

For the particular case discussed above, the reaction energy has already been calculated as Q = 22.2 MeV. Hence:

$$^{6}_{3}\mathrm{Li} + {}^{2}_{1}\mathrm{H} \rightarrow 2\,{}^{4}_{2}\mathrm{He} + 22.2\,\mathrm{MeV}.$$

The reaction energy (the "Q-value") is positive for exothermal reactions and negative for endothermal reactions, opposite to the similar expression in chemistry. On the one hand, it is the difference between the sums of kinetic energies on the final side and on the initial side. But on the other hand, it is also the difference between the nuclear rest masses on the initial side and on the final side (in this way, we have calculated the Q-value above).

Reaction Rates

If the reaction equation is balanced, that does not mean that the reaction really occurs. The rate at which reactions occur depends on the particle energy, the particle flux and the reaction cross section. An example of a large repository of reaction rates is the REACLIB database, as maintained by the Joint Institute for Nuclear Astrophysics.

Neutrons vs. Ions

In the initial collision which begins the reaction, the particles must approach closely enough so that the short range strong force can affect them. As most common nuclear particles are positively charged, this means they must overcome considerable electrostatic repulsion before the reaction can begin. Even if the target nucleus is part of a neutral atom, the other particle must penetrate well beyond the electron cloud and closely approach the nucleus, which is positively charged. Thus, such particles must be first accelerated to high energy, for example by:

- particle accelerators;
- nuclear decay (alpha particles are the main type of interest here, since beta and gamma rays are rarely involved in nuclear reactions);
- very high temperatures, on the order of millions of degrees, producing thermonuclear reactions;
- cosmic rays.

Also, since the force of repulsion is proportional to the product of the two charges, reactions between heavy nuclei are rarer, and require higher initiating energy, than those between a heavy and light nucleus; while reactions between two light nuclei are the most common ones.

Neutrons, on the other hand, have no electric charge to cause repulsion, and are able to initiate a nuclear reaction at very low energies. In fact, at extremely low particle energies (corresponding, say, to thermal equilibrium at room temperature), the neutron's de Broglie wavelength is greatly increased, possibly greatly increasing its capture cross section, at energies close to resonances of the nuclei involved. Thus low-energy neutrons *may* be even more reactive than high energy neutrons.

Notable Types

While the number of possible nuclear reactions is immense, there are several types which are more common, or otherwise notable. Some examples include:

- Fusion reactions — two light nuclei join to form a heavier one, with additional particles (usually protons or neutrons) emitted subsequently.
- Spallation — a nucleus is hit by a particle with sufficient energy and momentum to knock out several small fragments or smash it into many fragments.
- Induced gamma emission belongs to a class in which only photons were involved in creating and destroying states of nuclear excitation.
- Alpha decay — Though driven by the same underlying forces as spontaneous fission, α decay is usually considered to be separate from the latter. The often-quoted idea that "nuclear reactions" are confined to induced processes is incorrect. "Radioactive decays" are a subgroup of "nuclear reactions" that are spontaneous rather than induced. For example, so-called "hot alpha particles" with unusually high energies may actually be produced in induced ternary fission, which is an induced nuclear reaction (contrasting with spontaneous fission). Such alphas occur from spontaneous ternary fission as well.
- Fission reactions — a very heavy nucleus, after absorbing additional light particles (usually neutrons), splits into two or sometimes three pieces. This is an induced nuclear reaction. Spontaneous fission, which occurs without assistance of a neutron, is usually not considered a nuclear reaction. At most, it is not an *induced* nuclear reaction.

Direct Reactions

An intermediate energy projectile transfers energy or picks up or loses nucleons to the nucleus in a single quick (10^{-21} second) event. Energy and momentum transfer are relatively small. These are particularly useful in experimental nuclear physics, because the reaction mechanisms are often simple enough to calculate with sufficient accuracy to probe the structure of the target nucleus.

Inelastic Scattering

Only energy and momentum are transferred.

- (p,p') tests differences between nuclear states.
- (α,α') measures nuclear surface shapes and sizes. Since α particles that hit the nucleus react more violently, elastic and shallow inelastic α scattering are sensitive to the shapes and sizes of the targets, like light scattered from a small black object.
- (e,e') is useful for probing the interior structure. Since electrons interact less strongly than do protons and neutrons, they reach to the centers of the targets and their wave functions are less distorted by passing through the nucleus.

Transfer Reactions

Usually at moderately low energy, one or more nucleons are transferred between the projectile and target. These are useful in studying outer shell structure of nuclei.

- (α,n) and (α,p) reactions. Some of the earliest nuclear reactions studied involved an alpha particle produced by alpha decay, knocking a nucleon from a target nucleus.
- (d,n) and (d,p) reactions. A deuteron beam impinges on a target; the target nuclei absorb either the neutron or proton from the deuteron. The deuteron is so loosely bound that this is almost the same as proton or neutron capture. A compound nucleus may be formed, leading to additional neutrons being emitted more slowly. (d,n) reactions are used to generate energetic neutrons.
- The strangeness exchange reaction (K, π) has been used to study hypernuclei.
- The reaction $^{14}N(\alpha,p)^{17}O$ performed by Rutherford in 1917 (reported 1919), is generally regarded as the first nuclear transmutation experiment.

Reactions with Neutrons

	→ T	→ ^{7}Li	→ ^{14}C		
(n,α)	$^{6}Li + n \rightarrow T + \alpha$	$^{10}B + n \rightarrow {}^{7}Li + \alpha$	$^{17}O + n \rightarrow {}^{14}C + \alpha$	$^{21}Ne + n \rightarrow {}^{18}O + \alpha$	$^{37}Ar + n \rightarrow {}^{34}S + \alpha$
(n,p)	$^{3}He + n \rightarrow T + p$	$^{7}Be + n \rightarrow {}^{7}Li + p$	$^{14}N + n \rightarrow {}^{14}C + p$	$^{22}Na + n \rightarrow {}^{22}Ne + p$	

Reactions with neutrons are important in nuclear reactors and nuclear weapons. While the best-known neutron reactions are neutron scattering, neutron capture, and nuclear fission, for some light nuclei (especially odd-odd nuclei) the most probable reaction with a thermal neutron is a transfer reaction:

Some reactions are only possible with fast neutrons:

- (n,2n) reactions produce small amounts of protactinium-231 and uranium-232 in the thorium cycle which is otherwise relatively free of highly radioactive actinide products.
- 9Be + n → 2α + 2n can contribute some additional neutrons in the beryllium neutron reflector of a nuclear weapon.
- 7Li + n → T + α + n unexpectedly contributed additional yield in Castle Bravo, Castle Romeo, and Castle Yankee, the three highest-yield nuclear tests conducted by the U.S.

Compound Nuclear Reactions

Either a low-energy projectile is absorbed or a higher energy particle transfers energy to the nucleus, leaving it with too much energy to be fully bound together. On a time scale of about 10^{-19} seconds, particles, usually neutrons, are "boiled" off. That is, it remains together until enough energy happens to be concentrated in one neutron to escape the mutual attraction. Charged particles rarely boil off because of the coulomb barrier. The excited quasi-bound nucleus is called a compound nucleus.

- Low energy (e, e' xn), (γ, xn) (the xn indicating one or more neutrons), where the gamma or virtual gamma energy is near the giant dipole resonance. These increase the need for radiation shielding around electron accelerators.

Nuclear Chain Reaction

Nuclear chain reactions are reactions where nuclear energy is obtained, generally through nuclear fission. These chain reactions are what provide nuclear power plants with the energy that is then turned into electricity for use by people. In these reactions, neutrons generated by the fission process continue on to initiate fission in other atoms. These reactions generally occur with heavier isotopes such as uranium-235 where there is a continuous release and absorption of neutrons. Figure 1 below is a visual representation of what a nuclear fission chain reaction looks like. If at least one neutron from each fission strikes another U-235 nucleus and initiates fission, the chain reaction is sustained and it is said to be critical.

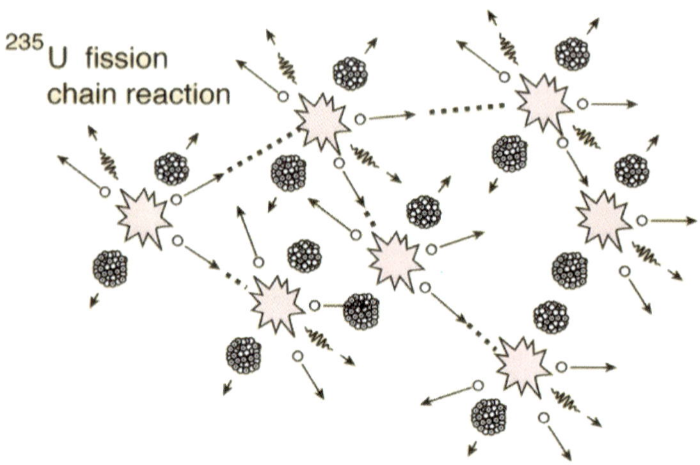

A nuclear fission chain reaction of uranium-253 atoms.

The mass of uranium-235 that is required to produce a reaction that is self-sustaining is said to be the "critical mass". This critical chain reaction can be accomplished at relatively low uranium-235 concentrations if the neutrons are moderated to lower their speed. A moderator is necessary because the probability for fission is greater with slow neutrons. The general equation for a nuclear chain reaction of uranium-235 is:

$$Uranium - 235 + {}^{1}_{0}n \rightarrow \sim 3{}^{1}_{0}n + \text{energy}$$

If for every one neutron inputted into the equation releases two or three more neutrons, then the number of fission events increases dramatically each generation. However, in reality not all of the released neutrons actually *cause* more fissions. Only 1.1 neutron per reaction actually goes on to cause more fissions and continue the chain, however the number of fission events still grows quickly. The process of a nuclear chain reaction releases large amounts of energy, but this energy can be utilized in different ways. On average, there is about 200 MeV of energy released during fission. To put this into context, burning coal provides only a couple eV, while 200 MeV is equal to 200 *million* electron volts. The difference in these energies is enormous. In nuclear reactors, the reaction is moderated and progresses at a slow pace to release its energy over a period of time so it can be harnessed and used for peaceful purposes. An atomic bomb utilizes this fission chain reaction as well, however it is designed to release its energy all at once—which is much more damaging. In either case, the release of the energy is controlled, but the time period taken to release the energy differs.

Fission Chain Reaction

Fission chain reactions occur because of interactions between neutrons and fissile isotopes (such as ^{235}U). The chain reaction requires both the release of neutrons from fissile isotopes undergoing nuclear fission and the subsequent absorption of some of these neutrons in fissile isotopes. When an atom undergoes nuclear fission, a few neutrons (the exact number depends on several factors) are ejected from the reaction. These free neutrons will then interact with the surrounding medium, and if more fissile fuel is present, some may be absorbed and cause more fissions. Thus, the cycle repeats to give a reaction that is self-sustaining.

Nuclear power plants operate by precisely controlling the rate at which nuclear reactions occur, and that control is maintained through the use of several redundant layers of safety measures. Moreover, the materials in a nuclear reactor core and the uranium enrichment level make a nuclear explosion impossible, even if all safety measures failed. On the other hand, nuclear weapons are specifically engineered to produce a reaction that is so fast and intense it cannot be controlled after it has started. When properly designed, this uncontrolled reaction can lead to an explosive energy release.

Nuclear Fission Fuel

Nuclear weapons employ high quality, highly enriched fuel exceeding the critical size and geometry (critical mass) necessary in order to obtain an explosive chain reaction. The fuel for energy purposes, such as in a nuclear fission reactor, is very different, usually consisting of a low-enriched oxide material (e.g. UO_2).

Fission Reaction Products

When a heavy atom undergoes nuclear fission, it breaks into two or more fission fragments. Also, several free neutrons, gamma rays, and neutrinos are emitted, and a large amount of energy is released. The sum of the rest masses of the fission fragments and ejected neutrons is less than the sum of the rest masses of the original atom and incident neutron (of course the fission fragments are not at rest). The mass difference is accounted for in the release of energy according to the equation $E=\Delta mc^2$:

$$\text{mass of released energy} = \frac{E}{c^2} = m_{\text{original}} - m_{\text{final}}$$

Due to the extremely large value of the speed of light, c, a small decrease in mass is associated with a tremendous release of active energy (for example, the kinetic energy of the fission fragments). This energy (in the form of radiation and heat) carries the missing mass, when it leaves the reaction system (total mass, like total energy, is always conserved). While typical chemical reactions release energies on the order of a few eVs (e.g. the binding energy of the electron to hydrogen is 13.6 eV), nuclear fission reactions typically release energies on the order of hundreds of millions of eVs.

Two typical fission reactions are shown below with average values of energy released and number of neutrons ejected:

$$^{235}U + neutron -> \text{fission fragments} + 2.4 \text{ neutrons} + 192.9 \text{ MeV}$$

$$^{239}Pu + neutron -> \text{fission fragments} + 2.9 \text{ neutrons} + 198.5 \text{ MeV}$$

Note that these equations are for fissions caused by slow-moving (thermal) neutrons. The average energy released and number of neutrons ejected is a function of the incident neutron speed. Also, note that these equations exclude energy from neutrinos since these subatomic particles are extremely non-reactive and, therefore, rarely deposit their energy in the system.

Timescales of Nuclear Chain Reactions

Prompt Neutron Lifetime

The prompt neutron lifetime, *l*, is the average time between the emission of neutrons and either their absorption in the system or their escape from the system. The neutrons that occur directly from fission are called "prompt neutrons," and the ones that are a result of radioactive decay of fission fragments are called "delayed neutrons." The term lifetime is used because the emission of a neutron is often considered its "birth," and the subsequent absorption is considered its "death." For thermal (slow-neutron) fission reactors, the typical prompt neutron lifetime is on the order of 10^{-4} seconds, and for fast fission reactors, the prompt neutron lifetime is on the order of 10^{-7} seconds. These extremely short lifetimes mean that in 1 second, 10,000 to 10,000,000 neutron lifetimes can pass. The *average* (also referred to as the *adjoint unweighted*) prompt neutron lifetime takes into account all prompt neutrons regardless of their importance in the reactor core; the *effective* prompt neutron lifetime (referred to as the *adjoint weighted* over space, energy, and angle) refers to a neutron with average importance.

Mean Generation Time

The mean generation time, Λ, is the average time from a neutron emission to a capture that results in fission. The mean generation time is different from the prompt neutron lifetime because the mean generation time only includes neutron absorptions that lead to fission reactions (not other absorption reactions). The two times are related by the following formula:

$$\Lambda = \frac{l}{k}$$

In this formula, k is the effective neutron multiplication factor, described below.

Effective Neutron Multiplication Factor

The effective neutron multiplication factor, *k*, is the average number of neutrons from one fission that cause another fission. The remaining neutrons either are absorbed in non-fission reactions or leave the system without being absorbed. The value of k determines how a nuclear chain reaction proceeds:

- $k < 1$ (subcriticality): The system cannot sustain a chain reaction, and any beginning of a chain reaction dies out over time. For every fission that is induced in the system, an average *total* of $1/(1 - k)$ fissions occur.
- $k = 1$ (criticality): Every fission causes an average of one more fission, leading to a fission (and power) level that is constant. Nuclear power plants operate with $k = 1$ unless the power level is being increased or decreased.
- $k > 1$ (supercriticality): For every fission in the material, it is likely that there will be "k" fissions after the next *mean generation time* (Λ). The result is that the number of fission reactions increases exponentially, according to the equation $e^{(k-1)t/\Lambda}$, where t is the elapsed time. Nuclear weapons are designed to operate under this state. There are two subdivisions of supercriticality: prompt and delayed.

When describing kinetics and dynamics of nuclear reactors, and also in the practice of reactor operation, the concept of reactivity is used, which characterizes the deflection of reactor from the critical state. $\rho=(k-1)/k$. InHour is a unit of reactivity of a nuclear reactor.

In a nuclear reactor, k will actually oscillate from slightly less than 1 to slightly more than 1, due primarily to thermal effects (as more power is produced, the fuel rods warm and thus expand, lowering their capture ratio, and thus driving k lower). This leaves the average value of k at exactly 1. Delayed neutrons play an important role in the timing of these oscillations.

In an infinite medium, the multiplication factor may be described by the four factor formula; in a non-infinite medium, the multiplication factor may be described by the six factor formula.

Prompt and Delayed Supercriticality

Not all neutrons are emitted as a direct product of fission; some are instead due to the radioactive decay of some of the fission fragments. The neutrons that occur directly from fission are called "prompt neutrons," and the ones that are a result of radioactive decay of fission fragments are called "delayed neutrons." The fraction of neutrons that are delayed is called β, and this fraction is typically less than 1% of all the neutrons in the chain reaction.

The delayed neutrons allow a nuclear reactor to respond several orders of magnitude more slowly than just prompt neutrons would alone. Without delayed neutrons, changes in reaction rates in nuclear reactors would occur at speeds that are too fast for humans to control.

The region of supercriticality between $k = 1$ and $k = 1/(1-\beta)$ is known as delayed supercriticality (or delayed criticality). It is in this region that all nuclear power reactors operate. The region of supercriticality for $k > 1/(1-\beta)$ is known as prompt supercriticality (or prompt criticality), which is the region in which nuclear weapons operate.

The change in k needed to go from critical to prompt critical is defined as a dollar.

Nuclear Weapons Application of Neutron Multiplication

Nuclear fission weapons require a mass of fissile fuel that is prompt supercritical.

For a given mass of fissile material the value of k can be increased by increasing the density. Since the probability per distance traveled for a neutron to collide with a nucleus is proportional to the material density, increasing the density of a fissile material can increase k. This concept is utilized in the implosion method for nuclear weapons. In these devices, the nuclear chain reaction begins after increasing the density of the fissile material with a conventional explosive.

In the gun-type fission weapon two subcritical pieces of fuel are rapidly brought together. The value of k for a combination of two masses is always greater than that of its components. The magnitude of the difference depends on distance, as well as the physical orientation.

The value of k can also be increased by using a neutron reflector surrounding the fissile material.

Once the mass of fuel is prompt supercritical, the power increases exponentially. However, the exponential power increase cannot continue for long since k decreases when the amount of fission material that is left decreases (i.e. it is consumed by fissions). Also, the geometry and density are expected to change during detonation since the remaining fission material is torn apart from the explosion.

Predetonation

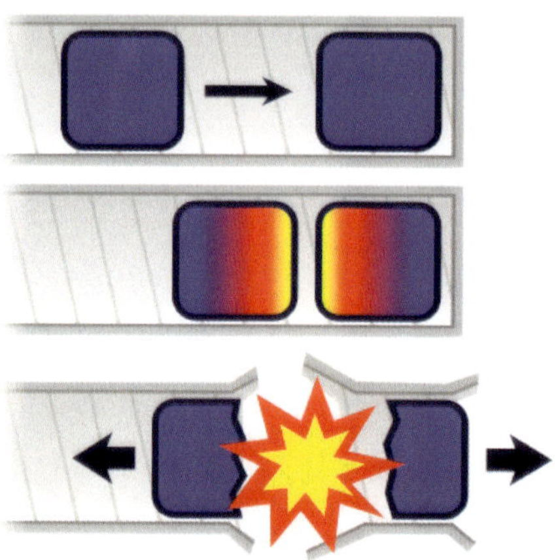

If two pieces of subcritical material are not brought together fast enough, nuclear predetonation can occur, whereby a smaller explosion than expected will blow the bulk of the material apart.

Detonation of a nuclear weapon involves bringing fissile material into its optimal supercritical state very rapidly. During part of this process, the assembly is supercritical, but not yet in an optimal state for a chain reaction. Free neutrons, in particular from spontaneous fissions, can cause the device to undergo a preliminary chain reaction that destroys the fissile material before it is ready to produce a large explosion, which is known as predetonation. To keep the probability of predetonation low, the duration of the non-optimal assembly period is minimized and fissile and other materials are used which have low spontaneous fission rates. In fact, the combination of materials has to be such that it is unlikely that there is even a single spontaneous fission during the period of supercritical assembly. In particular, the gun method cannot be used with plutonium.

Nuclear Power Plants and Control of Chain Reactions

Chain reactions naturally give rise to reaction rates that grow (or shrink) exponentially, whereas a nuclear power reactor needs to be able to hold the reaction rate reasonably constant. To maintain this control, the chain reaction criticality must have a slow enough time-scale to permit intervention by additional effects (e.g., mechanical control rods or thermal expansion). Consequently, all nuclear power reactors (even fast-neutron reactors) rely on delayed neutrons for their criticality. An operating nuclear power reactor fluctuates between being slightly subcritical and slightly delayed-supercritical, but must always remain below prompt-critical.

It is impossible for a nuclear power plant to undergo a nuclear chain reaction that results in an explosion of power comparable with a nuclear weapon, but even low-powered explosions due to uncontrolled chain reactions, that would be considered "fizzles" in a bomb, may still cause considerable damage and meltdown in a reactor. For example, the Chernobyl disaster involved a runaway chain reaction but the result was a low-powered steam explosion from the relatively small release of heat, as compared with a bomb. However, the reactor complex was destroyed by the heat, as well as by ordinary burning of the graphite exposed to air. Such steam explosions would be typical of the very diffuse assembly of materials in a nuclear reactor, even under the worst conditions.

In addition, other steps can be taken for safety. For example, power plants licensed in the United States require a negative void coefficient of reactivity (this means that if water is removed from the reactor core, the nuclear reaction will tend to shut down, not increase). This eliminates the possibility of the type of accident that occurred at Chernobyl (which was due to a positive void coefficient). However, nuclear reactors are still capable of causing smaller explosions even after complete shutdown, such as was the case of the Fukushima Daiichi nuclear disaster. In such cases, residual decay heat from the core may cause high temperatures if there is loss of coolant flow, even a day after the chain reaction has been shut down. This may cause a chemical reaction between water and fuel that produces hydrogen gas which can explode after mixing with air, with severe contamination consequences, since fuel rod material may still be exposed to the atmosphere from this process. However, such explosions do not happen during a chain reaction, but rather as a result of energy from radioactive beta decay, after the fission chain reaction has been stopped.

Four Factor Formula

A group of fast neutrons produced by fission can enter into several reactions. Some of these reactions reduce the size of the neutron group while other reactions allow the group to increase in size or produce a second generation. There are four factors that are completely independent of the size and shape of the reactor that give the inherent multiplication ability of the fuel and moderator materials without regard to leakage. This four factor formula accurately represents the infinite multiplication factor as shown in the equation below.

$$k = e \, p f h$$

where:

e = fast fission factor

p = resonance escape probability

f = thermal utilization factor

η= reproduction factor

There are four factors that are completely independent of the size and shape of the reactor that describe the inherent multiplication ability of the fuel and moderator materials without regard to leakage:

1. Fast Fission Factor

The fast fission process is in the multiplication factor characterized by the fast fission factor, ε, which increases the fast neutron population in one neutron generation. The fast fission factor is defined as the ratio of the fast neutrons produced by fissions at all energies to the number of fast neutrons produced in thermal fission.

2. Resonance Escape Probability

The resonance escape probability, symbolized by p, is the probability that a neutron will be slowed to thermal energy and will escape resonance capture. This probability is defined as the ratio of the number of neutrons that reach thermal energies to the number of fast neutrons that start to slow down.

3. Thermal Utilization Factor

The thermal utilisation factor, f, is the fraction of the thermal neutrons that are absorbed in the nuclear fuel, in all isotopes of the nuclear fuel. Itdescribes how effectively (how well are utilized) are thermal neutrons absorbed in the fuel.The value of the thermal utilization factor is given by the ratio of the number of thermal neutrons absorbed in the fuel (all nuclides) to the number of thermal neutrons absorbed in all the material that makes up the core.

4. Reproduction Factor

The number of neutrons created in the new generation is determined by the neutron reproduction factor. The reproduction factor, η, is defined as the ratio of the number of fast neutrons produced by thermal fission to the number of thermal neutrons absorbed in the fuel.

Infinite Multiplication Factor

The required condition for a stable, self-sustained fission chain reaction in a multiplying system (in a nuclear reactor) is that exactly every fission initiate another fission. The minimum condition is for each nucleus undergoing fission to produce, on the average, at least one neutron that causes fission of another nucleus. Also the number of fissions occurring per unit time (the reaction rate) within the system must be constant.

This condition can be expressed conveniently in terms of the multiplication factor. The infinite multiplication factor is the ratio of the neutrons produced by fission in one neutron generation to the number of neutrons lost through absorption in the preceding neutron generation. This can be expressed mathematically as shown below.

$$k_\infty = \frac{\text{neutron production from fission in one neutron generation}}{\text{neutron absorption in the preceding neutron generation}}$$

Six Factor Formula

A formula known as the six-factor formula for the neutron multiplication factor will be derived from a flow chart illustrating the various processes in the life cycle. This formula allows us to examine the effect of operational changes to the reactor, for example, the movement of control absorbers, build-up of fission products or replacement of fuel the neutron multiplication factor as the product of the six factors:

$$k = \varepsilon\, p \eta f \wedge_f \wedge_t$$

Now, the definition of each factor:

Fast fission factor denoted by the symbol e(epsilon). This is the factor by which the fast neutron population increases due to fast fission in U-238 (typically about 1.03 for natural uranium).

$$\varepsilon = \frac{No.\,of\,neutrons\,from\;thermal\;\;fission + No.\,of\;\;neutrons\,from\;\;fast\;\;fission}{No.\,of\;\;neutrons\,from\;\;fast\;\;fission}$$

Resonance escape probability (P). This is the probability that a neutron will not undergo resonance capture in U-238 while slowing down. The typical value is 0.90 for natural uranium fuel.

$$p = \frac{Number\;\;of\;\;neutrons\;leaving\;resonance\;energy\;range}{Number\;\;of\;\;neutrons\;\;entering\;\;resonance\;\;energy\;\;range}$$

Reproduction factor, denoted by the symbol η (eta). This is the number of fission neutrons produced per thermal neutron absorbed in the fuel. Note that in the present course the term "fuel" is taken to include all contents of the fuel bundle with the exception of the cladding (e.g., fissile material, U-238, fission products, Pu-240, etc.).

$$\eta = \nu \frac{\sum_f (fuel)}{\sum_a (fuel)} = \nu \frac{\sum_f (fuel)}{\sum_f (fuel) + \sum_{n.\gamma} (fuel)}$$

A typical value of η for natural uranium fuel is about 1.2. Note carefully the distinction between η, which is the number of fission neutrons per thermal neutron absorbed in the fuel and v, which is the average number of neutrons produced per fission.

Thermal utilization (f). This is the fraction of thermal neutrons absorbed by the fuel compared to the total thermal neutrons absorbed in the whole reactor. (It is essential that "fuel" be defined as for η above). A typical value of f is about 0.94.

$$f = \frac{\sum_a (fuel)\phi(fuel)}{\sum_a (total\ reactor)\phi(total\ reactor)}$$

where $\phi(\text{fuel})$ and $\phi(\text{total reactor})$ are the average thermal neutron fluxes in the fuel and in the total reactor.

Fast non-leakage probability denoted by $\wedge_f$, (lambda-f). This is the probability that a fast neutron will not leak out of the reactor. A typical value is about 0.995.

Thermal non-leakage probability, denoted by $\wedge_f$, (lambda-t). The probability that a thermal neutron will not leak out of the reactor. A typical value is about 0.98.

The first four factors depend essentially on the materials in the reactor and not on its size. The two non-leakage probabilities depend on the size and shape of the reactor and may also be increased by adding a reflector, as we will discuss later.

Let's redraw the cycle diagram using the symbols as defined above instead of the numbers we used before. For convenience, we'll also start at a different point in the cycle and, to make it more general, we won't make the assumption that the system is necessarily a critical one. Instead we'll use the diagram to derive an expression for the neutron multiplication factor for the cycle illustrated.

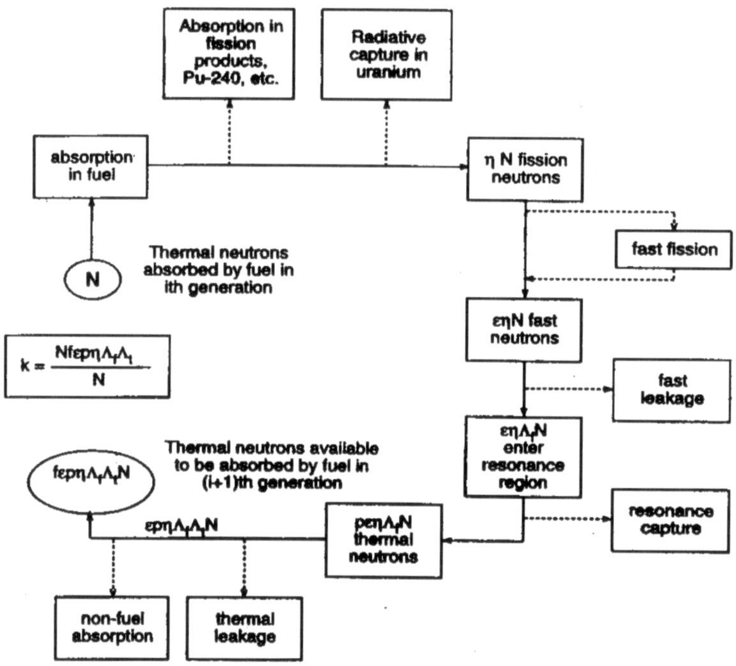

Multiplication factor in neutron cycle

The starting point in our diagram.this time is the number of thermal neutrons (N) absorbed in the fuel at the beginning of what we'll call the ith generation. From the definition of η given above, the number of fission neutrons resulting from these fuel absorptions will equal ηN. (Remember that the way η is defined takes care of all absorptions in the fuel that do not lead to fissions). On their way down to thermal energies, the fission neutrons will be augmented by fast fission in U-238, so that their number increases to $\varepsilon\eta$ N.

We take account of fast leakage by multiplying by the fast non-leakage probability, and of resonance capture in U-238 by multiplying by the resonance escape probability. The number of neutrons that actually reach thermal energies is therefore $\varepsilon p \eta \wedge_f N$.

We now allow for the thermal neutrons lost by leakage by multiplying by the thermal non-leakage probability; we allow for the loss by absorption in non-fuel materials by multiplying by f, the thermal utilization. We have accounted for all these potential losses, and the expression we have left (the product $N\varepsilon p \eta$ f$\wedge_f \wedge_t$), is the number of neutrons absorbed by the fuel at the start of the $(i+1)^{th}$ generation.

We can now write the expression for the neutron multiplication factor, k, which is simply the number of neutrons in the $(i+1)^{th}$ generation divided by the number in the ith generation, that is

$$k = \frac{N\varepsilon p\eta f \wedge_f \wedge_t}{N} = N\varepsilon p\eta f \wedge_f \wedge_t$$

If the reactor happens to be critical, of course. the . product of the six factors is equal to 1. In fact, what we do in operating a reactor at steady power is to adjust one or more of the six-factors to achieve a k-value of 1.

Of the six factors, the only ones which can be varied conveniently to make short-term changes in reactivity are the thermal utilization, f, and the leakage terms. Thermal utilization can be changed by varying the amount of absorption in the core, for example, by adjusting the liquid control zones or removing boron from the moderator. In some of the earlier CANDUs, the leakage terms were varied by changing the level of the moderator (and thereby the effective size of the core). The factors E and p are fixed by the initial design of the core, but 'll can be varied by removing exhausted fuel bundles and replacing them with fresh fuel, as is done in on-load refuelling.

Before we leave this topic, let's go back to the statement made earlier that the first four factors in the six-factor formula depend on the composition of the reactor and the last two on its size. For an infinitely large reactor, of course, the non-leakage probabilities would both be equal to one, since no neutrons could escape from an infinite system. In that case, k would simply equal the product of the first four terms, and we could write

$$k_\infty = \varepsilon p \eta f$$

where the terminology k_∞ is used to indicate that the k-value here is that for an infinitely large reactor of the composition specified by the appropriate values of ε, p, η and f. Equation ($k_\infty = \varepsilon p \eta f$) is known as the four-factor formula.

For a reactor of finite size, the relation between k_∞; as defined by the above equation, and the multiplication factor k, is

$$k = k_\infty \wedge_f \wedge_t$$

where, as before, the quantities $\wedge_f$ and $\wedge_t$ are the fast and thermal non-leakage probabilities.

Once a reactor has operated for some time, there will be a wide· variation in the composition ofthe fuel bundles, since each will have a different degree of burn up of U-235 and buildup of Pu-239 and fission product poisons.

References

- “Histoire secrète de la bombe atomique française”. Documents (in French). Paris. Cherche Midi. 2000. ISBN 978-2-862-74794-1. OCLC 45842105
- Jogalekar, Ashutosh. “Leo Szil rd, a traffic light and a slice of nuclear history”. Scientific American. Retrieved 4 January 2016
- Infinite-Multiplication-Factor-K-26: nuclearpowertraining.tpub.com, Retrieved 01 July 2018
- Shinn, E.; Et., al. (2013). “Nuclear energy conversion with stacks of graphene nanocapacitors”. Complexity. Bibcode:2013Cmplx..18c..24S. doi:10.1002/cplx.21427
- Tilley, R. J. D. (2004). Understanding Solids: The Science of Materials. John Wiley and Sons. p. 495. ISBN 0-470-85275-5
- Nuclear-power, reactor-physics, four-factor-formula-infinite-multiplication-factor, nuclear-fission-chain-reaction: nuclear-power.net, Retrieved 21 June 2018
- Carey Sublette (February 20, 1999). “4.1.5.3 Predetonation”. 4. Engineering and Design of Nuclear Weapons: 4.1 Elements of Fission Weapon Design. nuclearweaponarchive.org. Retrieved June 29, 2014

Nuclear Fusion and Fission

In nuclear physics and chemistry, a nuclear reaction is the process by which two nuclei undergo a collision to produce one or more nuclides. Nuclear reactions are of two types, nuclear fission and nuclear fusion. The topics elaborated in this chapter explore the concepts central to the understanding of nuclear fission and fusion such as thermonuclear fusion, inertial confinement fusion, nuclear fission product, delayed neutron, fissile material, etc.

Nuclear Fusion

In nuclear physics, nuclear fusion is a nuclear reaction in which two or more atomic nuclei collide at a very high energy and fuse together into a new nucleus, e.g. helium. If light nuclei are forced together, they will fuse with a yield of energy because the mass of the combination will be less than the sum of the masses of the individual nuclei. If the combined nuclear mass is less than that of iron at the peak of the binding energy curve, then the nuclear particles will be more tightly bound than they were in the lighter nuclei, and that decrease in mass comes off in the form of energy according to the Albert Einstein relationship. For elements like the uranium and thorium, fission will yield energy. Fusion reactions have an energy density many times greater than nuclear fission and fusion reactions are themselves millions of times more energetic than chemical reactions.

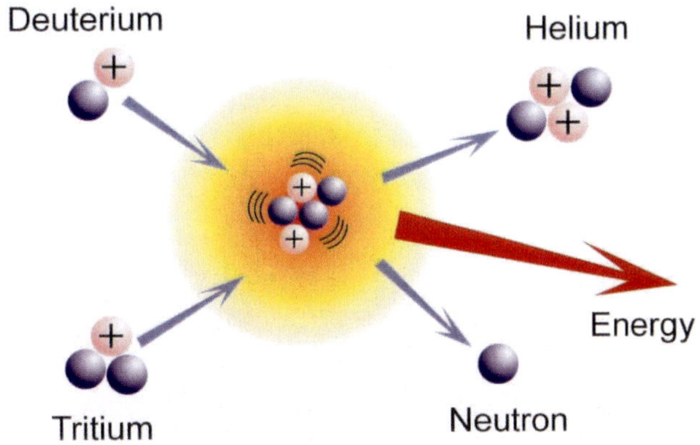

Nuclear fusion is a nuclear reaction in which two or more atomic nuclei (e.g. D+T) collide at a very high energy and fuse together.

Requirements

A substantial energy barrier must be overcome before fusion can occur. At large distances two naked nuclei repel one another because of the repulsive electrostatic force between their positively charged protons. If two nuclei can be brought close enough together, however, the electrostatic repulsion can be overcome by the nuclear force which is stronger at close distances.

When a nucleon such as a proton or neutron is added to a nucleus, the nuclear force attracts it to other nucleons, but primarily to its immediate neighbors due to the short range of the force. The nucleons in the interior of a nucleus have more neighboring nucleons than those on the surface. Since smaller nuclei have a larger surface area-to-volume ratio, the binding energy per nucleon due to the strong force generally increases with the size of the nucleus but approaches a limiting value corresponding to that of a fully surrounded nucleon.

The electrostatic force, on the other hand, is an inverse-square force, so a proton added to a nucleus will feel an electrostatic repulsion from *all* the other protons in the nucleus. The electrostatic energy per nucleon due to the electrostatic force thus increases without limit as nuclei get larger.

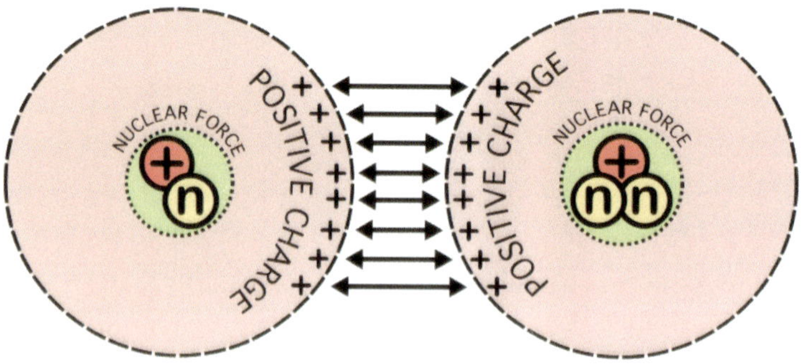

The electrostatic force caused by positively charged nuclei are very strong over long distances, but at short distances the nuclear force is stronger. As such, the main technical difficulty for fusion is getting the nuclei close enough to fuse (distances not to scale).

The net result of these opposing forces is that the binding energy per nucleon generally increases with increasing size, up to the elements iron and nickel, and then decreases for heavier nuclei. Eventually, the binding energy becomes negative and very heavy nuclei are not stable. The four most tightly bound nuclei, in decreasing order of binding energy, are ^{62}Ni, ^{58}Fe, ^{56}Fe, and ^{60}Ni. Even though the nickel isotope ^{62}Ni is more stable, the iron isotope ^{56}Fe is an order of magnitude more common. This is due to a greater disintegration rate for ^{62}Ni in the interior of stars driven by photon absorption.

A notable exception to this general trend is the helium-4 nucleus, whose binding energy is higher than that of lithium, the next heavier element. The Pauli exclusion principle provides an explanation for this exceptional behavior—it says that because protons and neutrons are fermions, they cannot exist in exactly the same state. Each proton or neutron energy state in a nucleus can accommodate both a spin up particle and a spin down particle. Helium-4 has an anomalously large binding energy because its nucleus consists of two protons and two neutrons; so all four of its nucleons can be in the ground state. Any additional nucleons would have to go into higher energy states.

The situation is similar if two nuclei are brought together. As they approach each other, all the protons in one nucleus repel all the protons in the other. Not until the two nuclei actually come in contact can the strong nuclear force take over. Consequently, even when the final energy state is lower, there is a large energy barrier that must first be overcome. It is called the Coulomb barrier.

The Coulomb barrier is smallest for isotopes of hydrogen—they contain only a single positive charge in the nucleus. A bi-proton is not stable, so neutrons must also be involved, ideally in such a way that a helium nucleus, with its extremely tight binding, is one of the products.

Using deuterium-tritium fuel, the resulting energy barrier is about 0.01 MeV. In comparison, the energy needed to remove an electronfrom hydrogen is 13.6 eV, about 750 times less energy. The (intermediate) result of the fusion is an unstable ^{5}He nucleus, which immediately ejects a neutron with 14.1 MeV. The recoil energy of the remaining ^{4}He nucleus is 3.5 MeV, so the total energy liberated is 17.6 MeV. This is many times more than what was needed to overcome the energy barrier.

If the energy to initiate the reaction comes from accelerating one of the nuclei, the process is called *beam-target* fusion; if both nuclei are accelerated, it is *beam-beam* fusion. If the nuclei are part of a plasma near thermal equilibrium, one speaks of *thermonuclear* fusion. Temperature is a measure of the average kinetic energy of particles, so by heating the nuclei they will gain energy and eventually have enough to overcome this 0.01 MeV. Converting the units between electron-volts and Kelvin shows that the barrier would be overcome at a temperature in excess of 120 million Kelvin—a very high temperature.

There are two effects that lower the actual temperature needed. One is the fact that temperature is the *average* kinetic energy, implying that some nuclei at this temperature would actually have much higher energy than 0.01 MeV, while others would be much lower. It is the nuclei in the high-energy tail of the velocity distribution that account for most of the fusion reactions. The other effect is quantum tunneling. The nuclei do not actually have to have enough energy to overcome the Coulomb barrier completely. If they have nearly enough energy, they can tunnel through the remaining barrier. For this reason fuel at lower temperatures will still undergo fusion events at a lower rate.

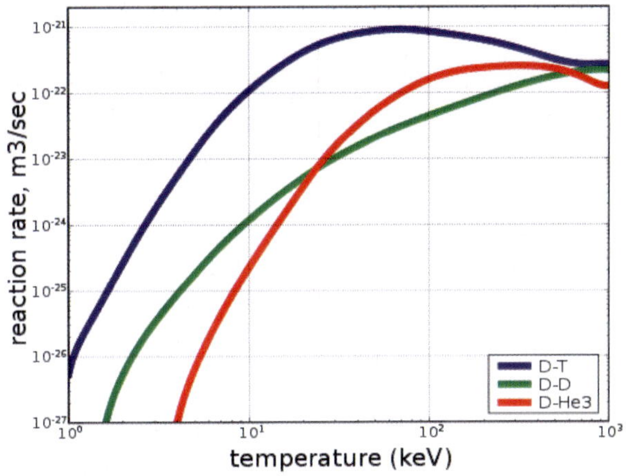

The fusion reaction rate increases rapidly with temperature until it maximizes and then gradually drops off. The DT rate peaks at a lower temperature (about 70 keV, or 800 million Kelvin) and at a higher value than other reactions commonly considered for fusion energy.

The reaction cross section σ is a measure of the probability of a fusion reaction as a function of the relative velocity of the two reactant nuclei. If the reactants have a distribution of velocities, e.g. a thermal distribution with thermonuclear fusion, then it is useful to perform an average over the distributions of the product of cross section and velocity. The reaction rate (fusions per volume per time) is $\langle\sigma v\rangle$ times the product of the reactant number densities:

$$f = n_1 n_2 \langle \sigma v \rangle$$

If a species of nuclei is reacting with itself, such as the DD reaction, then the product $n_1 n_2$ must be replaced by $(1/2)n^2$.

$\langle \sigma v \rangle$ increases from virtually zero at room temperatures up to meaningful magnitudes at temperatures of 10–100 keV. At these temperatures, well above typical ionization energies (13.6 eV in the hydrogen case), the fusion reactants exist in a plasma state.

The significance of $\langle \sigma v \rangle$ as a function of temperature in a device with a particular energy confinement time is found by considering the Lawson criterion.

Fuel Confinement Methods

Gravitational

One force capable of confining the fuel well enough to satisfy the Lawson criterion is gravity. The mass needed, however, is so great that gravitational confinement is only found in stars (the smallest of which are brown dwarfs). Even if the more reactive fuel deuterium were used, a mass greater than that of the planet Jupiter would be needed.

Magnetic

Since plasmas are very good electrical conductors, magnetic fields can also confine fusion fuel. A variety of magnetic configurations can be used, the most basic distinction being between mirror confinement and toroidal confinement, especially tokamaks and stellarators.

Inertial

A third confinement principle is to apply a rapid pulse of energy to a large part of the surface of a pellet of fusion fuel, causing it to simultaneously "implode" and heat to very high pressure and temperature. If the fuel is dense enough and hot enough, the fusion reaction rate will be high enough to burn a significant fraction of the fuel before it has dissipated. To achieve these extreme conditions, the initially cold fuel must be explosively compressed. Inertial confinement is used in the hydrogen bomb, where the driver is x-rayscreated by a fission bomb. Inertial confinement is also attempted in "controlled" nuclear fusion, where the driver is a laser, ion, or electron beam, or a Z-pinch.

Some other confinement principles have been investigated, such as muon-catalyzed fusion, the Farnsworth-Hirsch fusor and Polywell (inertial electrostatic confinement), and bubble fusion.

Production Methods

A variety of methods are known to effect nuclear fusion. Some are "cold" in the strict sense that no part of the material is hot (except for the reaction products), some are "cold" in the limited sense that the bulk of the material is at a relatively low temperature and pressure but the reactants are not, and some are "hot" fusion methods that create macroscopic regions of very high temperature and pressure.

Locally Cold Fusion

- Muon-catalyzed fusion is a well-established and reproducible fusion process that occurs at ordinary temperatures. It was studied in detail by Steven Jones in the early 1980s. It has not been reported to produce net energy. Net energy production from this reaction is not believed to be possible because of the energy required to create muons, their 2.2 μs half-life, and the chance that a muon will bind to the new alpha particle and thus stop catalyzing fusion.

Generally Cold, Locally Hot Fusion

- Accelerator based light-ion fusion. Using particle accelerators it is possible to achieve particle kinetic energies sufficient to induce many light ion fusion reactions. Of particular relevance into this discussion are devices referred to as sealed-tube neutron generators. These small devices are miniature particle accelerators filled with deuterium and tritium gas in an arrangement which allows ions of these nuclei to be accelerated against hydride targets, also containing deuterium and tritium, where fusion takes place. Hundreds of neutron generators are produced annually for use in the petroleum industry where they are used in measurement equipment for locating and mapping oil reserves. Despite periodic reports in the popular press by scientists claiming to have invented "table-top" fusion machines, neutron generators have been around for half a century. The sizes of these devices vary but the smallest instruments are often packaged in sizes smaller than a loaf of bread. These devices do not produce a net power output.
- In sonoluminescence, acoustic shock waves create temporary bubbles that collapse shortly after creation, producing very high temperatures and pressures. In 2002, Rusi P. Taleyarkhan reported the possibility that bubble fusion occurs in those collapsing bubbles (sonofusion). As of 2005, experiments to determine whether fusion is occurring give conflicting results. If fusion is occurring, it is because the local temperature and pressure are sufficiently high to produce hot fusion.
- The Farnsworth-Hirsch Fusor is a tabletop device in which fusion occurs. This fusion comes from high effective temperatures produced by electrostatic acceleration of ions. The device can be built inexpensively, but it too is unable to produce a net power output.
- Antimatter-initialized fusion uses small amounts of antimatter to trigger a tiny fusion explosion. This has been studied primarily in the context of making nuclear pulse propulsion feasible. This is not near becoming a practical power source, due to the cost of manufacturing antimatter alone.
- Pyroelectric fusion was reported in April 2005 by a team at UCLA. The scientists used a pyroelectric crystal heated from −34 to 7 °C (−30 to 45 °F), combined with a tungsten needle to produce an electric field of about 25 gigavolts per meter to ionize and accelerate deuterium nuclei into an erbium deuteride target. Though the energy of the deuterium ions generated by the crystal has not been directly measured, the authors used 100 keV (a temperature of about 10^9 K) as an estimate in their modeling. At these energy levels, two deuterium nuclei can fuse together to produce a helium-3 nucleus, a 2.45 MeV neutron and bremsstrahlung. Although it makes a useful neutron generator, the apparatus is not intended for power generation since it requires far more energy than it produces.

Hot Fusion

- "Standard" "hot" fusion, in which the fuel reaches tremendous temperature and pressure inside a fusion reactor or nuclear weapon.

The methods in the second group are examples of non-equilibrium systems, in which very high temperatures and pressures are produced in a relatively small region adjacent to material of much lower temperature. In his doctoral thesis for MIT, Todd Rider did a theoretical study of all quasineutral, isotropic, non-equilibrium fusion systems. He demonstrated that all such systems will leak energy at a rapid rate due to bremsstrahlung, radiation produced when electrons in the plasma hit other electrons or ions at a cooler temperature and suddenly decelerate. The problem is not as pronounced in a hot plasma because the range of temperatures, and thus the magnitude of the deceleration, is much lower. Note that Rider's work does not apply to non-neutral and/or anisotropic non-equilibrium plasmas.

Important Reactions

Astrophysical Reaction Chains

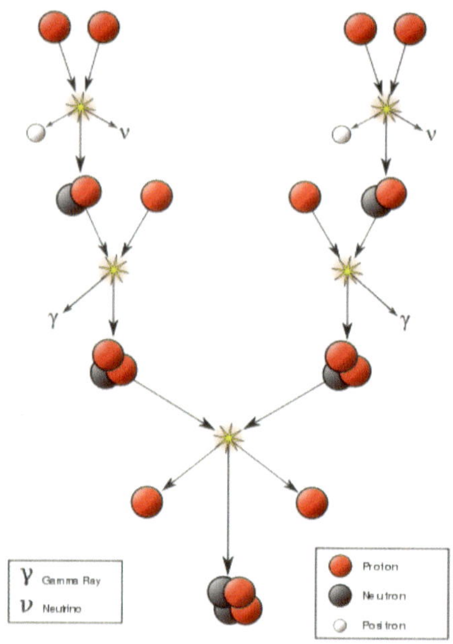

The proton-proton chain dominates in stars the size of the sun or smaller

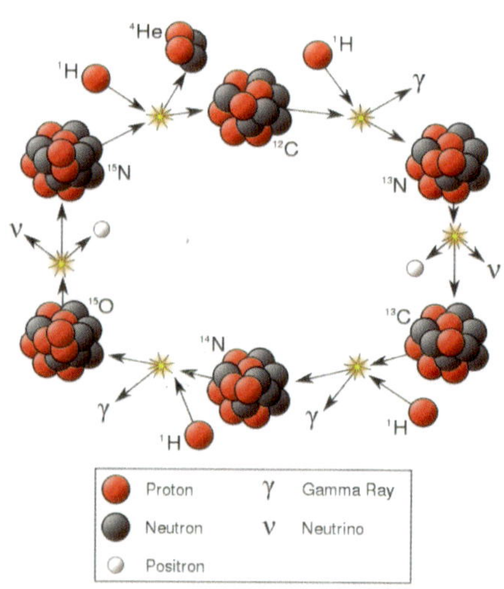

The CNO cycle dominates in stars heavier than the sun

The most important fusion process in nature is that which powers the stars. The net result is the fusion of four protons into one alpha particle, with the release of two positrons, two neutrinos (which changes two of the protons into neutrons), and energy, but several individual reactions are involved, depending on the mass of the star. For stars the size of the sun or smaller, the proton-proton chain dominates. In heavier stars, the CNO cycle is more important. Both types of processes are responsible for the creation of new elements as part of stellar nucleosynthesis.

At the temperatures and densities in stellar cores the rates of fusion reactions are notoriously slow. For example, at solar core temperature (T ~ 15 MK) and density (~120 g/cm^3), the energy release rate is only ~0.1 microwatt/cm^3—millions of times less than the rate of energy release of ordinary candela and thousands of times less than the rate at which a human body generates heat. Thus, reproduction of stellar core conditions in a lab for nuclear fusion power production is completely impractical. Because nuclear reaction rates strongly depend on temperature (~exp(-E/kT)), then in order to achieve reasonable rates of energy production in terrestrial fusion reactors 10–100 times higher temperatures (compared to stellar interiors) are required.

Criteria and Candidates for Terrestrial Reactions

In man-made fusion, the primary fuel is not constrained to be protons and higher temperatures can be used, so reactions with larger cross-sections are chosen. This implies a lower Lawson criterion, and therefore less startup effort. Another concern is the production of neutrons, which activate the reactor structure radiologically, but also have the advantages of allowing volumetric extraction of the fusion energy and tritium breeding. Reactions that release no neutrons are referred to as *aneutronic*.

In order to be useful as a source of energy, a fusion reaction must satisfy several criteria. It must,

- be exothermic: This may be obvious, but it limits the reactants to the low Z (number of protons) side of the curve of binding energy. It also makes helium-4 the most common product because of its extraordinarily tight binding, although He-3 and H-3 also show up;
- involve low Z nuclei: This is because the electrostatic repulsion must be overcome before the nuclei are close enough to fuse;
- have two reactants: At anything less than stellar densities, three body collisions are too improbable. It should be noted that in inertial confinement, both stellar densities and temperatures are exceeded to compensate for the shortcomings of the third parameter of the Lawson criterion, ICF's very short confinement time;
- have two or more products: This allows simultaneous conservation of energy and momentum without relying on the electromagnetic force;
- conserve both protons and neutrons: The cross sections for the weak interaction are too small.

Few reactions meet these criteria. The following are those with the largest cross sections:

(1)	D	+	T	→		4He	(3.5 MeV)	+		n	(14.1 MeV)					
(2i)	D	+	D	→		T	(1.01 MeV)	+		p	(3.02 MeV)					50%
(2ii)				→		3He	(0.82 MeV)	+		n	(2.45 MeV)					50%
(3)	D	+	3He	→		4He	(3.6 MeV)	+		p	(14.7 MeV)					
(4)	T	+	T	→		4He		+	2	n	+ 11.3 MeV					
(5)	3He	+	3He	→		4He		+	2	p	+ 12.9 MeV					
(6i)	3He	+	T	→		4He		+		p		+	n	+ 12.1 MeV		51%
(6ii)				→		4He	(4.8 MeV)	+		D	(9.5 MeV)					43%
(6iii)				→		4He	(0.5 MeV)	+		n	(1.9 MeV)	+	p	(11.9 MeV)		6%
(7i)	D	+	6Li	→	2	4He	+ 22.4 MeV									__%
(7ii)				→		3He		+		4He		+	n	+ 2.56 MeV		__%
(7iii)				→		7Li		+		p	+ 5.0 MeV					__%
(7iv)				→		7Be		+		n	+ 3.4 MeV					__%
(8)	p	+	6Li	→		4He	(1.7 MeV)	+		3He	(2.3 MeV)					
(9)	3He	+	6Li	→	2	4He		+		p	+ 16.9 MeV					
(10)	p	+	11B	→	3	4He	+	8.7 MeV								

p (protium), D (deuterium), and T (tritium) are shorthand notation for the main three isotopes of hydrogen.

For reactions with two products, the energy is divided between them in inverse proportion to their masses, as shown. In most reactions with three products, the distribution of energy varies. For reactions that can result in more than one set of products, the branching ratios are given.

Some reaction candidates can be eliminated at once. The D-^{6}Li reaction has no advantage compared to p-^{11}B because it is roughly as difficult to burn but produces substantially more neutrons through D-D side reactions. There is also a p-^{7}Li reaction, but the cross section is far too low, except possibly when T_i > 1 MeV, but at such high temperatures an endothermic, direct neutron-producing reaction also becomes very significant. Finally there is also a p-^{9}Be reaction, which is not only difficult to burn, but ^{9}Be can be easily induced to split into two alphas and a neutron.

In addition to the fusion reactions, the following reactions with neutrons are important in order to "breed" tritium in "dry" fusion bombs and some proposed fusion reactors:

$$n + {}^{6}Li \rightarrow T + {}^{4}He$$

$$n + {}^{7}Li \rightarrow T + {}^{4}He + n$$

To evaluate the usefulness of these reactions, in addition to the reactants, the products, and the energy released, one needs to know something about the cross section. Any given fusion device will have a maximum plasma pressure that it can sustain, and an economical device will always operate near this maximum. Given this pressure, the largest fusion output is obtained when the temperature is chosen so that $<\sigma v>/T^2$ is a maximum. This is also the temperature at which the value of the triple product $nT\tau$ required for ignition is a minimum (a plasma is "ignited" if the fusion reactions produce enough power to maintain the temperature without external heating). This optimum temperature and the value of $<\sigma v>/T^2$ at that temperature is given for a few of these reactions in the following table.

fuel	T [keV]	$<\sigma v>/T^2$ [$m^3/s/keV^2$]
D-T	13.6	1.24×10^{-24}
D-D	15	1.28×10^{-26}
D-^{3}He	58	2.24×10^{-26}
p-^{6}Li	66	1.46×10^{-27}
p-^{11}B	123	3.01×10^{-27}

Note that many of the reactions form chains. For instance, a reactor fueled with T and ^{3}He will create some D, which is then possible to use in the D + ^{3}He reaction if the energies are "right." An elegant idea is to combine the reactions (8) and (9). The ^{3}He from reaction (8) can react with ^{6}Li in reaction (9) before completely thermalizing. This produces an energetic proton which in turn undergoes reaction (8) before thermalizing. A detailed analysis shows that this idea will not really work well, but it is a good example of a case where the usual assumption of a Maxwellian plasma is not appropriate.

Neutronicity, Confinement Requirement, and Power Density

The only fusion reactions thus far produced by humans to achieve ignition are those which have been created in hydrogen bombs; the first of which, shot Ivy Mike, is shown here.

Any of the reactions above can in principle be the basis of fusion power production. In addition to the temperature and cross section discussed above, we must consider the total energy of the fusion products E_{fus}, the energy of the charged fusion products E_{ch}, and the atomic number Z of the non-hydrogenic reactant.

Specification of the D-D reaction entails some difficulties, though. To begin with, one must average over the two branches (2) and (3). More difficult is to decide how to treat the T and ^{3}He products. T burns so well in a deuterium plasma that it is almost impossible to extract from the plasma. The D-^{3}He reaction is optimized at a much higher temperature, so the burnup at the optimum D-D temperature may be low, so it seems reasonable to assume the T but not the ^{3}He gets burned up and adds its energy to the net reaction. Thus we will count the DD fusion energy as E_{fus} = (4.03+17.6+3.27)/2 = 12.5 MeV and the energy in charged particles as E_{ch} = (4.03+3.5+0.82)/2 = 4.2 MeV.

Another unique aspect of the D-D reaction is that there is only one reactant, which must be taken into account when calculating the reaction rate.

With this choice, we tabulate parameters for four of the most important reactions.

fuel	***Z***	**E_{fus} [MeV]**	**E_{ch} [MeV]**	**neutronicity**
D-T	1	17.6	3.5	0.80
D-D	1	12.5	4.2	0.66
D-^{3}He	2	18.3	18.3	~0.05
p-^{11}B	5	8.7	8.7	~0.001

The last column is the neutronicity of the reaction, the fraction of the fusion energy released as neutrons. This is an important indicator of the magnitude of the problems associated with neutrons like radiation damage, biological shielding, remote handling, and safety. For the first two reactions it is calculated as $(E_{fus}-E_{ch})/E_{fus}$. For the last two reactions, where this calculation would give zero, the values quoted are rough estimates based on side reactions that produce neutrons in a plasma in thermal equilibrium.

Of course, the reactants should also be mixed in the optimal proportions. This is the case when each reactant ion plus its associated electrons accounts for half the pressure. Assuming that the total pressure is fixed, this means that density of the non-hydrogenic ion is smaller than that of the hydrogenic ion by a factor $2/(Z+1)$. Therefore the rate for these reactions is reduced by the same factor, on top of any differences in the values of $<\sigma v>/T^2$. On the other hand, because the D-D reaction has only one reactant, the rate is twice as high as if the fuel were divided between two hydrogenic species.

Thus there is a "penalty" of (2/(Z+1)) for non-hydrogenic fuels arising from the fact that they require more electrons, which take up pressure without participating in the fusion reaction (It is usually a good assumption that the electron temperature will be nearly equal to the ion temperature. Some authors, however, discuss the possibility that the electrons could be maintained substantially colder than the ions. In such a case, known as a "hot ion mode," the "penalty" would not apply.

There is at the same time a "bonus" of a factor 2 for D-D due to the fact that each ion can react with any of the other ions, not just a fraction of them.

We can now compare these reactions in the following table:

fuel	<σv>/T²	penalty/bonus	reactivity	Lawson criterion	power density
D-T	1.24×10^{-24}	1	1	1	1
D-D	1.28×10^{-26}	2	48	30	68
D-^{3}He	2.24×10^{-26}	2/3	83	16	80
p-^{11}B	3.01×10^{-27}	1/3	1240	500	2500

The maximum value of $<\sigma v>/T^2$ is taken from a previous table. The "penalty/bonus" factor is that related to a non-hydrogenic reactant or a single-species reaction. The values in the column "reactivity" are found by dividing 1.24×10^{-24} by the product of the second and third columns. It indicates the factor by which the other reactions occur more slowly than the D-T reaction under comparable conditions. The column "Lawson criterion" weights these results with E_{ch} and gives an indication of how much more difficult it is to achieve ignition with these reactions, relative to the difficulty for the D-T reaction. The last column is labeled "power density" and weights the practical reactivity with E_{fus}. It indicates how much lower the fusion power density of the other reactions is compared to the D-T reaction and can be considered a measure of the economic potential.

Bremsstrahlung Losses in Quasineutral, Isotropic Plasmas

The ions undergoing fusion in many systems will essentially never occur alone but will be mixed with electrons that in aggregate neutralize the ions' bulk electrical charge and form a plasma. The electrons will generally have a temperature comparable to or greater than that of the ions, so they will collide with the ions and emit x-ray radiation of 10–30 keV energy (Bremsstrahlung). The sun and stars are opaque to x-rays, but essentially any terrestrial fusion reactor will be optically thin for x-rays of this energy range. X-rays are difficult to reflect but they are effectively absorbed (and converted into heat) in less than mm thickness of stainless steel (which is part of reactor shield). The ratio of fusion power produced to x-ray radiation lost to walls is an important figure of merit. This ratio is generally maximized at a much higher temperature than that which maximizes the power density. The following table shows the rough optimum temperature and the power ratio at that temperature for several reactions.

fuel	T_i (keV)	$P_{fusion}/P_{Bremsstrahlung}$
D-T	50	140
D-D	500	2.9
D-^{3}He	100	5.3
^{3}He-^{3}He	1000	0.72
p-^{6}Li	800	0.21
p-^{11}B	300	0.57

The actual ratios of fusion to Bremsstrahlung power will likely be significantly lower for several reasons. For one, the calculation assumes that the energy of the fusion products is transmitted completely to the fuel ions, which then lose energy to the electrons by collisions, which in turn lose energy by Bremsstrahlung. However because the fusion products move much faster than the fuel ions, they will give up a significant fraction of their energy directly to the electrons. Secondly, the plasma is assumed to be composed purely of fuel ions. In practice, there will be a significant proportion of impurity ions, which will lower the ratio. In particular, the fusion products themselves *must* remain in the plasma until they have given up their energy, and *will* remain some time after that in any proposed confinement scheme. Finally, all channels of energy loss other than Bremsstrahlung have been neglected. The last two factors are related. On theoretical and experimental grounds, particle and energy confinement seem to be closely related. In a confinement scheme that does a good job of retaining energy, fusion products will build up. If the fusion products are efficiently ejected, then energy confinement will be poor, too.

The temperatures maximizing the fusion power compared to the Bremsstrahlung are in every case higher than the temperature that maximizes the power density and minimizes the required value of the fusion triple product. This will not change the optimum operating point for D-T very much because the Bremsstrahlung fraction is low, but it will push the other fuels into regimes where the power density relative to D-T is even lower and the required confinement even more difficult to achieve. For D-D and D-^{3}He, Bremsstrahlung losses will be a serious, possibly prohibitive problem. For ^{3}He-^{3}He, p-^{6}Li and p-^{11}B the Bremsstrahlung losses appear to make a fusion reactor using these fuels with a quasineutral, anisotropic plasma impossible. Some ways out of this dilemma are considered—and rejected—in "Fundamental limitations on plasma fusion systems not in thermodynamic equilibrium" by Todd Rider. This limitation does not apply to non-neutral and anisotropic plasmas; however, these have their own challenges to contend with.

Mathematical Description

To determine the rate of fusion reactions, the value of most interest is the cross section, which describes the probability that particle with fuse by giving a characteristic area of interaction. In a classical picture, the cross section is the radius of the nuclei (on order of a femtometer) and can only occur when the particles have enough energy to overcome the Coulomb barrier and approach within this radius (Needing energy on the order of a MeV). This greatly overestimates the energy needed for fusion reactions to occur because it doesn't include quantum effects of the smeared radius as the DeBroglie wavelength as well as quantum tunnelling. An estimation of the fusion cross section is often broken into three pieces:

$$\sigma \approx \sigma_{geometry} \times T \times R$$

Where $\sigma_{geometry}$ is the geometric cross section, T is the barrier transparency and R is the reaction characteristics of the reaction.

$\sigma_{geometry}$ is of the order of the square of the de-Broglie wavelength $\sigma_{geometry} \approx \lambda^2 = \left(\frac{\hbar^2}{m_r v}\right)^2 \propto \frac{1}{\epsilon}$ where m_r is the reduced mass of the system and ϵ is the center of mass energy of the system.

T can be approximated by the Gamow transparency, which has the form: $T \approx e^{-\sqrt{\epsilon_G/\epsilon}}$ where $\epsilon_G = (\pi\alpha Z_1 Z_2)^2 \times 2m_r c^2$ is the Gamow factor.

R contains all the nuclear physics of the specific reaction and takes very different values depending on the nature of the interaction. However, for most reactions, the variation of $R(\epsilon)$ is small compared to the variation from the Gamow factor and so is approximated by a function called the Astrophysical S-factor, $S(\epsilon)$, which is weakly varying in energy. Putting these dependencies together, one estimation for the fusion cross section as a function of energy takes the form:

$$\sigma(\epsilon) \approx \frac{S(\epsilon)}{\epsilon} e^{-\sqrt{\epsilon_G/\epsilon}}$$

Thermonuclear Fusion

Thermonuclear fusion refers to nuclear fusion reactions which take place at extremely high temperatures (example: reactions in the sun). The energy produced here is extremely high but such reactions cannot be controlled. If we are able to achieve conditions where such a reaction can take place while controlling its rate, then we can achieve controlled thermonuclear fusion.

The hydrogen bomb is an example of an uncontrolled fusion reaction. Another differentiating factor between controlled and uncontrolled fusion reactions is that, since conditions are unpredictable in an uncontrolled reaction, they can't be tapped for any use.

Another issue to consider in controlled thermonuclear reaction is about containing it. The temperature of the hot plasma is so high that it cannot be in contact with any material. Therefore it has to be checked , but in such a way that there is no contact.

There are a few ways to do this

Magnetic Confinement – used as particles are charged

Inertial Confinement – rapid pulse dispensed to achieve required conditions

Gravitational Confinement – found only in stars

Electrostatic Confinement – confine ions using electrostatic field

Since there aren't any possibilities for chain reactions to occur, it's easier to control and stop them than fission reactions. This is why people are trying to tap into this source rather than using fission reactions. Along with the fact that it can produce virtually an abundant source of fuel with very little and easily available fuel, it also has almost no nuclear waste!

Unfortunately, there is no possibility of harnessing this energy in the near future, till perhaps about 20 – 25 years from now and no economical feasibility until then as well. There is a common thought that, in case this is not a foreseeable future, the amount of resources spent on research could have been diverted for other renewable sources of energy.

The utilization of thermonuclear fusion belongs among the most promising ways of electricity generation. In this process at a very high temperature, a synthesis (fusion) of light elements into heavier ones occurs with the release of a significant energy. At the level of current knowledge about the problems in question, the synthesis of isotopes of hydrogen (deuterium and tritium) seems promising. The output of this process is helium that itself is not radioactive and waste from the thermonuclear synthesis thus will not be radioactive. If the process taking place in reactor is sufficiently shielded and insulated, the material itself from which the reactor is built becomes radioactive. From the above mentioned it is obvious that even during thermonuclear synthesis certain amount of radwaste will be produced, however, it will be a significantly less amount than in the existing nuclear power plants.

The technology of thermonuclear synthesis is mastered experimentally at present and the future will show if it will be utilized for electric energy production in the industrial scale.

Temperature Requirements

Temperature is a measure of the average kinetic energy of particles, so by heating the material it will gain energy. After reaching sufficient temperature, given by the Lawson criterion, the energy of accidental collisions within the plasma is high enough to overcome the Coulomb barrier and the particles may fuse together.

In a deuterium–tritium fusion reaction, for example, the energy necessary to overcome the Coulomb barrier is 0.1 MeV. Converting between energy and temperature shows that the 0.1 MeV barrier would be overcome at a temperature in excess of 1.2 billion kelvins.

There are two effects that lower the actual temperature needed. One is the fact that temperature is the *average* kinetic energy, implying that some nuclei at this temperature would actually have much higher energy than 0.1 MeV, while others would be much lower. It is the nuclei in the high-energy tail of the velocity distribution that account for most of the fusion reactions. The other effect is quantum tunnelling. The nuclei do not actually have to have enough energy to overcome the Coulomb barrier completely. If they have nearly enough energy, they can tunnel through the remaining barrier. For these reasons fuel at lower temperatures will still undergo fusion events, at a lower rate.

Thermonuclear fusion is one of the methods being researched in the attempts to produce fusion power. If Thermonuclear fusion becomes favorable to use, it would significantly reduce the world's carbon footprint.

Confinement

The key problem in achieving thermonuclear fusion is how to confine the hot plasma. Due to the high temperature, the plasma can not be in direct contact with any solid material, so it has to be located in a vacuum. Also, high temperatures imply high pressures. The plasma tends to expand immediately and some force is necessary to act against it. This force can take one of three forms: gravitation in stars, magnetic forces in magnetic confinement fusion reactors, or inertial as the fusion reaction may occur before the plasma starts to expand, so the plasma's inertia is keeping the material together.

Gravitational Confinement

One force capable of confining the fuel well enough to satisfy the Lawson criterion is gravity. The mass needed, however, is so great that gravitational confinement is only found in stars—the least massive stars capable of sustained fusion are red dwarfs, while brown dwarfs are able to fuse deuterium and lithium if they are of sufficient mass. In stars heavy enough, after the supply of hydrogen is exhausted in their cores, their cores (or a shell around the core) start fusing helium to carbon. In the most massive stars (at least 8–11 solar masses), the process is continued until some of their energy is produced by fusing lighter elements to iron. As iron has one of the highest binding energies, reactions producing heavier elements are generally endothermic. Therefore significant amounts of heavier elements are not formed during stable periods of massive star evolution, but are formed in supernova explosions. Some lighter stars also form these elements in the outer parts of the stars over long periods of time, by absorbing energy from fusion in the inside of the star, by absorbing neutrons that are emitted from the fusion process.

All of the elements heavier than iron have some potential energy to release, in theory. At the extremely heavy end of element production, these heavier elements can produce energyin the process of being split again back toward the size of iron, in the process of nuclear fission. Nuclear fission thus releases energy which has been stored, sometimes billions of years before, during stellar nucleosynthesis.

Magnetic Confinement

Electrically charged particles (such as fuel ions) will follow magnetic field lines. The fusion fuel can therefore be trapped using a strong magnetic field. A variety of magnetic configurations exist, including the toroidal geometries of tokamaks and stellarators and open-ended mirror confinement systems.

Inertial Confinement

A third confinement principle is to apply a rapid pulse of energy to a large part of the surface of a pellet of fusion fuel, causing it to simultaneously "implode" and heat to very high pressure and temperature. If the fuel is dense enough and hot enough, the fusion reaction rate will be high enough to burn a significant fraction of the fuel before it has dissipated. To achieve these extreme conditions, the initially cold fuel must be explosively compressed. Inertial confinement is used in the hydrogen bomb, where the driver is x-rays created by a fission bomb. Inertial confinement is also attempted in "controlled" nuclear fusion, where the driver is a laser, ion, or electron beam, or a Z-pinch. Another method is to use conventional high explosive material to compress a fuel to fusion conditions. The UTIAS explosive-driven-implosion facility was used to produce stable, centred and focused hemispherical implosions to generate neutrons from D-D reactions. The simplest and most direct method proved to be in a predetonated stoichiometric mixture of deuterium-oxygen. The other successful method was using a miniature Voitenko compressor, where a plane diaphragm was driven by the implosion wave into a secondary small spherical cavity that contained pure deuterium gas at one atmosphere.

Electrostatic Confinement

There are also electrostatic confinement fusion devices. These devices confine ions using electrostatic fields. The best known is the Fusor. This device has a cathode inside an anode wire cage. Positive ions fly towards the negative inner cage, and are heated by the electric field in the process. If they miss the inner cage they can collide and fuse. Ions typically hit the cathode, however, creating prohibitory high conduction losses. Also, fusion rates in fusors are very low due to competing physical effects, such as energy loss in the form of light radiation. Designs have been proposed to avoid the problems associated with the cage, by generating the field using a non-neutral cloud. These include a plasma oscillating device, a penning trap and the polywell. The technology is relatively immature, however, and many scientific and engineering questions remain.

Inertial Confinement Fusion

Laser Fusion, usually known as Inertial Confinement Fusion (ICF), is one of two major routes to releasing energy by the process of nuclear fusion for energy generation purposes.

Laser fusion attempts to force nuclear fusion in tiny pellets or microballoons of a deuterium-tritium mixture by zapping them with such a high energy density that they will fuse before they have time to move away from each other. This is an example of inertial confinement.

Two experimental laser fusion devices have been developed at Lawrence Livermore Laboratory, called Shiva and Nova. They deliver high power bursts of laser light from multiple lasers onto a small deuterium-tritium target. These lasers are neodymium glass lasers which are capable of extremely high power pulses.

In this approach, a fuel mass is compressed rapidly to densities 1,000 to 10,000 times greater than normal by generating a pressure as high as 10^{17} pascals (10^{12} atmospheres) for periods as short as a nanosecond (10^{-9}second). Near the end of this time period, the implosion speed exceeds about 3×10^5 metres per second. At maximum compression of the fuel, which is now in a cool plasma state, the energy in converging shock waves is sufficient to heat the very centre of the fuel to temperatures high enough to induce fusion reactions (greater than an equivalent energy of about 4,400 eV). If the mass of this highly compressed fuel material is large enough, energy will be generated through fusion reactions before this hot plasma ball disassembles. Under proper conditions, much more energy can be released than is required to compress and shock heat the fuel to thermonuclear burning conditions.

The physical processes in ICF bear a relationship to those in thermonuclear weapons and in star formation—namely, collapse, compression heating, and the onset of nuclear fusion. The situation in star formation differs in one respect: gravity is the cause of the collapse, and a collapsed star begins to expand again due to heat from exoergic nuclear fusion reactions. The expansion is ultimately arrested by the gravitational force associated with the enormous mass of the star, at which point a state of equilibrium in both size and temperature is achieved. In contrast, the fuel in a thermonuclear weapon or ICF completely disassembles. In the ideal ICF case, however, this does not occur until about 30 percent of the fusion fuel has burned.

Over the decades, very significant progress has been made in developing the technology and systems for high-energy, short-time-pulse drivers that are necessary to implode the fusion fuel. The most common driver is a high-power laser, though particle accelerators capable of producing beams of high-energy ions are also used. Lasers that produce more than 100,000 joules in pulses of about one nanosecond are now used in experiments, and the power available in short bursts exceeds 10^{14} watts.

Two lasers capable of delivering up to 5,000,000 joules in equally short bursts, generating a power level on the fusion targets in excess of 5×10^{14}watts, are operational.

Principles of Inertial Confinement

In an inertial confinement fusion (ICF) reactor, a tiny solid pellet of fuel—such as deuterium-tritium (D-T)—would be compressed to tremendous density and temperature so that fusion power is produced in the few nanoseconds before the pellet blows apart. The compression is accomplished by focusing an intense laser beam or a charged particle beam, referred to as the driver, upon the small pellet (typically 1 to 10 mm in diameter). For efficient thermonuclear burn, the time allotted for the pellet to burn must be less than the disassembly time. This means that, in the compressed state, the product of the pellet mass density and the pellet radius must exceed about 3 grams per square centimetre. A high mass density will hasten the burn, and a large radius will slow the disassembly time. The ratio of fusion energy produced in the pellet explosion to the driver energy is called the pellet gain. High pellet gains of 100 or more are required for an ICF reactor.

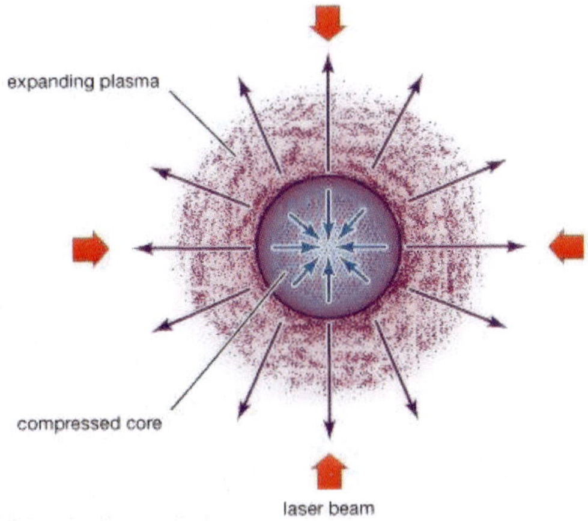

Laser fusion

Pellets are multilayered, consisting of several concentric spheres. The surface of the pellet is ionized by the driver beam, and ablation of the ionized material generates a large inward force on the pellet. Recoil from the ablation implodes the inner layer, producing a shock wave that compresses the inner layers of the D-T fuel. The implosion speed is several hundred kilometres per second, produced by a force equivalent to some 10 billion atmospheres. The target layers are designed such that the laser or particle-beam energy provides compression, not heat (entropy), during this stage. At the final stage of compression, the pellet is compressed to 1,000 to 10,000 times the density of typical solids.

In conventional ICF (usually referred to as shock-heated ICF), the laser or particle-beam energy pulses are accurately set such that the shocks produced during the implosion phase converge in the centre of the pellet, heating it to fusion temperatures. The burn initiates in the central D-T layer and spreads outward as the alpha particles collide with and heat the rest of the pellet to a value sufficient to produce fusion reactions. Ignition occurs, and the pellet, now a dense plasma, is burned up in a small microexplosion. An alternative method for ICF, known as fast ignition, has emerged in recent years because of rapid progress in generating intense picosecond (10^{-12} second) laser systems. Fast ignition can reduce the driver energy considerably. In this scheme, the main laser or particle beam is used to compress the fuel similar to shock heating. Then a short, intense laser pulse heats a small portion of the compressed fuel to fusion temperature, initiating the fusion burn. This approach may lead to target gains that are 3 to 10 times larger than in shock heating.

It is essential that the implosion of the outer layer of the target be symmetric and uniform to a high degree of accuracy. Any asymmetry can grow during compression, and, more important, the shocks may not precisely converge on the centre, which would prevent ignition of the fuel. Thus, the pellet must be manufactured with a high degree of smoothness, with tolerances of less than a thousandth of a millimetre. The driver should also deposit its energy on the pellet uniformly, with a variation of less than 1 percent. There are two methods to achieve this uniformity. In the first method, known as indirect drive, the pellet is located inside a hollow cylindrical shell known as a hohlraum, and the driver is aimed at the walls of the hohlraum. The hohlraum absorbs the driver's energy and then radiates the target with intense X-rays, which cause the pellet to heat and implode. Because a hohlraum is effectively a resonant cavity, the X-ray intensity on the target will be quite uniform even if few driver beams are used. The drawback of this technique is that the target gain is reduced because of inefficiency in converting the driver energy to X-rays. A higher gain is seen in the second method, known as direct drive, but here the driver system is much more complex, as many driver beams and special optical elements are needed to achieve the necessary uniform delivery of energy to the target.

ICF Mechanism of Action

In a hydrogen bomb, the fusion fuel is compressed and heated with a separate fission bomb. A variety of mechanisms transfers the energy of the fission "trigger"'s explosion into the fusion fuel. The requirement of a fission bomb makes the method impractical for power generation. Not only would the triggers be prohibitively expensive to produce, but there is a minimum size that such a bomb can be built, defined roughly by the critical mass of the plutonium fuel used. Generally it seems difficult to build nuclear devices smaller than about 1 kiloton in yield, which would make it a difficult engineering problem to extract power from the resulting explosions.

As the explosion size is scaled down, so too is the amount of energy needed to start the reaction off. Studies from the late 1950s and early 1960s suggested that scaling down into the megajoule energy range would require energy levels that could be delivered by any number of means. This led to the idea of using a device that would "beam" the energy at the fusion fuel, ensuring mechanical separation. By the mid-1960s, it appeared that the laser would develop to the point where the required energy levels would be available.

Generally ICF systems use a single laser, the *driver*, whose beam is split up into a number of beams which are subsequently individually amplified by a trillion times or more. These are sent into the reaction chamber (called a target chamber) by a number of mirrors, positioned in order to illuminate the target evenly over its whole surface. The heat applied by the driver causes the outer layer of the target to explode, just as the outer layers of an H-bomb's fuel cylinder do when illuminated by the X-rays of the fission device.

The material exploding off the surface causes the remaining material on the inside to be driven inwards with great force, eventually collapsing into a tiny near-spherical ball. In modern ICF devices the density of the resulting fuel mixture is as much as one-hundred times the density of lead, around 1000 g/cm³. This density is not high enough to create any useful rate of fusion on its own. However, during the collapse of the fuel, shock waves also form and travel into the center of the fuel at high speed. When they meet their counterparts moving in from the other sides of the fuel in the center, the density of that spot is raised much further.

Given the correct conditions, the fusion rate in the region highly compressed by the shock wave can give off significant amounts of highly energetic alpha particles. Due to the high density of the surrounding fuel, they move only a short distance before being "thermalised", losing their energy to the fuel as heat. This additional energy will cause additional fusion reactions in the heated fuel, giving off more high-energy particles. This process spreads outward from the centre, leading to a kind of self-sustaining burn known as *ignition*.

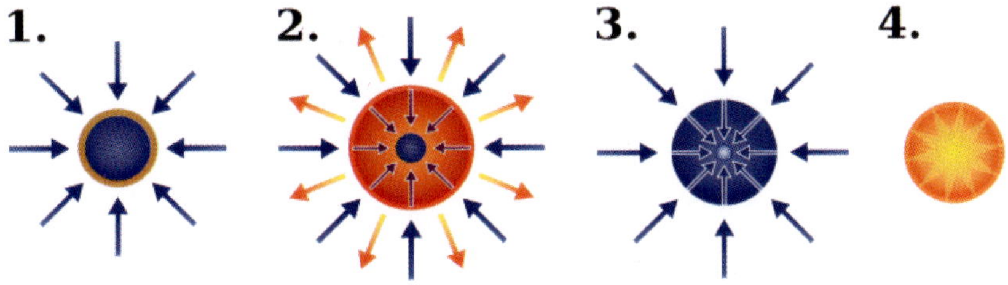

Schematic of the stages of inertial confinement fusion using lasers. The blue arrows represent radiation; orange is blowoff; purple is inwardly transported thermal energy.

1. Laser beams or laser-produced X-rays rapidly heat the surface of the fusion target, forming a surrounding plasma envelope.
2. Fuel is compressed by the rocket-like blowoff of the hot surface material.
3. During the final part of the capsule implosion, the fuel core reaches 20 times the density of lead and ignites at 100,000,000 °C.
4. Thermonuclear burn spreads rapidly through the compressed fuel, yielding many times the input energy.

Issues with Successful Achievement

The primary problems with increasing ICF performance since the early experiments in the 1970s have been of energy delivery to the target, controlling symmetry of the imploding fuel, preventing premature heating of the fuel (before maximum density is achieved), preventing premature mixing of hot and cool fuel by hydrodynamic instabilities and the formation of a ‹tight' shock-wave convergence at the compressed fuel center.

In order to focus the shock wave on the center of the target, the target must be made with extremely high precision and sphericity with aberrations of no more than a few micrometres over its surface (inner and outer). Likewise the aiming of the laser beams must be extremely precise and the beams must arrive at the same time at all points on the target. Beam timing is a relatively simple issue though and is solved by using delay lines in the beams' optical path to achieve picosecond levels of timing accuracy. The other major problem plaguing the achievement of high symmetry and high temperatures/densities of the imploding target are so called "beam-beam" imbalance and beam anisotropy. These problems are, respectively, where the energy delivered by one beam may be higher or lower than other beams impinging on the target and of "hot spots" within a beam diameter hitting a target which induces uneven compression on the target surface, thereby forming Rayleigh–Taylor instabilities in the fuel, prematurely mixing it and reducing heating efficacy at the time of maximum compression. The Richtmyer-Meshkov instability is also formed during the process due to shock waves being formed.

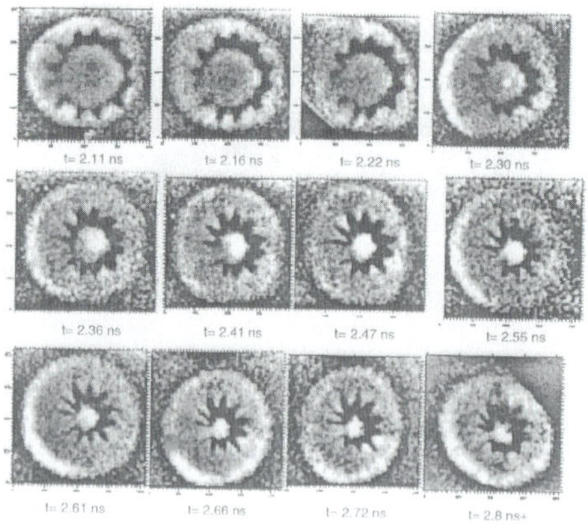

An Inertial confinement fusion target, which was a foam filled cylindrical target with machined perturbations, being compressed by the Nova Laser. This shot was done in 1995. The image shows the compression of the target, as well as the growth of the Rayleigh-Taylor instabilities.

All of these problems have been substantially mitigated to varying degrees in the past two decades of research by using various beam smoothing techniques and beam energy diagnostics to balance beam to beam energy; however, RT instability remains a major issue. Target design has also improved tremendously over the years. Modern cryogenic hydrogen ice targets tend to freeze a thin layer of deuterium just on the inside of a plastic sphere while irradiating it with a low power IR laser to smooth its inner surface while monitoring it with a microscope equipped camera, thereby allowing the layer to be closely monitored ensuring its "smoothness". Cryogenic targets filled with a deuterium tritium (D-T) mixture are "self-smoothing" due to the small amount of heat created by the decay of the radioactive tritium isotope. This is often referred to as "beta-layering".

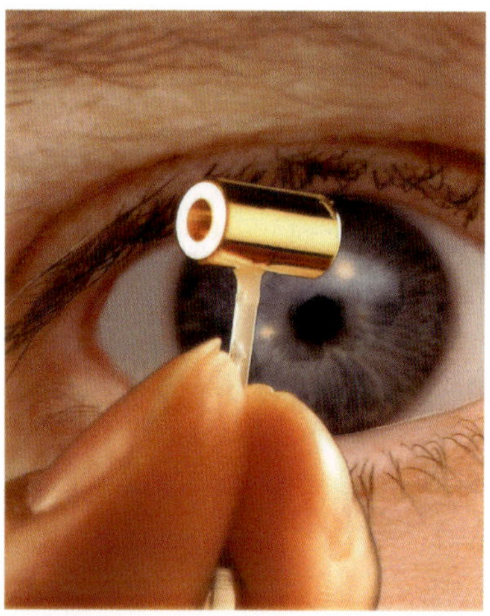

Mockup of a gold plated NIF hohlraum.

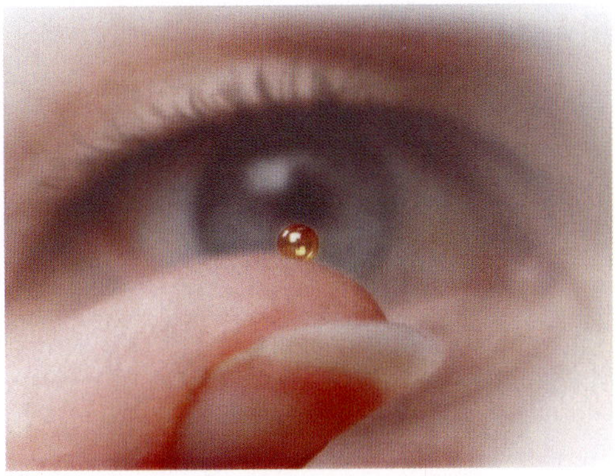

An inertial confinement fusion fuel microcapsule (sometimes called a “microballoon”) of the size to be used on the NIF which can be filled with either deuteriumand tritium gas or DT ice.

The capsule can be either inserted in a hohlraum (as above) and imploded in the indirect drive mode or irradiated directly with laser energy in the direct drive configuration. Microcapsules used on previous laser systems were significantly smaller owing to the less powerful irradiation earlier lasers were capable of delivering to the target.

Certain targets are surrounded by a small metal cylinder which is irradiated by the laser beams instead of the target itself, an approach known as “*indirect drive*”. In this approach the lasers are focused on the inner side of the cylinder, heating it to a superhot plasma which radiates mostly in X-rays. The X-rays from this plasma are then absorbed by the target surface, imploding it in the same way as if it had been hit with the lasers directly. The absorption of thermal x-rays by the target is more efficient than the direct absorption of laser light, however these *hohlraums* or “burning chambers” also take up considerable energy to heat on their own thus significantly reducing the overall efficiency of laser-to-target energy transfer. They are thus a debated feature even today; the equally numerous “*direct-drive*” design does not use them. Most often, indirect drive hohlraum targets are used to simulate thermonuclear weapons tests due to the fact that the fusion fuel in them is also imploded mainly by X-ray radiation.

A variety of ICF drivers are being explored. Lasers have improved dramatically since the 1970s, scaling up in energy and power from a few joules and kilowatts to megajoules and hundreds of terawatts, using mostly frequency doubled or tripled light from neodymium glass amplifiers.

Heavy ion beams are particularly interesting for commercial generation, as they are easy to create, control, and focus. On the downside, it is very difficult to achieve the very high energy densities required to implode a target efficiently, and most ion-beam systems require the use of a hohlraum surrounding the target to smooth out the irradiation, reducing the overall efficiency of the coupling of the ion beam’s energy to that of the imploding target further.

As an Energy Source

Practical power plants built using ICF have been studied since the late 1970s when ICF experiments were beginning to ramp up to higher powers; they are known as inertial fusion energy, or IFE plants. These devices would deliver a successive stream of targets to the reaction chamber, several a second typically, and capture the resulting heat and neutron radiation from their implosion and fusion to drive a conventional steam turbine.

Technical Challenges

IFE faces continued technical challenges in reaching the conditions needed for ignition. But even if these were all to be solved, there are a significant number of practical problems that seem just as difficult to overcome. Laser-driven systems were initially believed to be able to generate commercially useful amounts of energy. However, as estimates of the energy required to reach ignition grew dramatically during the 1970s and '80s, these hopes were abandoned. Given the low efficiency of the laser amplification process (about 1 to 1.5%), and the losses in generation (steam-driven turbine systems are typically about 35% efficient), fusion gains would have to be on the order of 350 just to energetically break even. These sorts of gains appeared to be impossible to generate, and ICF work turned primarily to weapons research.

With the recent introduction of fast ignition and similar approaches, things have changed dramatically. In this approach gains of 100 are predicted in the first experimental device, HiPER. Given a gain of about 100 and a laser efficiency of about 1%, HiPER produces about the same amount of *fusion* energy as electrical energy was needed to create it. It also appears that an order of magnitude improvement in laser efficiency may be possible through the use of newer designs that replace the flash lamps with laser diodes that are tuned to produce most of their energy in a frequency range that is strongly absorbed. Initial experimental devices offer efficiencies of about 10%, and it is suggested that 20% is a real possibility with some additional development.

With "classical" devices like NIF about 330 MJ of electrical power are used to produce the driver beams, producing an expected yield of about 20 MJ, with the maximum credible yield of 45 MJ. Using the same sorts of numbers in a reactor combining fast ignition with newer lasers would offer dramatically improved performance. HiPER requires about 270 kJ of laser energy, so assuming a first-generation diode laser driver at 10% the reactor would require about 3 MJ of electrical power. This is expected to produce about 30 MJ of fusion power. Even a very poor conversion to electrical energy appears to offer real-world power output, and incremental improvements in yield and laser efficiency appear to be able to offer a commercially useful output.

Practical Problems

ICF systems face some of the same secondary power extraction problems as magnetic systems in generating useful power from their reactions. One of the primary concerns is how to successfully remove heat from the reaction chamber without interfering with the targets and driver beams. Another serious concern is that the huge number of neutrons released in the fusion reactions react with the plant, causing them to become intensely radioactive themselves, as well as mechanically weakening metals. Fusion plants built of conventional metals like steel would have a fairly short lifetime and the core containment vessels will have to be replaced frequently.

One current concept in dealing with both of these problems, as shown in the HYLIFE-II baseline design, is to use a "waterfall" of FLiBe, a molten mix of fluoride salts of lithium and beryllium, which both protect the chamber from neutrons and carry away heat. The FLiBe is then passed into a heat exchanger where it heats water for use in the turbines. The tritium produced by fissioning lithium nuclei can also be extracted in order to close the power plant's thermonuclear fuel cycle, a necessity for perpetual operation because tritium does not exist in quantity naturally and must be manufactured. Another concept, Sombrero, uses a reaction chamber built of Carbon-fiber-reinforced polymer which has a very low neutron cross section. Cooling is provided by a molten ceramic, chosen because of its ability to stop the neutrons from traveling any further, while at the same time being an efficient heat transfer agent.

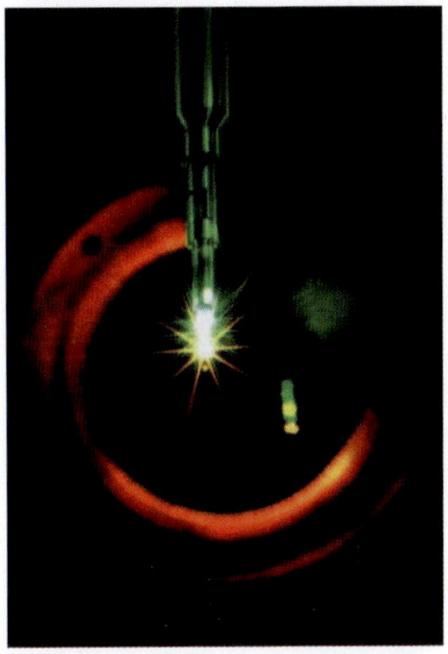

An inertial confinement fusion implosion in Nova, creating "micro sun" conditions of tremendously high density and temperature rivalling even those found at the core of our Sun.

Economic Viability

Even if these technical advances solve the considerable problems in IFE, another factor working against IFE is the cost of the fuel. Even as Nuckolls was developing his earliest detailed calculations on the idea, co-workers pointed this out: if an IFE machine produces 50 MJ of fusion energy, one might expect that a shot could produce perhaps 10 MJ of power for export. Converted to better known units, this is the equivalent of 2.8 kWh of electrical power. Wholesale rates for electrical power on the grid were about 0.3 cents/kWh at the time, which meant the monetary value of the shot was perhaps one cent. In the intervening 50 years the price of power has remained about even with the rate of inflation, and the rate in 2012 in Ontario, Canada was about 2.8 cents/kWh.

Thus, in order for an IFE plant to be economically viable, fuel shots would have to cost considerably less than ten cents in year 2012 dollars. At the time this objection was first noted, Nuckolls suggested using liquid droplets sprayed into the hohlraum from an eye-dropper-like apparatus. Given the ever-increasing demands for higher uniformity of the targets, this approach does not appear practical, as even the inner ablator and fuel itself currently costs several orders of magnitude more than this. Moreover, Nuckolls' solution had the fuel dropped into a fixed hohlraum that would be re-used in a continual cycle, but at current energy levels the hohlraum is destroyed with every shot.

Direct-drive systems avoid the use of a hohlraum and thereby may be less expensive in fuel terms. However, these systems still require an ablator, and the accuracy and geometrical considerations are even more important. They are also far less developed than the indirect-drive systems, and face considerably more technical problems in terms of implosion physics. Currently there is no strong consensus whether a direct-drive system would actually be less expensive to operate.

Nuclear Fission

Nuclear fission is the splitting of the nucleus of an atom into parts (lighter nuclei), often producing photons (in the form of gamma rays), free neutrons, and other subatomic particles as by-products. Fission of heavy elements is an exothermic reaction which can release large amounts of energy both as electromagnetic radiation and as kinetic energy of the fragments (heating the bulk material where fission takes place). Fission is a form of elemental transmutation because the resulting fragments are not the same element as the original atom.

Nuclear fission produces energy for nuclear power and to drive the explosion of nuclear weapons. Both uses are made possible because certain substances, called nuclear fuels, undergo fission when struck by free neutrons and in turn generate neutrons when they break apart. This makes possible a self-sustaining chain reaction that releases energy at a controlled rate in a nuclear reactor or at a very rapid uncontrolled rate in a nuclear weapon.

The amount of free energy contained in nuclear fuel is millions of times the amount of free energy contained in a similar mass of chemical fuel such as gasoline, making nuclear fission a very attractive source of energy. However, the products of nuclear fission are radioactive and remain so for significantly long periods of time, leading to a nuclear waste problem. Concerns over nuclear waste accumulation and over the destructive potential of nuclear weapons may counterbalance the desirable qualities of fission as an energy source, and give rise to ongoing political debate over nuclear power.

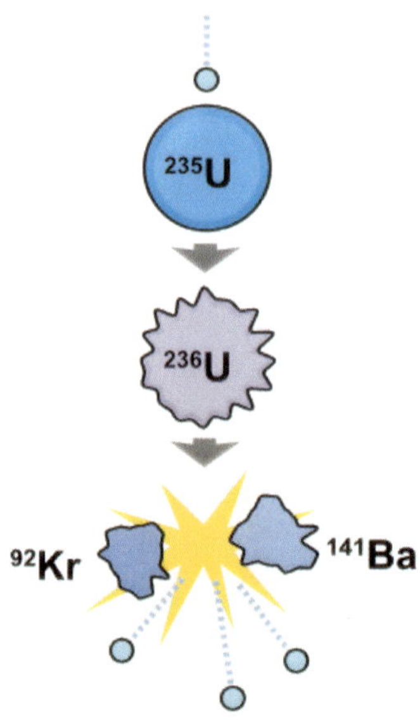

An induced nuclear fission event. A slow-moving neutron is absorbed by the nucleus of a uranium-235 atom, which in turn splits into fast-moving lighter elements (fission products) and free neutrons.

Physical Overview

Nuclear fission differs from other forms of radioactive decay in that it can be harnessed and controlled via a chain reaction: Free neutrons released by each fission event can trigger yet more events, which in turn release more neutrons and cause more fissions. Chemical isotopes that can sustain a fission chain reaction are called nuclear fuels, and are said to be fissile. The most common nuclear fuels are ^{235}U (the isotope of uranium with an atomic mass of 235 and of use in nuclear reactors, 238) and ^{239}Pu (the isotope of plutonium with an atomic mass of 239). These fuels break apart into a range of chemical elements with atomic masses near 100 (fission products). Most nuclear fuels undergo spontaneous fission only very slowly, decaying mainly via an alpha/beta decay chain over periods of millennia to eons. In a nuclear reactor or nuclear weapon, most fission events are induced by bombardment with another particle, such as a neutron.

Typical fission events release several hundred million eV of energy for each fission event. By contrast, most chemical oxidation reactions (such as burning coal or TNT) release at most a few eV per event, so nuclear fuel contains at least ten million times more usable energy than does chemical fuel. The energy of nuclear fission is released as kinetic energy of the fission products and fragments, and as electromagnetic radiation in the form of gamma rays; in a nuclear reactor, the energy is converted to heat as the particles and gamma rays collide with the atoms that make up the reactor and its working fluid, usually water or occasionally heavy water.

Nuclear fission of heavy elements produces energy because the specific binding energy (binding energy per mass) of intermediate-mass nuclei with atomic numbers and atomic masses close to ^{61}Ni and ^{56}Fe is greater than the specific binding energy of very heavy nuclei, so that energy is released when heavy nuclei are broken apart.

The total rest masses of the fission products (Mp) from a single reaction is less than the mass of the original fuel nucleus (M). The excess mass $\Delta m = M - Mp$ is the invariant mass of the energy that is released as photons (gamma rays) and kinetic energy of the fission fragments, according to the mass-energy equivalence formula, $E = mc^2$.

In nuclear fission events, the nuclei may break into any combination of lighter nuclei, but the most common event is not fission to equal mass nuclei of about mass 120; the most common event (depending on isotope and process) is a slightly unequal fission in which one daughter nucleus has a mass of about 90 to 100 u and the other, the remaining 130 to 140 u. Unequal fissions are energetically more favorable because this allows one product to be closer to the energetic minimum near mass 60 u (only a quarter of the average fissionable mass), while the other nucleus with mass 135 u is still not far out of the range of the most tightly bound nuclei (another statement of this is that the atomic binding energy curve is slightly steeper to the left of mass 120 u than to the right of it).

The variation in specific binding energy with atomic number is due to the interplay of the two fundamental forces acting on the component nucleons (protons and neutrons) that make up the nucleus. Nuclei are bound by an attractive strong nuclear force between nucleons, which overcomes the electrostatic repulsion between protons. However, the strong nuclear force acts only over extremely short ranges, since it follows a Yukawa potential. For this reason, large nuclei are less tightly bound per unit mass than small nuclei, and breaking a very large nucleus into two or more intermediate-sized nuclei releases energy.

Because of the short range of the strong binding force, large nuclei must contain proportionally more neutrons than light elements do, which are most stable with a 1-1 ratio of protons and neutrons. Extra neutrons stabilize heavy elements because they add to strong-force binding without adding to proton-proton repulsion. Fission products have, on average, about the same ratio of neutrons and protons as their parent nucleus, and are therefore usually unstable because they have proportionally too many neutrons compared to stable isotopes of similar mass. This is the fundamental cause of the problem of radioactive high level waste from nuclear reactors. Fission products tend to be beta emitters, emitting fast-moving electrons to conserve electric charge as excess neutrons convert to protons inside the nucleus of the fission product atoms.

The most common nuclear fuels, ^{235}U and ^{239}Pu, are not major radiologic hazards by themselves: ^{235}U has a half-life of approximately 700 million years, and although ^{239}Pu has a half-life of only about 24,000 years, it is a pure alpha particle emitter and, hence, not particularly dangerous unless ingested. Once a fuel element has been used, the remaining fuel material is intimately mixed with highly radioactive fission products that emit energetic beta particles and gamma rays. Some fission products have half-lives as short as seconds; others have half-lives of tens of thousands of years, requiring long-term storage in facilities such as Yucca mountain until the fission products decay into non-radioactive stable isotopes.

Chain Reactions

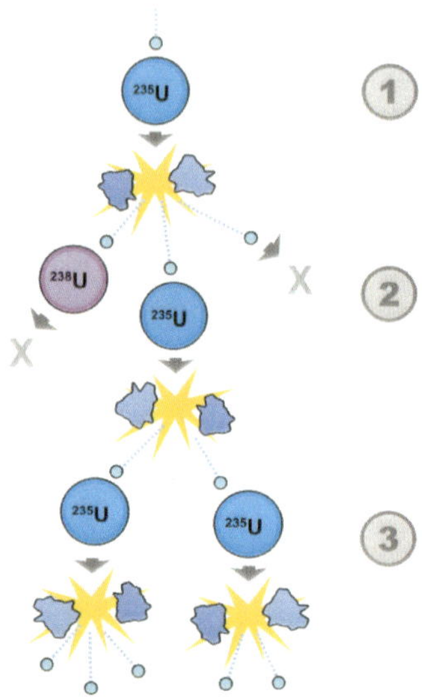

A schematic nuclear fission chain reaction.

1. A uranium-235 atom absorbs a neutron and fissions into two new atoms (fission fragments), releasing three new neutrons and some binding energy. 2. One of those neutrons is absorbed by an atom of uranium-238 and does not continue the reaction. Another neutron is simply lost and does not collide with anything, also not continuing the reaction. However one neutron does collide with an atom of uranium-235, which then fissions and releases two neutrons and some binding energy. 3. Both of those neutrons collide with uranium-235 atoms, each of which fissions and releases between one and three neutrons, which can then continue the reaction.

Many heavy elements, such as uranium, thorium, and plutonium, undergo both spontaneous fission, a form of radioactive decay, and *induced fission*, a form of nuclear reaction. Elemental isotopes that undergo induced fission when struck by a free neutron are called fissionable; isotopes that undergo fission when struck by a thermal, slow moving neutron are also called fissile. A few particularly fissile and readily obtainable isotopes (notably ^{235}U and ^{239}Pu) are called nuclear fuels because they can sustain a chain reaction and can be obtained in large enough quantities to be useful.

All fissionable and fissile isotopes undergo a small amount of spontaneous fission which releases a few free neutrons into any sample of nuclear fuel. Such neutrons escape rapidly from the fuel and become known as free neutrons, with a half-life of about 15 minutes before they decayed to protons and beta particles. However, neutrons almost invariably impact and are absorbed by other nuclei in the vicinity long before this happens (newly-created fission neutrons are moving at about 7 percent of the speed of light, and even moderated neutrons are moving at about 8 times the speed of sound). Some neutrons will impact fuel nuclei and induce further fissions, releasing yet more neutrons. If enough nuclear fuel is assembled into one place, or if the escaping neutrons are sufficiently contained, then these freshly generated neutrons outnumber the neutrons that escape from the assembly, and a *sustained nuclear chain reaction* will take place.

An assembly that supports a sustained nuclear chain reaction is called a critical assembly or, if the assembly is almost entirely made of a nuclear fuel, a critical mass. The word "critical" refers to a cusp in the behavior of the differential equation that governs the number of free neutrons present in the fuel: If less than a critical mass is present, then the amount of neutrons is determined by radioactive decay, but if a critical mass or more is present, then the amount of neutrons is controlled instead by the physics of the chain reaction. The actual mass of a *critical mass* of nuclear fuel depends strongly on the geometry and surrounding materials.

Not all fissionable isotopes can sustain a chain reaction. For example, ^{238}U, the most abundant form of uranium, is fissionable but not fissile: It undergoes induced fission when impacted by an energetic neutron with over 1 MeV of kinetic energy. But too few of the neutrons produced by ^{238}U fission are energetic enough to induce further fissions in ^{238}U, so no chain reaction is possible with this isotope. Instead, bombarding ^{238}U with slow neutrons causes it to absorb them (becoming ^{239}U) and decay by beta emission to ^{239}Np which then decays again by the same process to ^{239}Pu; that process is used to manufacture ^{239}Pu in breeder reactors, but does not contribute to a neutron chain reaction.

Fissionable, non-fissile isotopes can be used as fission energy source even without a chain reaction. Bombarding ^{238}U with fast neutrons induces fissions, releasing energy as long as the external neutron source is present. That effect is used to augment the energy released by modern thermonuclear weapons, by jacketing the weapon with ^{238}U to react with neutrons released by nuclear fusion at the center of the device.

Fission Reactors

Critical fission reactors are the most common type of nuclear reactor. In a critical fission reactor, neutrons produced by fission of fuel atoms are used to induce yet more fissions, to sustain a controllable amount of energy release. Devices that produce engineered but non-self-sustaining fission reactions are *subcritical fission reactors*. Such devices use radioactive decay or particle accelerators to trigger fissions.

Critical fission reactors are built for three primary purposes, which typically involve different engineering trade-offs to take advantage of either the heat or the neutrons produced by the fission chain reaction:

- *Power reactors* are intended to produce heat for nuclear power, either as part of a generating station or a local power system such as in a nuclear submarine.
- *Research reactors* are intended to produce neutrons and/or activate radioactive sources for scientific, medical, engineering, or other research purposes.
- *Breeder reactors* are intended to produce nuclear fuels in bulk from more abundant isotopes. The better known fast breeder reactor makes ^{239}Pu (a nuclear fuel) from the naturally very abundant ^{238}U (not a nuclear fuel). Thermal breeder reactors previously tested using ^{232}Th continue to be studied and developed.

While, in principle, all fission reactors can act in all three capacities, in practice the tasks lead to conflicting engineering goals and most reactors have been built with only one of the above tasks in mind. (There are several early counter-examples, such as the Hanford N reactor, now decommissioned.) Power reactors generally convert the kinetic energy of fission products into heat, which is used to heat a working fluid and drive a heat engine that generates mechanical or electrical power. The working fluid is usually water with a steam turbine, but some designs use other materials, such as gaseous helium. Research reactors produce neutrons that are used in various ways, with the heat of fission being treated as an unavoidable waste product. Breeder reactors are a specialized form of research reactor, with the caveat that the sample being irradiated is usually the fuel itself, a mixture of ^{238}U and ^{235}U.

Fission Bombs

One class of nuclear weapon, a *fission bomb*, otherwise known as an *atomic bomb* or *atom bomb,* is a fission reactor designed to liberate as much energy as possible as rapidly as possible, before the released energy causes the reactor to explode (and the chain reaction to stop). Development of nuclear weapons was the motivation behind early research into nuclear fission: The Manhattan Project of the U.S. military during World War II carried out most of the early scientific work on fission chain reactions, culminating in the Little Boy and Fat Man and Trinity bombs that were exploded over test sites, the cities Hiroshima, and Nagasaki, Japan, in August of 1945.

Even the first fission bombs were thousands of times more explosive than a comparable mass of chemical explosive. For example, Little Boy weighed a total of about four tons (of which 60 kg was nuclear fuel) and was 11 feet long; it also yielded an explosion equivalent to about 15,000 tons of TNT, destroying a large part of the city of Hiroshima. Modern nuclear weapons (which include a thermonuclear *fusion* as well as one or more fission stages) are literally hundreds of times more energetic for their weight than the first pure fission atomic bombs, so that a modern single missile warhead bomb weighing less than 1/8th as much as Little Boy has a yield of 475,000 tons of TNT, and could bring destruction to 10 times the city area.

While the fundamental physics of the fission chain reaction in a nuclear weapon is similar to the physics of a controlled nuclear reactor, the two types of device must be engineered quite differently. It would be extremely difficult to convert a nuclear reactor to cause a true nuclear explosion (though partial fuel meltdowns and steam explosions have occurred), and similarly difficult to extract useful power from a nuclear explosive (though at least one rocket propulsion system, Project Orion, was intended to work by exploding fission bombs behind a massively padded vehicle).

The strategic importance of nuclear weapons is a major reason why the technology of nuclear fission is politically sensitive. Viable fission bomb designs are within the capabilities of bright undergraduates being incredibly simple, but nuclear fuel to realize the designs is thought to be difficult to obtain being rare.

Ternary Fission

Ternary fission is defined as a nuclear breakup into three fragments with usually the third fragment being an α-particle. Though several other third fragments have been reported, the break up into three equal-mass fragments has not yet been detected (normally referred to as true ternary fission, TTF). However, theoretically, it has been envisaged that such heavy third fragment accompanied fission is energetically a viable process. Further, this ternary fission process carries nuclear structural information as is evident from the experimental signature. The favourable ternary breakups are found to be associated with either spherical or deformed closed shells. Thus, the study of this exotic decay mode, the ternary fission, will give structural information of nuclear species.

The branching ratio of ternary fission and/or light charged particle accompanied fission with respect to binary fission is small and being an exotic process, theoretical as well as experimental understanding of this process is not completely exploited. Hence the study on this process is important for understanding fission mechanism. Such a study would help us to understand the conditions of the nucleus at scission, for example, velocity of the fragments, dynamics, trajectories, etc.

Methodology

A model, called the three-cluster model (TCM) has been put forth. This accounts for the energy minimization of all possible ternary breakups of a heavy radioactive nucleus. Further, within the TCM we have analysed the competition between different geometries as well as different positioning of the fragments. Also, an attempt was made to calculate the mass distribution of ternary fission process within the statistical approach. The TCM and the statistical approach used to calculate the mass distribution are discussed as follows:

The Three-cluster Model

The TCM is based on the dynamical or quantum mechanical fragmentation theory of cold fusion phenomenon in heavy-ion reactions and fission dynamics, including the prediction of cluster radioactivity (CR).

In TCM the available Q-value to the kinetic energies Ei of the three fragments, i.e., $Q = E_1 + E_2 + E_3$, is defined as:

$$Q = M - \sum_{i=1}^{3} mi \ ,$$

where M and mi are the mass excesses of the decaying nucleus and the product nuclei, respectively, which are expressed in MeV and taken from.

The ternary fragmentation potential between the three (spherical) fragments (referred to as PES), within the TCM [1–6], is defined as the sum of the total Coulomb potential, total nuclear potential, ℓ-dependent potential and the sum of the mass excesses of ternary fragments. It can be written as

$$V_{tot}(\theta) = \sum_{i=1}^{3}\sum_{j>i}^{3}\left(m_i + V_{ij}(\theta)\right) + V_l(\theta),$$

$$V_{ij}(\theta) = V_{Cij}(\theta) + V_{Nij}(\theta).$$

The Coulomb interaction energy V_{Cij} describes the force of repulsion between the two interacting charges and we consider the pairwise interaction of the three fragments. The Coulomb energy expression is defined as

$$V_{Cij}(\theta) = \begin{cases} \dfrac{Z_i Z_j e^2}{R_{ij}}, & R_{ij} \geq R_i + R_j, \\ \dfrac{Z_i Z_j e^2}{R_{ij}} \Delta, & R_i - R_j \leq R_{ij} \leq R_i + R_j, \\ Z_i Z_j e^2 \Upsilon, & 0 \leq R_{ij} \leq R_i - R_j, \end{cases}$$

$$\Delta = \left[1 - \frac{\left(R_i + R_j - R_{ij}\right)^4}{160 R_i^3 R_j^3}\right],$$

With $\Delta_1 = \left(R_{ij}^2 + 4R_{ij}\left(R_i + R_j\right) + 20R_iR_j - 5R_i^2 - 5R_j^2\right),$

$$\Upsilon = \frac{15R_i^2 - 3R_j^2 - 5R_{ij}^2}{10R_i^3},$$

where R_{ij} is the centre-to-centre distance between the fragments i and j. For the nuclear part of the potential V_{Nij}, one can use either the proximity potential V_{Pij} or the short-range Yukawa plus exponential nuclear attractive potential V_{Yij} amongst the three fragments. The proximity potential V_{Pij} is defined as $V_{Pij} = 4\pi R_Y b\varphi(\xi)$.

The universal function $\varphi(\xi)$ depends only on the distance between two nuclei and is independent of the atomic numbers of the two nuclei. This is given as

$$\varphi(\xi) = \begin{cases} -\dfrac{1}{2}(\xi - 2.54)^2 - 0.0852(\xi - 2.54)^3, & \xi \langle 1.2511, \\ -3.437\exp(-\xi/0.75), & \xi \geq 1.2511. \end{cases}$$

Here $\xi = s/b$. The function $\varphi(\xi)$ is defined for negative (the overlap region), zero (touching configuration) and positive (separated configuration) values of the surface separation $s = R_{ij} - (R_i + R_j)$ and $b \approx 1$ fm is the diffusivity parameter.

The specific nuclear surface tension γ is given by

$$\gamma = 0.9517\left[1 - 1.7826\left(\frac{N - Z}{A}\right)^2\right] \; MeV\, fm^{-2} \,,$$

where R^{-} is the mean curvature radius. The potential as defined in eq. ($V_{tot}(\theta) = \sum_{i=1}^{3}\sum_{j>i}^{3}\left(m_i + V_{ij}(\theta)\right) + V_l(\theta),$) can be calculated as a function of angle θ between the line passing through the centre of the fixed fragment along the fission direction denoted as x-axis and the line passing through the centres of the fixed fragment and the other two fragments. The value of the angle θ = 0 refers to a collinear arrangement as shown in figure . Varying the angle of the end fragments would lead to a triangular configuration as shown in figure where all the surfaces of the three fragments touch each other. The distance between the centres of the interacting fragments R_{ij} can be denoted as

$$R_{32} = R_3 + R_2; \qquad R_{31} = R_1 + R_3,$$

$$R12 = \sqrt{R_{23}^2 + R_{13}^2 + 2R_{23}R_{13}\cos 2\theta}.$$

The radius of each fragment is defined as $R_x \quad roA_i^{1\ 3}$, with r_0 = 1.16 fm (i taking the values of 1, 2 and 3 corresponding to fragments A_1, A_2 and A_3).

The distance vector R_{ij} given as the angle components between I th and j th nuclei are given as

$$R_{31} = \left(-R_{31}\cos\theta\right)\hat{i} + \left(R_{31}\sin\theta\right)\hat{j},$$

$$R_{32} = R_{32}\cos\theta\,\hat{i} + R_{32}\sin\theta\,\hat{j}.$$

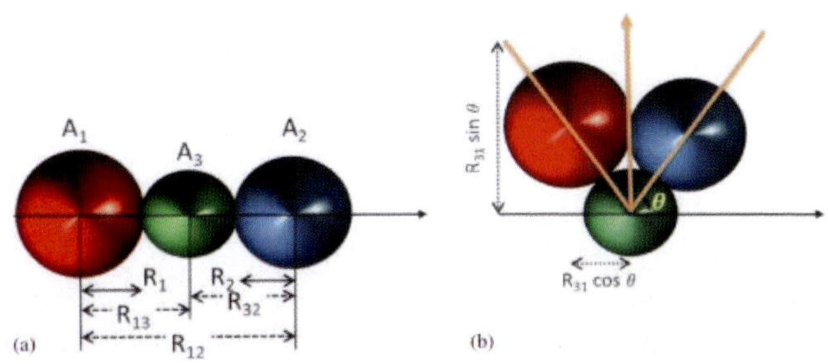

Figure . Schematic configurations of three nuclei in (a) a collinear arrangement and (b) triangular arrangement, with the parameters as labelled.

A Level Density Approach to Ternary Fission

In the work of Rajasekaran and Devanathan, the mass distribution for the binary fission of a parent nucleus was obtained by considering the charge-to-mass ratios of the fissioning fragments. The same idea is extended for the ternary fission as,

$$\frac{Z_1}{A_1} \approx \frac{Z_2}{A_2} \approx \frac{Z_3}{A_3} \approx \frac{Z_p}{A_p},$$

where A_p, Z_p and A_i, Z_i (i = 1, 2 and 3) correspond to the mass and charge numbers of the parent nucleus and the three fission fragments. Further, we impose a condition $A_1 \geq A_2 \geq A_3$ to avoid repetition in fragment combinations. The ternary fission yield, the ratio between the probability of a given ternary fragmentation and the sum of the probabilities of all the possible ternary fragmentations, is given by,

$$Y\left(A_j, Z_j\right) = \frac{P\left(A_j, Z_j\right)}{\sum P\left(A_j, Z_j\right)}.$$

Here A_j and Z_j refer to a ternary fragmentation involving three fragments with mass and charge numbers as A_1, A_2, A_3 and Z_1, Z_2, Z_3 which are obtained from eq.

In the statistical theory of Fong, the fission probability at the scissioning stage is considered to be proportional to the product of nuclear level densities of the fission fragments as,

$$P\left(A_j, Z_j\right) \propto \rho_1\left(A_1, Z_1\right)\rho_2\left(A_2, Z_2\right)\rho_3\left(A_3, Z_3\right).$$

The nuclear level density is

$$\rho = \frac{1}{12}\left(\pi^2 / a\right)^{1/4} E^{-5/4} \exp\left(2\sqrt{aE}\right),$$

where the level density parameter $a = E/T^2$ and the excitation energy $E = E_{tot} - E_0$. The total energy of the system $E_{tot} = \sum_k n_k^Z \varepsilon_k^Z + \sum_k n_k^N \varepsilon_k^N$, where n_k^Z and n_k^N are the occupation probabilities of Z protons and N neutrons of a particular fragment and the summation is for all the single-particle energies considered. The ground-state energy $E_0 = \sum_{k=1}^{Z} \varepsilon_k^Z + \sum_{k=1}^{N} \varepsilon_k^N$. The particle number equations

$$Z = \sum_k n_k^Z = \frac{1}{1 + \exp\left(-\alpha^Z + \beta\varepsilon_k^Z\right)},$$

$$N = \sum_k n_k^N = \frac{1}{1 + \exp\left(-\alpha^N + \beta\varepsilon_k^N\right)}$$

are numerically solved to determine the Lagrangian multipliers α^Z and α^N at a given temperature, $T = 1/\beta$.

Nuclear Fission Product

Fission product, in physics, any of the lighter atomic nuclei formed by splitting heavier nuclei (nuclear fission), including both the primary nuclei directly produced (fission fragments) and the nuclei subsequently generated by their radioactive decay. The fission fragments are highly unstable because of their abnormally large number of neutrons compared with protons; consequently they undergo successive radioactive decays by emitting neutrons, by converting neutrons into protons, antineutrinos, and ejected electrons (beta decay), and by radiating energy (gamma decay).

One of the many known fission reactions of uranium-235 induced by absorbing a neutron results, for example, in two extremely unstable fission fragments, a barium and a krypton nucleus. These fragments almost instantaneously release three neutrons between themselves, becoming barium-144 and krypton-89. By repeated beta decay, the barium-144 in turn is converted step by step to other fission products, lanthanum-144, cerium-144, praseodymium-144, and eventually relatively stable neodymium-144; and krypton-89 is similarly transformed to stable yttrium-89 by way of rubidium-89 and strontium-89. Fission products are identified by their chemical properties and by their radioactive properties, such as their half-lives and the kinds of particles they emit.

Prompt energy is released immediately, and delayed energy is released by fission products after the fission has occurred, this delay can vary from milliseconds to minutes.

Prompt energy:

- The fission fragments fly apart at high speed; their kinetic energy is ≈ 170 MeV. This energy will be deposited locally as heat in the fuel.
- The prompt neutrons will also have a kinetic energy of ≈ 2 MeV. Due to their high energy, these neutrons are also called fast neutrons. On average 2.4 prompt neutrons are released in a U-235 fission, and hence the total energy of prompt neutrons is ≈ 5 MeV. The neutrons will lose this energy within the moderator.
- Prompt gamma rays are emitted from the fission fragments, with an energy ≈ 7 MeV. This energy will be absorbed somewhere within the reactor.

Delayed energy:

- Most fission fragments are neutron-rich and will beta decay after some time has passed, this is the source of delayed energy.
- Beta particles (fast electrons) are emitted, with an energy of ≈ 8 MeV. This energy is deposited in the fuel.
- Beta decay will also produce neutrinos, with an energy of ≈ 10 MeV. These neutrinos and hence their energy will escape the reactor (and our solar system).
- Gamma rays will then be emitted after these beta decays. These delayed gamma rays carry an energy of ≈ 7 MeV. Like the prompt gamma rays, this energy is absorbed somewhere within the reactor.

Formation and Decay

The sum of the atomic mass of the two atoms produced by the fission of one fissile atom is always less than the atomic mass of the original atom. This is because some of the mass is lost as free neutrons, and once kinetic energy of the fission products has been removed (i.e., the products have been cooled to extract the heat provided by the reaction), then the mass associated with this energy is lost to the system also, and thus appears to be “missing” from the cooled fission products.

Since the nuclei that can readily undergo fission are particularly neutron-rich (e.g. 61% of the nucleons in uranium-235 are neutrons), the initial fission products are often more neutron-rich than stable nuclei of the same mass as the fission product (e.g. stable zirconium-90 is 56% neutrons compared to unstable strontium-90 at 58%). The initial fission products therefore may be unstable and typically undergo beta decay to move towards a stable configuration, converting a neutron to a proton with each beta emission. (Fission products do not decay via alpha decay.)

A few neutron-rich and short-lived initial fission products decay by ordinary beta decay (this is the source of perceptible half life, typically a few tenths of a second to a few seconds), followed by immediate emission of a neutron by the excited daughter-product. This process is the source of so-called delayed neutrons, which play an important role in control of a nuclear reactor.

The first beta decays are rapid and may release high energy beta particles or gamma radiation. However, as the fission products approach stable nuclear conditions, the last one or two decays may have a long half-life and release less energy.

Radioactivity Over Time

Fission products have half-lives of 90 years (samarium-151) or less, except for seven long-lived fission products that have half lives of 211,100 years (technetium-99) and more. Therefore, the total radioactivity of a mixture of pure fission products decreases rapidly for the first several hundred years (controlled by the short-lived products) before stabilizing at a low level that changes little for hundreds of thousands of years (controlled by the seven long-lived products).

This behavior of pure fission products with actinides removed, contrasts with the decay of fuel that still contains actinides. This fuel is produced in the so-called "open" (i.e., no nuclear reprocessing) nuclear fuel cycle. A number of these actinides have half lives in the missing range of about 100 to 200,000 years, causing some difficulty with storage plans in this time-range for open cycle non-reprocessed fuels.

Proponents of nuclear fuel cycles which aim to consume all their actinides by fission, such as the Integral Fast Reactor and molten salt reactor, use this fact to claim that within 200 years, their fuel wastes are no more radioactive than the original uranium ore.

Fission products emit beta radiation, while actinides primarily emit alpha radiation. Many of each also emit gamma radiation.

Yield

Each fission of a parent atom produces a different set of fission product atoms. However, while an individual fission is not predictable, the fission products are statistically predictable. The amount of any particular isotope produced per fission is called its yield, typically expressed as percent per parent fission; therefore, yields total to just over 200% (because of ternary fissions), not 100%.

While fission products include every element from zinc through the lanthanides, the majority of the fission products occur in two peaks. One peak occurs at about (expressed by atomic number) strontium to ruthenium while the other peak is at about tellurium to neodymium. The yield is somewhat dependent on the parent atom and also on the energy of the initiating neutron.

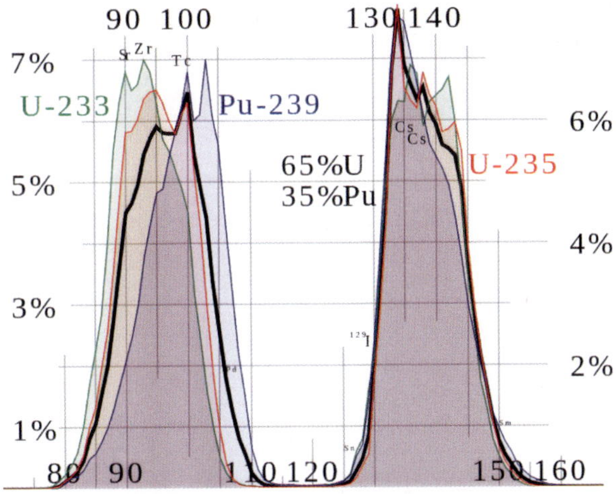

Fission product yields by mass for thermal neutron fission of U-235, Pu-239, a combination of the two typical of current nuclear power reactors, and U-233 used in the thorium cycle.

In general the higher the energy of the state that undergoes nuclear fission, the more likely that the two fission products have similar mass. Hence as the neutron energy increases and/or the energy of the fissile atom increases, the valley between the two peaks becomes more shallow. For instance, the curve of yield against mass for Pu-239 has a more shallow valley than that observed for U-235 when the neutrons are thermal neutrons. The curves for the fission of the later actinides tend to make even more shallow valleys. In extreme cases such as ^{259}Fm, only one peak is seen.

The adjacent figure shows a typical fission product distribution from the fission of uranium. Note that in the calculations used to make this graph, the activation of fission products was ignored and the fission was assumed to occur in a single moment rather than a length of time. In this bar chart results are shown for different cooling times — time after fission. Because of the stability of nuclei with even numbers of protons and/or neutrons, the curve of yield against element is not a smooth curve but tends to alternate. Note that the curve against mass number is smooth.

Production

Small amounts of fission products are naturally formed as the result of either spontaneous fission of natural uranium, which occurs at a low rate, or as a result of neutrons from radioactive decay or reactions with cosmic ray particles. The microscopic tracks left by these fission products in some natural minerals (mainly apatite and zircon) are used in fission track dating to provide the cooling (crystallization) ages of natural rocks. The technique has an effective dating range of 0.1 Ma to >1.0 Ga depending on the mineral used and the concentration of uranium in that mineral.

About 1.5 billion years ago in a uranium ore body in Africa, a natural nuclear fission reactor operated for a few hundred thousand years and produced approximately 5 tonnes of fission products. These fission products were important in providing proof that the natural reactor had occurred. Fission products are produced in nuclear weapon explosions, with the amount depending on the type of weapon. The largest source of fission products is from nuclear reactors. In current nuclear power reactors, about 3% of the uranium in the fuel is converted into fission products as a by-product of energy generation. Most of these fission products remain in the fuel unless there is fuel element failure or a nuclear accident, or the fuel is reprocessed.

Power Reactors

In commercial nuclear fission reactors, the system is operated in the otherwise self-extinguishing prompt subcritical state. The reactor specific physical phenomena that nonetheless maintains the temperature above the decay heat level, are the predictably delayed, and therefore easily controlled, transformations or movements of a vital class of fission product, or reaction ember, as they decay, with Bromine-87 being one such long-lived "ember", with a half-life of about a minute and thus it emits a *delayed* neutron upon decay. Operating in this delayed critical state, the dependence on the inherently delayed transformation or movement of fission products/embers to maintain the temperature, is a process that occurs slow enough to permit human feedback on the temperature control. In an analogous manner to fire dampers varying the opening to control the movement of wood embers towards new fuel, control rods are comparatively varied up or down, as the nuclear fuel burns up over time.

In a nuclear power reactor, the main sources of radioactivity are fission products, alongside actinides and activation products. Fission products are the largest source of radioactivity for the first several hundred years, while actinides are dominant roughly 10^3 to 10^5 years after fuel use.

Fission occurs in the nuclear fuel, and the fission products are primarily retained within the fuel close to where they are produced. These fission products are important to the operation of the reactor because some fission products contribute delayed neutrons that are useful for reactor control while others are neutron poisons that tend to inhibit the nuclear reaction. The buildup of the fission product poisons is a key factor in determining the maximum duration a given fuel element can be kept within the reactor. The decay of short-lived fission products also provide a source of heat within the fuel that continues even after the reactor has been shut down and the fission reactions stopped. It is this decay heat that sets the requirements for cooling of a reactor after shutdown.

If the fuel cladding around the fuel develops holes, then fission products can leak into the primary coolant. Depending on the fission product chemistry, it may settle within the reactor core or travel through the coolant system. Coolant systems include chemistry control systems that tend to remove such fission products. In a well-designed power reactor running under normal conditions, the radioactivity of the coolant is very low.

It is known that the isotope responsible for the majority of the gamma exposure in fuel reprocessing plants (and the Chernobyl site in 2005) is Cs-137. Iodine-129 is one of the major radioactive elements released from reprocessing plants. In nuclear reactors both Cs-137 and strontium-90 are found in locations remote from the fuel. This is because these isotopes are formed by the beta decay of noble gases (xenon-137 {halflife of 3.8 minutes} and krypton-90 {halflife 32 seconds}) which enable these isotopes to be deposited in locations remote from the fuel (e.g. on control rods).

Nuclear Reactor Poisons

Some fission products decay with the release of a neutron. Since there may be a short delay in time between the original fission event (which releases its own prompt neutronsimmediately) and the release of these neutrons, the latter are termed "delayed neutrons". These delayed neutrons are important to nuclear reactor control.

Some of the fission products, such as xenon-135 and samarium-149, have a high neutron absorption cross section. Since a nuclear reactor depends on a balance in the neutron production and absorption rates, those fission products that remove neutrons from the reaction will tend to shut the reactor down or "poison" the reactor. Nuclear fuels and reactors are designed to address this phenomenon through such features as burnable poisons and control rods. Build-up of xenon-135 during shutdown or low-power operation may poison the reactor enough to impede restart or to interfere with normal control of the reaction during restart or restoration of full power, possibly causing or contributing to an accident scenario.

Nuclear Weapons

Nuclear weapons use fission as either the partial or the main energy source. Depending on the weapon design and where it is exploded, the relative importance of the fission product radioactivity will vary compared to the activation product radioactivity in the total fallout radioactivity.

The immediate fission products from nuclear weapon fission are essentially the same as those from any other fission source, depending slightly on the particular nuclide that is fissioning. However, the very short time scale for the reaction makes a difference in the particular mix of isotopes produced from an atomic bomb.

For example, the $^{134}Cs/^{137}Cs$ ratio provides an easy method of distinguishing between fallout from a bomb and the fission products from a power reactor. Almost no Cs-134 is formed by nuclear fission (because xenon-134 is stable). The ^{134}Cs is formed by the neutron activation of the stable ^{133}Cs which is formed by the decay of isotopes in the isobar (A = 133). So in a momentary criticality by the time that the neutron flux becomes zero too little time will have passed for any ^{133}Cs to be present. While in a power reactor plenty of time exists for the decay of the isotopes in the isobar to form ^{133}Cs, the ^{133}Cs thus formed can then be activated to form ^{134}Cs only if the time between the start and the end of the criticality is long.

According to Jiri Hala's textbook, the radioactivity in the fission product mixture in an atom bomb is mostly caused by short-lived isotopes such as I-131 and Ba-140. After about four months Ce-141, Zr-95/Nb-95, and Sr-89 represent the largest share of radioactive material. After two to three years, Ce-144/Pr-144, Ru-106/Rh-106, and Promethium-147 are the bulk of the radioactivity. After a few years, the radiation is dominated by strontium-90 and caesium-137, whereas in the period between 10,000 and a million years it is technetium-99 that dominates.

Application

Some fission products (such as Cs-137) are used in medical and industrial radioactive sources. $^{99}TcO_4^-$ ion can react with steel surfaces to form a corrosion resistant layer. In this way these metaloxo anions act as anodic corrosion inhibitors - it renders the steel surface passive. The formation of $^{99}TcO_2$ on steel surfaces is one effect which will retard the release of ^{99}Tc from nuclear waste drums and nuclear equipment which has become lost prior to decontamination (e.g. nuclea r submarine reactors which have been lost at sea).

In a similar way the release of radio-iodine in a serious power reactor accident could be retarded by adsorption on metal surfaces within the nuclear plant. Much of the other work on the iodine chemistry which would occur during a bad accident has been done.

Fissile Material

Fissile material, also called Fissionable Material, in nuclear physics, any species of atomic nucleus that can undergo the fission reaction. The principal fissile materials are uranium-235 (0.7 percent of naturally occurring uranium), plutonium-239, and uranium-233, the last two being artificially produced from the fertile materials uranium-238 and thorium-232, respectively. A fertile material, not itself capable of undergoing fission with low-energy neutrons, is one that decays into fissile material after neutron absorption within a reactor. Thorium-232 and uranium-238 are the only two naturally occurring fertile materials.

For heavy nuclides with atomic number of higher than 90, most of fissile isotopes meet the fissile rule:

Fissile isotopes have $2 \times Z - N = 43 \pm 2$ (example for ^{235}U: $2 \times 92 - 143 = 41$)

where Z is number of protons and N is number of neutrons.

Distinction between Fissionable, Fissile and Fertile

Fissile materials undergoes fission reaction after absorption of the binding energy of thermal neutron. They do not require additional kinetic energy for fission. If the neutron has higher kinetic energy, this energy will be transformed into additional excitation energy of the compound nucleus. On the other hand, the binding energy released by compound nucleus of (^{238}U + n) after absorption of thermal neutron is less than the critical energy, so the fission reaction cannot occur. The distinction is described in the following points:

- Fissile materials are a subset of fissionable materials.
- Fissionable material consist of isotopes that are capable of undergoing nuclear fission after capturing either fast neutron (high energy neutron – let say >1 MeV) or thermal neutron (low energy neutron – let say 0.025 eV). Typical fissionable materials: 238U, 240Pu, but also 235U, 233U, 239Pu, 241Pu.
- Fissile material consist of fissionable isotopes that are capable of undergoing nuclear fission only after capturing a thermal neutron. 238U is not fissile isotope, because 238U cannot be fissioned by thermal neutron. 238U does not meet also alternative requirement to fissile materials. 238U is not capable of sustaining a nuclear fission chain reaction, because neutrons produced by fission of 238U have lower energies than

original neutron (usually below the threshold energy of 1 MeV). Typical fissile materials: 235U, 233U, 239Pu, 241Pu.

- Fertile material consist of isotopes that are not fissionable by thermal neutrons, but can be converted into fissile isotopes (after neutron absorption and subsequent nuclear decay). Typical fertile materials: 238U, 232Th.

Fissile Nuclides

In general, most actinide isotopes with an odd neutron number are fissile. Most nuclear fuels have an odd atomic mass number ($A = Z + N$ = the total number of nucleons), and an even atomic number Z. This implies an odd number of neutrons. Isotopes with an odd number of neutrons gain an extra 1 to 2 MeV of energy from absorbing an extra neutron, from the pairing effect which favors even numbers of both neutrons and protons. This energy is enough to supply the needed extra energy for fission by slower neutrons, which is important for making fissionable isotopes also fissile.

More generally, nuclides with an even number of protons and an even number of neutrons, and located near a well-known curve in nuclear physics of atomic number vs. atomic mass number are more stable than others; hence, they are less likely to undergo fission. They are more likely to "ignore" the neutron and let it go on its way, or else to absorb the neutron but without gaining enough energy from the process to deform the nucleus enough for it to fission. These "even-even" isotopes are also less likely to undergo spontaneous fission, and they also have relatively much longer partial half-lives for alpha or beta decay. Examples of these isotopes are uranium-238 and thorium-232. On the other hand, other than the lightest nuclides, nuclides with an odd number of protons and an odd number of neutrons (odd Z, odd N) are usually short-lived (a notable exception is neptunium-236 with a half-life of 154,000 years) because they readily decay by beta-particle emission to their isobars with an even number of protons and an even number of neutrons (even Z, even N) becoming much more stable. The physical basis for this phenomenon also comes from the pairing effect in nuclear binding energy, but this time from both proton–proton and neutron–neutron pairing. The relatively short half-life of such odd-odd heavy isotopes means that they are not available in quantity and are highly radioactive.

Nuclear Fuel

To be a useful fuel for nuclear fission chain reactions, the material must:

- Be in the region of the binding energy curve where a fission chain reaction is possible (i.e., above radium).
- Have a high probability of fission on neutron capture.
- Release two or more neutrons on average per neutron capture (which means a higher average number of them on each fission, to compensate for nonfissions, and absorptions in the moderator).
- Have a reasonably long half-life.
- Be available in suitable quantities.

Capture-fission ratios of fissile nuclides						
Thermal neutrons				Epithermal neutrons		
σ_F (b)	σ_γ (b)	%		σ_F (b)	σ_γ (b)	%
531	46	8.0%	^{233}U	760	140	16%
585	99	14.5%	^{235}U	275	140	34%
750	271	26.5%	^{239}Pu	300	200	40%
1010	361	26.3%	^{241}Pu	570	160	22%

Fissile nuclides in nuclear fuels include:

- Uranium-235 which occurs in natural uranium and enriched uranium.
- Plutonium-239 bred from uranium-238 by neutron capture.
- Plutonium-241 bred from plutonium-240 by neutron capture. The ^{240}Pu comes from ^{239}Pu by the same process.
- Uranium-233 bred from thorium-232 by neutron capture.

Fissile nuclides do not have a 100% chance of undergoing fission on absorption of a neutron. The chance is dependent on the nuclide as well as neutron energy. For low and medium-energy neutrons, the neutron capture cross sections for fission (σ_F), the cross section for neutron capture with emission of a gamma ray (σ_γ), and the percentage of non-fissions are in the table at right.

Prompt Neutron

Prompt fission neutrons and prompt gamma rays are emitted when excited primary fission fragments release their energy to reach a more stable configuration. This happens shortly after the fission of the nucleus on two fission fragments. Studying prompt neutrons and gamma rays is of importance not only for deepening our understanding of the nuclear fission process, but also in core design calculations or in radiation shielding calculations.

Most of the neutrons produced in fission are prompt neutrons. Usually more than 99 percent of the fission neutrons are the prompt neutrons, but the exact fraction is dependent on the nuclide to be fissioned and is also dependent on an incident neutron energy (usually increases with energy). For example a fission of ^{235}U by thermal neutron yields 2.43 neutrons, of which 2.42 neutrons are the prompt neutrons and 0.01585 neutrons (0.01585/2.43=0.0065=ß) are the delayed neutrons. The production of prompt neutrons slightly increase with incident neutron energy.

Prompt neutrons are emitted within 10^{-14} second. This is very important, because it is very very fast and causes very very fast response of the reactor power in case of prompt criticality. Therefore nuclear reactors must operate in a prompt subcritical, delayed critical condition. All power reactors are designed to operate in a delayed critical conditions and are provided with safety systems to prevent them from ever achieving prompt criticality.

It must be noted the response of the reactor on the reactivity insertion is determined by a neutron generation time (not by emission time), which is the average time from a prompt neutron emission to a capture that results only in fission.

Isotope	fission cross-section 0.025eV / 2MeV	prompt neutrons 0.025eV / 2MeV	delayed neutrons 0.025eV / 2MeV
235U	585 / 1.27	2.42 / 2.63	0.0162 / 0.0165
238U	0.000027 / 0.57	2.36 / 2.60	0.0478 / 0.0478
233U	531 / 1.98	2.48 / 2.63	0.0067 / 0.0077
239Pu	747 / 1.93	2.87 / 3.16	0.0065 / 0.0067
241Pu	1 012 / 1.76	2.92 / 3.21	0.0160 / 0.0160

Table of key prompt and delayed neutrons characteristics. Thermal vs. Fast Fission

Key Characteristics of Prompt Neutrons

- Prompt neutrons are emitted directly from fission and they are emitted within very short time of about 10^{-14} second.
- Most of the neutrons produced in fission are prompt neutrons – about 99.9%.
- For example a fission of ^{235}U by thermal neutron yields 2.43 neutrons, of which 2.42 neutrons are prompt neutrons and 0.01585 neutrons are the delayed neutrons.
- The production of prompt neutrons slightly increase with incident neutron energy.
- Almost all prompt fission neutrons have energies between 0.1 MeV and 10 MeV.
- The mean neutron energy is about 2 MeV. The most probable neutron energy is about 0.7 MeV.
- In reactor design the prompt neutron lifetime (PNL) belongs to key neutron-physical characteristics of reactor core.
- Its value depends especially on the type of the moderator and on the energy of the neutrons causing fission.
- In an infinite reactor (without escape) prompt neutron lifetime is the sum of the slowing down time and the diffusion time.
- In LWRs the PNL increases with the fuel burnup.
- The typical prompt neutron lifetime in thermal reactors is on the order of 10^{-4} second.
- The typical prompt neutron lifetime in fast reactors is on the order of 10^{-7} second.

Prompt Gamma Rays

With the prompt neutrons prompt gamma rays are associated. Most of prompt gamma rays are emitted after prompt neutrons. The fission reaction releases approximately ~7 MeV in prompt gamma rays and additional ~7 MeV (for ^{235}U) in delayed gamma rays. This is a significant portion of energy (~7 % of fission energy released) and it must be considered in many fields of reactor design or in the design of nuclear reactor shields. A particular challenge is the calculation of the gamma heat deposition in core baffles (reflectors), pressure vessels or excore detector channels.

In past the comparisons of various benchmarks experiments with calculated gamma heating showed a systematic underestimation for the main fuel isotopes ^{235}U and ^{239}Pu. Discrepancies observed for C/E ratios in various benchmarks range from 10 to 28%, while required accuracy is 7.5%. Therefore requests for new measurements of prompt gamma rays in the reactions 235U(n,f) and 239Pu(n,f) have been formulated in the Nuclear Data High Priority Request List of the Nuclear Energy Agency.

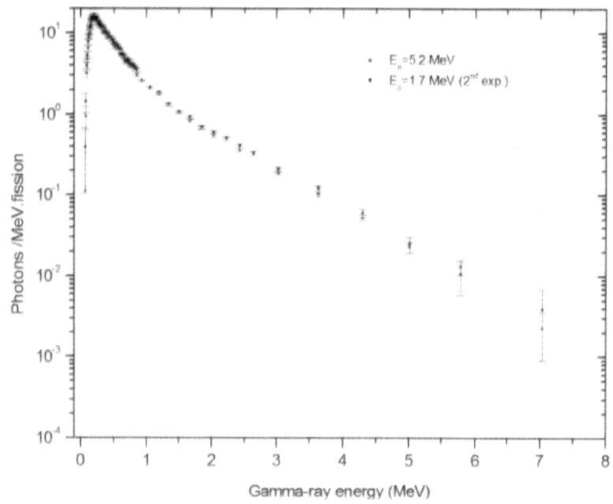

Energy Spectra of Prompt Gamma Rays – New Measurements.

Prompt Neutron Lifetime

Prompt neutron lifetime, l, is the average time from a prompt neutron emission to either its absorbtion(fission or radiative capture) or to its escape from the system. This parameter is defined in multiplying or also in nonmultiplying systems. In both systems the prompt neutron lifetimes depend strongly on:

- Material composition of the system
  - Multiplying – nonmultiplying system
  - System with or without thermalization
  - Isotopic composition of the system
- Geometric configuration of the system
  - Homogeneous or heterogeneous system
  - Shape of entire system
  - Size of the system

In an infinite reactor (without escape) prompt neutron lifetime is the sum of the slowing down time and the diffusion time.

$$l = t_s + t_d$$

In an infinite thermal reactor $t_s << t_d$ and therefore $l \approx t_d$. The typical prompt neutron lifetime in thermal reactors is on the order of 10^{-4} second. Generally, the longer neutron lifetimes take place in systems in which the neutrons must be thermalized in order to be absorbed.

Systems in which most of the neutrons are absorbed at higher energies and the neutron thermalization is suppressed (e.g. in fast reactors), have much shorter prompt neutron lifetimes. The typical prompt neutron lifetime in fast reactors is on the order of 10^{-7} second.

	Slowing Down Time [s]	Diffusion Time [s]
Water	7.1×10^{-4}	2.4×10^{-4}
Heavy water	5.0×10^{-5}	6.0×10^{-2}
Beryllium	5.7×10^{-5}	4.2×10^{-3}
Graphite	1.4×10^{-4}	1.6×10^{-2}

Slowing Down and Diffision Times for Thermal Neutrons in an Infinite Medium

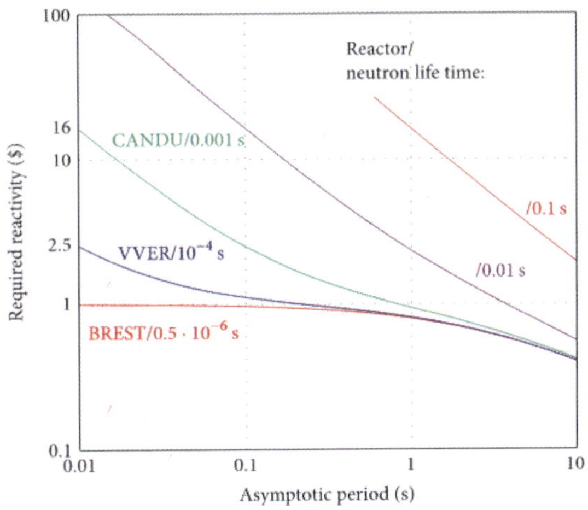

Dependencies of asymptotic time period on the reactivity required for different reactor types with different prompt neutron lifetimes.

	LWR	CANDU	Fast Reactor
Λ [s]	2×10^{-5}	1×10^{-3}	4×10^{-7}
β_{eff}	0.006	0.005	0.0035
λ [s^{-1}]	0.08	0.08	0.08

Table of main kinetic parameters.

Prompt Generation Time – Mean Generation Time

In multiplying systems, in which the absorption of a prompt fission neutron can initiate a fission reaction, l is equal to the average time between two generations of prompt neutrons (at k_{eff}=1). This time is known as the prompt neutron generation time.

Prompt Neutron Generation Time (or Mean Generation Time), Λ, is the average time from a prompt neutron emission to a capture that results only in fission. The prompt neutron generation time is designated as:

$$\Lambda = l/k_{eff}$$

In power reactors the prompt generation time changes with the fuel burnup. In LWRs increases with the fuel burnup. It is simple, fresh uranium fuel contains much fissile material (in case of uranium fuel about 4% of ^{235}U). This causes significant excess of reactivity and this excess must be compensated via chemical shim (in case of PWRs) or via burnable absorbers.

Owing to these factors (high probability of absorption in fuel and high probability of absorption in moderator) the prompt neutron lives much shorter and prompt neutron lifetime is low. With fuel burnup the amount of fissile material as well as the absorption in moderator decreases and therefore the prompt neutron is able to "live"much longer.

Effect of Prompt Neutron Lifetime on Nuclear Safety

The prompt neutron lifetime belongs to key neutron-physical characteristics of reactor core. Its value depends especially on the type of the moderator and on the energy of the neutronscausing fission. Its importance for nuclear reactor safety is well known for a long time.

The longer prompt neutron lifetimes can substantially improve kinetic response of reactor (the longer prompt neutron lifetime gives simply slower power increase). For example under RIA conditions (Reactivity-Initiated Accidents) reactors should withstand a jump-like insertion of relatively large (~1$ or even more) positive reactivity and the PNL (prompt neutron lifetime) plays here the key role. Therefore the PNL should be verified in a reload safety evaluation process.

In some cases (especially in some fast reactors) reactor cores or can be modified in order to increase the PNL and in order to improve nuclear safety.

Prompt Neutron Energy Spectrum

Studying prompt neutron energy spectra is of importance in many nuclear reactor applications (e.g. criticality calculations) and also constitute the most important component of the source term for nuclear reactor shielding calculations.

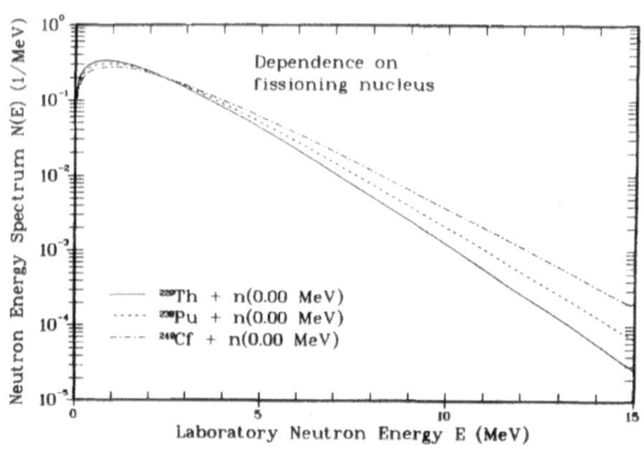

Prompt Neutron Energy Spectra – Dependence on fissioning nucleus.

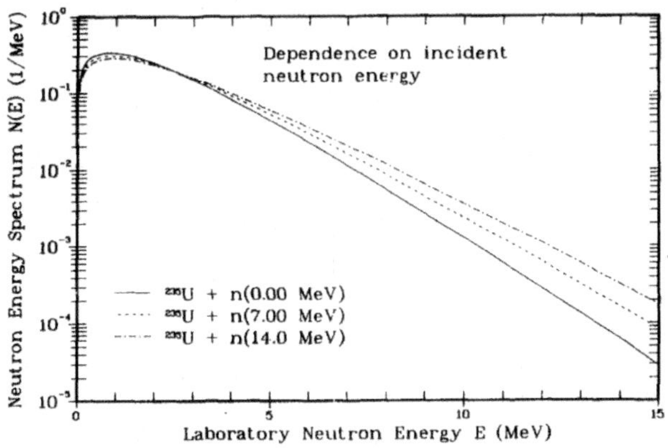

Prompt Neutron Energy Spectra – Dependence on incident neutron energy.

Basic features of prompt neutron energy spectra are summarized below:

- The neutrons produced by fission are high energy neutrons.
- Almost all fission neutrons have energies between 0.1 MeV and 10 MeV.
- The prompt neutron energy distribution, or spectrum, may be best described by dependence of the fraction of neutrons per MeV on neutron energy.
- The most probable neutron energy is about 0.7 MeV.
  The mean neutron energy is about 2 MeV.
- These values are for thermal fission of ^{235}U, but these values vary only slightly for other nuclides.

Prompt neutron fission spectra evaluation is one of the most interesting aspects of evaluation of actinides. Many experimental and theoretical researches have been carried out for the determination of prompt neutron spectra. There are several representations of prompt fission neutron spectra. Two early models of the prompt fission neutron spectrum, which are still used today, are the Maxwellian and Watt spectrum.

The modern spectrum representation of the prompt fission neutron spectrum and average prompt neutron multiplicity is called the Madland-Nix Spectrum (Los Alamos Model). This model is based upon the standard nuclear evaporation theory and utilizes an isospin-dependent optical potential for the inverse process of compound nucleus formation in neutron-rich fission fragments.

Fraction Definitions

The factor β is defined as:

$$\beta = \frac{\text{precursor atoms}}{\text{prompt neutrons} + \text{precursor atoms}}.$$

and it is equal to 0.0064 for U-235.

The delayed neutron fraction (DNF) is defined as:

$$DNF = \frac{\text{delayed neutrons}}{\text{prompt neutrons} + \text{delayed neutrons}}.$$

These two factors, β and *DNF*, are not the same thing in case of a rapid change in the number of neutrons in the reactor.

Another concept, is the *effective fraction of delayed neutrons*, which is the fraction of delayed neutrons weighted (over space, energy, and angle) on the adjoint neutron flux. This concept arises because delayed neutrons are emitted with an energy spectrum more thermalized relative to prompt neutrons. For low enriched uranium fuel working on a thermal neutron spectrum, the difference between the average and effective delayed neutron fractions can reach 50 pcm (1 pcm = 1e-5).

Delayed Neutron

While the most of the neutrons produced in fission are prompt neutrons, the delayed neutrons are of importance in the reactor control. In fact the presence of delayed neutrons is perhaps most important aspect of the fission process from the viewpoint of reactor control.

	Possible precursor nuclei	Mean energy (MeV)	Average half-life of the group [s]			Delayed neutron fraction [%]		
i			235U	239Pu	233U	235U	239Pu	233U
1	87Br, 142Cs	0.25	55.72	54.28	55.0	0.021	0.0072	0.0226
2	137I, 88Br	0.56	22.72	23.4	20.57	0.140	0.0626	0.0786
3	138I, 89Br, (93,94)Rb	0.43	6.22	5.60	5.00	0.126	0.0444	0.0658
4	139I, (93,94)Kr 143Xe, (90,92)Br	0.62	2.3	2.13	2.13	0.252	0.0685	0.0730
5	140I, 145Cs	0.42	0.61	0.618	0.615	0.074	0.018	0.0135
6	(Br, Rb, As etc.)	-	0.23	0.257	0.277	0.027	0.0093	0.0087
		Total				0.64	0.21	0.26

Delayed neutrons are traditionally represented by six delayed neutron groups, whose yields and decay constants (λ) are obtained from nonlinear least-squares fits to experimental measurements.

The term "delayed" in this context means, that the neutron is emitted with half-lifes, ranging from few milliseconds up to 55 s for the longest-lived precursor ^{87}Br. These neutrons have to be distinguished from the prompt neutrons which are emitted immediately (on the order of 10^{-14} s) after a fission event from a neutron-rich nucleus. Despite the fact the amount of delayed neutrons is only on the order of tenths of percent of the total amount, the timescale in seconds plays the extremely important role.

Key Characteristics of Delayed Neutrons

- The presence of delayed neutrons is perhaps most important aspect of the fission process from the viewpoint of reactor control.
- Delayed neutrons are emitted by neutron rich fission fragments that are called the delayed neutron precursors.

- These precursors usually undergo beta decay but a small fraction of them are excited enough to undergo neutron emission.
- The emission of neutron happens orders of magnitude later compared to the emission of the prompt neutrons.
- About 240 n-emitters are known between ^{8}He and ^{210}Tl, about 75 of them are in the non-fission region.
- In order to simplify reactor kinetic calculations it is suggested to group together the precursors based on their half-lives.
- Therefore delayed neutrons are traditionally represented by six delayed neutron groups.
- Neutrons can be produced also in (γ, n) reactions (especially in reactors with heavy water moderator) and therefore they are usually referred to as photoneutrons. Photoneutrons are usually treated no differently than regular delayed neutrons in the kinetic calculations.
- The total yield of delayed neutrons per fission, v_d, depends on:
    - Isotope, that is fissioned.
    - Energy of a neutron that induces fission.
- Variation among individual group yields is much greater than variation among group periods.
- In reactor kinetic calculations it is convenient to use relative units usually referred to as delayed neutron fraction (DNF).
- At the steady state condition of criticality, with $k_{eff} = 1$, the delayed neutron fraction is equal to the precursor yield fraction β.
- In LWRs the β decreases with fuel burnup. This is due to isotopic changes in the fuel.
- Delayed neutrons have initial energy between 0.3 and 0.9 MeV with an average energy of 0.4 MeV.
- Depending on the type of the reactor, and their spectrum, the delayed neutrons may be more (in thermal reactors) or less effective than prompt neutrons (in fast reactors). In order to include this effect into the reactor kinetic calculations the effective delayed neutron fraction – β_{eff} must be defined.
- The effective delayed neutron fraction is the product of the average delayed neutron fraction and the importance factor $\beta_{eff} = \beta \cdot I$.
- The weighted delayed generation time is given by $\tau = \sum_i \tau_i \cdot \beta_i / \beta = 13.05$ s, therefore the weighted decay constant $\lambda = 1 / \tau \approx 0.08\ s^{-1}$.
- The mean generation time with delayed neutrons is about ~0.1 s, rather than ~10^{-5} as in section Prompt Neutron Lifetime, where the delayed neutrons were omitted.

- Their presence completely changes the dynamic time response of a reactor to some reactivity change, making it controllable by control systems such as the control rods.

Precursors of Delayed Neutrons

Delayed neutrons originate from the radioactive decay of nuclei produced in fission and hence they are different for each fissile material. They are emitted by excited neutron rich fission fragments (so called the delayed neutron precursors) some appreciable time after the fission. How long afterward, is dependent on the half-life of the precursor, since the neutron emission itself occurs in a very short time. The precursors usually undergo beta decay without any neutron emission but a small fraction of them (highly excited nuclei) can undergo the neutron emission instead of the gamma emission. In addition current nuclear physics facilities can produce more neutron-rich isotopes that can emit multiple neutrons. Currently about 18 2n-emitters are experimentally confirmed [IAEA – INDC(NDS)-0599], but only two of them are also fission products.

As an example, the isotope ^{87}Br is the major component of the first group of precursor nuclei. This isotope has half-life of 55.6 seconds. It undergoes negative beta decay through its two main branches with emission of 2.6 MeV and 8 MeV beta particles. This decay leads to the formation of $^{87}Kr^*$ and ^{87}Kr (ground state) and the ^{87}Kr nucleus subsequently decays via two successive beta decays into the stable isotope ^{87}Sr. But there is also one possible way for the ^{87}Br nucleus to beta decay. The ^{87}Br nucleus can beta decay into an excited state of the ^{87}Kr nucleus at an energy of 5.5 MeV, which is larger than the binding energy of a neutron in the ^{87}Kr nucleus. In this case, the ^{87}Kr nucleus can undergo (with probability of 2.5%) a neutron emission leading to the formation of stable ^{87}Kr isotope.

According to the JEFF 3.1 database, about 240 n-emitters are known between ^{8}He and ^{210}Tl, about 75 of them are in the non-fission region. Furthermore 18 2n-emitter, and only four 3n-emitters (^{11}Li, ^{14}Be, ^{17}B, ^{31}Na) are experimentally confirmed. These numbers are not certainly final. Since new IAEA Co-ordinated Research Project (CRP) on Beta-delayed neutron emission evaluation has been started in 2013, it is expected these numbers will change significantly.

As can be seen it was identified many precursor nuclei. Not all of them are fission products (about 75 of them in the non-fission region A<70), but there are also many precursor nuclei, that are in the fission region between A=70-150. Their half-lives range between tenths of second (0.12 s) and tens of seconds (55.6 s), therefore their delayed neutrons appear with considerably differing delay times.

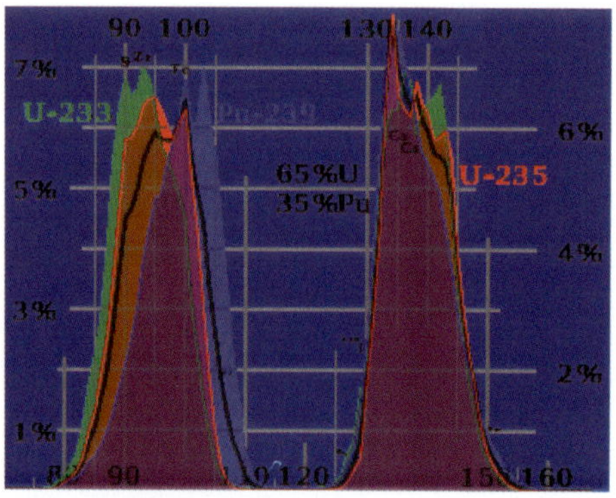

Fission fragment yield for different nuclei. The most probable fragment masses are around mass 95 (Krypton) and 137 (Barium).

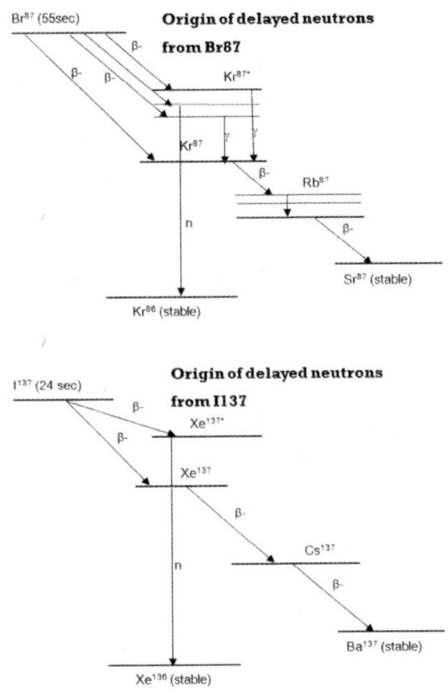

Precursors of Delayed Neutrons

Six Groups of Delayed Neutrons – Parameters

Reactor-kinetic calculations with considering of such a number of initial conditions would be correct, but it also would be very complicated. Therefore G. R. Keepin and his co-workers suggested to group together the precursors based on their half-lives.

Therefore delayed neutrons are traditionally represented by six delayed neutron groups, whose yields and decay constants (λ) are obtained from nonlinear least-squares fits to experimental measurements. This model has following disadvantages:

- All constants for each group of precursors are empirical fits to the data.
- They cannot be matched with decay constants of specific precursors.
- These constants are different for each fissionable nuclide.
- These constants change also with the neutron energy spectrum.

Although this six group parameterization still satisfies the requirements of commercial organizations, a higher accuracy of the delayed neutron yields and a better energy resolution in the delayed neutron spectra is desired. It was recognised that the half-lives in six-group structure do not accurately reproduce the asymptotic die-away time constants associated with the three longest-lived dominant precursors: ^{87}Br, ^{137}I and ^{88}Br.

This model may be insufficient especially in case of epithermal reactors, because virtually all delayed neutron activity measurements have been performed for fast or thermal-neutron-induced fission. In case of fast reactors, in which the nuclear fission of six fissionable isotopes of uranium and plutonium is important, the accuracy and energy resolution may play an important role.

U^{235}			
Group	Decay constant λ_i (sec^{-1})	Yield (neutrons per fission)	Fraction β_i
1	0.0124	0.00052	0.000215
2	0.0305	0.00346	0.001424
3	0.111	0.00310	0.001274
4	0.301	0.00624	0.002568
5	1.14	0.00182	0.000748
6	3.01	0.00066	0.000273
	Total yield: 0.0158		
	Total delayed fraction (β):		0.0065

Pu^{239}			
Group	Decay constant λ_i (sec^{-1})	Yield (neutrons per fission)	Fraction β_i
1	0.0128	0.00021	0.000073
2	0.0301	0.00182	0.000626
3	0.124	0.00129	0.000443
4	0.325	0.00199	0.000685
5	1.12	0.00052	0.000181
6	2.69	0.00027	0.000092
	Total yield: 0.0061		
	Total delayed fraction (β):		0.0021

U^{233}			
Group	Decay constant λ_i (sec^{-1})	Yield (neutrons per fission)	Fraction β_i
1	0.0126	0.00057	0.000224
2	0.0337	0.00197	0.000777
3	0.139	0.00166	0.000655
4	0.325	0.00184	0.000723
5	1.13	0.00034	0.000133
6	2.50	0.00022	0.000088
	Total yield: 0.0066		
	Total delayed fraction (β):		0.0026

Eight Groups of Delayed Neutrons – Parameters

In order to reduce discrepancies between measured and calculated values of the reactivity scale based on reactor kinetics (this discrepancy results in excessive conservatism in the design, because it must be covered by uncertainties during core design and safety analyses calculations), the NEA (Nuclear Energy Agency) and its NEANSC/WPEC Subgroup 6 recommends a new eight-group representation. It seems reasonable to increase the number of delayed neutron groups, because many studies have shown that most of delayed neutrons are produced approximately by twelve precursors that are common to all fissioning isotopes.

The eight-group representation uses a set of eight-group half-lives for all fissioning systems, with the half-lives adopted for thethree longest-lived groups corresponding to the three dominant long-lived precursors: ^{87}Br, ^{137}I and ^{88}Br (these precursors was separated into single groups).

Delayed Neutrons Energy Spectra

The key properties of delayed neutrons, which are very important for the nuclear reactor design, belong also delayed neutron energy spectra. The energy spectra of the delayed neutrons are the poorest known of all input data required, because it very difficult to measure it.

Depending on the type of the reactor, and their spectrum, the delayed neutrons may be more (in thermal reactors) or less effective than prompt neutrons (in fast reactors). In order to include this effect into the reactor kinetic calculations the effective delayed neutron fraction – β_{eff} must be defined.

Effective Delayed Neutron Fraction – β_{eff}

The delayed neutron fraction, β, is the fraction of delayed neutrons in the core at creation, that is, at high energies. But in case of thermal reactors the fission can be initiated mainly by thermal neutron. Thermal neutrons are of practical interest in study of thermal reactor behaviour. The effective delayed neutron fraction, usually referred to as β_{eff}, is the same fraction at thermal energies.

The effective delayed neutron fraction reflects the ability of the reactor to thermalize and utilize each neutron produced. The β is not the same as the β_{eff} due to the fact delayed neutrons do not have the same properties as prompt neutrons released directly from fission. In general, delayed neutrons have lower energies than prompt neutrons. Prompt neutrons have initial energy between 1 MeV and 10 MeV, with an average energy of 2 MeV. Delayed neutrons have initial energy between 0.3 and 0.9 MeV with an average energy of 0.4 MeV.

Therefore in thermal reactors a delayed neutron traverses a smaller energy range to become thermal and it is also less likely to be lost by leakage or by parasitic absorption than is the 2 MeV prompt neutron. On the other hand, delayed neutrons are also less likely to cause fast fission, because their average energy is less than the minimum required for fast fission to occur.

These two effects (lower fast fission factor and higher fast non-leakage probability for delayed neutrons) tend to counteract each other and forms a term called the importance factor (I). The importance factor relates the average delayed neutron fraction to the effective delayed neutron fraction. As a result, the effective delayed neutron fraction is the product of the average delayed neutron fraction and the importance factor.

$$\beta_{eff} = \beta \,.\, I$$

The delayed and prompt neutrons have a difference in their effectiveness in producing a subsequent fission event. Since the energy distribution of the delayed neutrons differs also from group to group, the different groups of delayed neutrons will also have a different effectiveness. Moreover, a nuclear reactor contains a mixture of fissionable isotopes. Therefore, in some cases, the importance factor is insufficient and an importance function must be defined.

For example:

In a small thermal reactor with highly enriched fuel, the increase in fast non-leakage probability will dominate the decrease in the fast fission factor, and the importance factor will be greater than one.

In a large thermal reactor with low enriched fuel, the decrease in the fast fission factor will dominate the increase in the fast non-leakage probability and the importance factor will be less than one (about 0.97 for a commercial PWR).

In large fast reactors, the decrease in the fast fission factor will also dominate the increase in the fast non-leakage probability and the β_{eff} is less than β by about 10%.

	LWR	CANDU	Fast Reactor
Λ [s]	2 x 10^{-5}	1 x 10^{-3}	4 x 10^{-7}
β_{eff}	0.006	0.005	0.0035
λ [s^{-1}]	0.08	0.08	0.08

Table of main kinetic parameters.

Mean Generation Time with Delayed Neutrons

Mean Generation Time with Delayed Neutrons, l_d, is the weighted average of the prompt generation times and a delayed neutron generation time. The delayed neutron generation time, τ, is the weighted average of mean precursor lifetimes of the six groups (or more groups) of delayed neutron precursors.

It must be noted, the true lifetime of delayed neutrons (the slowing down time and the diffusion time) is very short compared with the mean lifetime of their precursors ($t_s + t_d << \tau_i$). Therefore τ_i is also equal to the mean lifetime of a neutron from the ith group, that is, $\tau_i = l_i$ and the equation for mean generation time with delayed neutrons is the following:

$$l_d = (1 - \beta).l_p + \sum l_i \, . \, \beta_i \Rightarrow l_d = (1 - \beta).l_p + \sum \tau_i \, . \, \beta_i$$

where

- $(1 - \beta)$ is the fraction of all neutrons emitted as prompt neutrons
- l_p is the prompt neutron lifetime
- τ_i is the mean precursor lifetime, the inverse value of the decay constant $\tau_i = 1/\lambda_i$
- The weighted delayed generation time is given by $\tau = \sum \tau_i \, . \, \beta_i / \beta = 13.05$ s
- Therefore the weighted decay constant $\lambda = 1 / \tau \approx 0.08\ s^{-1}$

The number, 0.08 s^{-1}, is relatively high and have a dominating effect of reactor time response, although delayed neutrons are a small fraction of all neutrons in the core. This is best illustrated by calculating a weighted mean generation time with delayed neutrons:

$$l_d = (1 - \beta).l_p + \sum \tau_i \, . \, \beta_i = (1 - 0.0065).\ 2 \times 10^{-5} + 0.085 = 0.00001987 + 0.085 \approx 0.085$$

In short, the mean generation time with delayed neutrons is about ~0.1 s, rather than ~10-5 as in section Prompt Neutron Lifetime, where the delayed neutrons were omitted.

	LWR	CANDU	Fast Reactor
Λ [s]	2 x 10^{-5}	1 x 10^{-3}	4 x 10^{-7}
β_{eff}	0.006	0.005	0.0035
λ [s^{-1}]	0.08	0.08	0.08

Table of main kinetic parameters.

Example – Infinite Multiplying System without Source and with Delayed Neutrons

An simplest equation governing the neutron kinetics of the system with delayed neutrons is the point kinetics equation. This equation states that the time change of the neutron population is equal to the excess of neutron production (by fission) minus neutron loss by absorption in one mean generation time with delayed neutrons (ld). The role of ld is evident. Longer lifetimes give simply slower responses of multiplying systems.

If there are neutrons in the system at t=0, that is, if n(0) > 0, the solution of this equation gives the simplest point kinetics equation with delayed neutrons (similarly to the case without delayed neutrons):

$$\frac{d}{dt}n(t) = \frac{(k-1)}{l_d}n(t)$$

$$n(t) = n(0)exp\left[\frac{(k-1)}{l_d}t\right]$$

Let us consider that the mean generation time with delayed neutrons is ~0.085 and k (k_∞ – neutron multiplication factor) will be step increased by only 0.01% (i.e. 10 pcm or ~1.5 cents), that is k_∞=1.0000 will increase to k_∞=1.0001.

It must be noted such reactivity insertion (10pcm) is very small in case of LWRs. The reactivity insertions of the order of one pcm are for LWRs practically unrealizable. In this case the reactor period will be:

$$T = l_d / (k_\infty - 1) = 0.085 / (1.0001-1) = 850s$$

This is a very long period. In ~14 minutes the neutron flux (and power) in the reactor would increase by a factor of e = 2.718. This is completely different dimension of the response on reactivity insertion in comparison with the case without presence of delayed neutrons, where the reactor period was 1 second.

Reactors with such a kinetics would be quite easy to control. From this point of view it may seem that reactor control will be a quite boring affair. It will not! The presence of delayed neutrons entails many many specific phenomena.

References

- M. Kikuchi, K. Lackner & M. Q. Tran (2012). Fusion Physics. International Atomic Energy Agency. p. 22. ISBN 9789201304100
- Lawrence, J. E.; Lamm, S. H.; Pino, S.; Richman, K.; Braverman, L. E. (2000). "The Effect of Short-Term Low-Dose Perchlorate on Various Aspects of Thyroid Function". Thyroid. 10(8): 659–63. doi:10.1089/10507250050137734. PMID 11014310
- Controlled-thermonuclear-fusion, physics: byjus.com, Retrieved 29 June 2018
- Stone, Brad (3 April 2017). "Former Google Vice President Starts a Company Promising Clean and Safe Nuclear Energy". Bloomberg.com. Retrieved 2017-05-01

- Barzilai, D; Sheinfeld, M (1966). "Fatal complications following use of potassium perchlorate in thyrotoxicosis. Report of two cases and a review of the literature". Israel journal of medical sciences. 2 (4): 453–6. PMID 4290684
- Nuclear-fission: newworldencyclopedia.org, Retrieved 19 July 2018
- Hala, Jiri; James D. Navratil (2003). Radioactivity, Ionizing Radiation, and Nuclear Energy. Brno: Konvoj. ISBN 80-7302-053-X
- Woenckhaus, U.; Girlich, C. (2005). "Therapie und Prävention der Hyperthyreose" [Therapy and prevention of hyperthyroidism]. Der Internist (in German). 46 (12): 1318–23. doi:10.1007/s00108-005-1508-4. PMID 16231171
- Fission-product, science: britannica.com, Retrieved 19 March 2018
- Philip Ball (12 February 2014). "Laser fusion experiment extracts net energy from fuel". Nature. pp. 12–27. doi:10.1038/nature.2014.14710. Retrieved 2014-02-13
- Bartalena, L.; Brogioni, S; Grasso, L; Bogazzi, F; Burelli, A; Martino, E (1996). "Treatment of amiodarone-induced thyrotoxicosis, a difficult challenge: Results of a prospective study". Journal of Clinical Endocrinology & Metabolism. 81 (8): 2930–3. doi:10.1210/jc.81.8.2930. PMID 8768854
- Nuclear-Reactor-Physics: owlcation.com, Retrieved 22 May 2018
- John R. Lamarsh and Anthony John Baratta (Third Edition) (2001). Introduction to Nuclear Engineering. Prentice Hall. ISBN 0-201-82498-1

6

Understanding Nuclear Decay

An unstable nucleus loses energy in the form of radiation in a process called nuclear decay. The emission of radiation can be in the form of an alpha or beta particle or as gamma radiation. Therefore, an understanding of the different forms of nuclear decay, such as alpha, beta and gamma decay as well as spontaneous fission, proton emission, neutron emission, etc. is vital to the understanding of nuclear reactions in a nuclear reactor. These topics have been extensively discussed in the following chapter.

Nuclear decay (Radioactive decay) occurs when an unstable atom loses energy by emitting ionizing radiation. Radioactive decay is a random process at the level of single atoms, in that, according to quantum theory, it is impossible to predict when a particular atom will decay. In other words, a nucleus of a radionuclide has no "memory". A nucleus does not "age" with the passage of time. Thus, the probability of its breaking down does not increase with time, but stays constant no matter how long the nucleus has existed. During its unpredictable decay this unstable nucleus spontaneosly and randomly decomposes to form a different nucleus (or a different energy state – gamma decay), giving off radiation in the form of atomic partices or high energy rays. This decay occurs at a constant, predictable rate that is referred to as half-life. A stable nucleus will not undergo this kind of decay and is thus non-radioactive. There are many modes of radioactive decay:

- Alpha radioactivity. Alpha particles consist of two protons and two neutrons bound together into a particle identical to a helium nucleus. Because of its very large mass (more than 7000 times the mass of the beta particle) and its charge, it heavy ionizes material and has a very short range.
- Beta radioactivity. Consist of beta particles. Beta particles are high-energy, high-speed electrons or positrons emitted by certain types of radioactive nuclei such as potassium-40. The beta particles have greater range of penetration than alpha particles, but still much less than gamma rays.The beta particles emitted are a form of ionizing radiation also known as beta rays. The production of beta particles is termed beta decay.
- Gamma radioactivity. Gamma radioactivity consist of gamma rays. Gamma rays are electromagnetic radiation (high energy photons) of an very high frequency and of a high energy. They are produced by the decay of nuclei as they transition from a high energy state to a lower state known as gamma decay. Most of nuclear reactions are accompanied by gamma emission.
- Neutron emission. Neutron emission is a type of radioactive decay of nuclei containing excess neutrons (especially fission products), in which a neutron is simply ejected from the nucleus. This type of radiation plays key role in nuclear reactor control, because these neutrons are delayed neutrons.

Nature of Decay

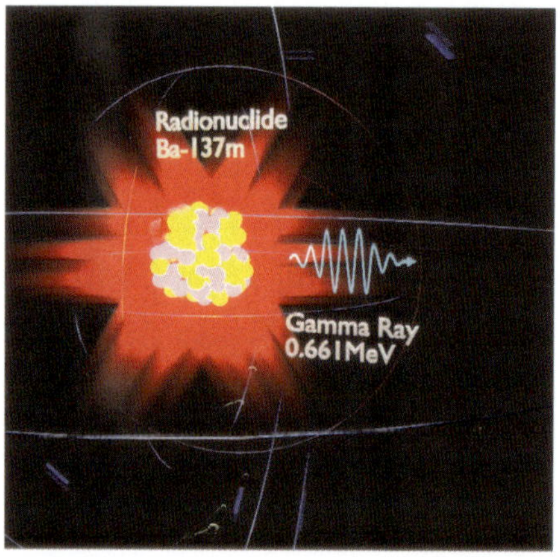

Barium-137m is a product of a common fission product – Caesium – 137.
The main gamma ray of Barium-137m is 661keV photon.

As was written, atomic nuclei consist of protons and neutrons, which attract each other through the nuclear force, while protons repel each other via the electromagnetic force due to their positive charge. These two forces compete, leading to various stability of nuclei. There are only certain combinations of neutrons and protons, which forms stable nuclei. Neutrons stabilize the nucleus, because they attract each other and protons , which helps offset the electrical repulsion between protons. As a result, as the number of protons increases, an increasing ratio of neutrons to protons is needed to form a stable nucleus. If there are too many (neutrons also obey the Pauli exclusion principle) or too few neutrons for a given number of protons, the resulting nucleus is not stable and it undergoes radioactive decay. Most atoms found in nature are stable and do not emit particles or energy that change form over time. Of the first 82 elements in the periodic table, 80 have isotopes considered to be stable. Technetium, promethium and all the elements with an atomic number over 82 are unstable and decompose through radioactive decay. Unstable isotopes decay spontaneously through various radioactive decay pathways, most commonly alpha decay, beta decay, gamma decay or electron capture. Many other rare types of decay, such as spontaneous fission or neutron emission are known.

Radioactive Series

In a simple, one-step radioactive decay, the new nucleus that emerges is stable. C-14 undergoing beta decay to N-14 and K-40 undergoing electron capture to Ar-40 are examples.

On the other hand, the daughter nuclide of a decay event can be unstable, sometimes even more unstable than the parent. If this is the case, it will proceed to decay again. A sequence of several decay events, producing in the end a stable nuclide, is a *decay chain*. Ultrapure uranium, for instance, is hardly radioactive at all. After a few weeks, however, the unstable daughter nucleides accumulate—such as radium—and it is their radioactivity that becomes noticeable.

Of the commonly occurring forms of radioactive decay, the only one that changes the number of aggregate protons and neutrons *(nucleons)* contained in the nucleus is alpha emission, which reduces it by four. Thus, the number of nucleons modulo 4 is preserved across any decay chain. This leads to the four radioactive decay series with atomic weights 4n+0, 4n+1, 4n+2, and 4n+3.

In an alpha decay, the atomic weight decreases by 4 and the atomic number decreases by 2. In a beta decay, the atomic weight stays the same and the atomic number increases by 1. In a gamma decay, both atomic weight and number stay the same. A branching path occurs when there are alternate routes to the same stable destination. One branch is usually highly favored over the other.

These are the four radioactive decay series.

Uranium-235 Series (4n+3)

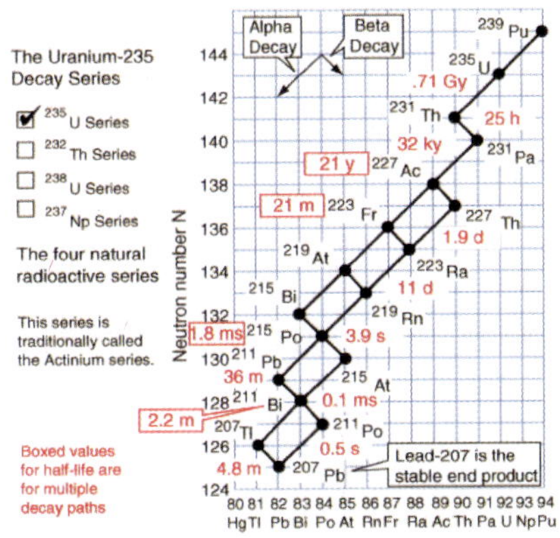

Thorium-232 Series (4n+0)

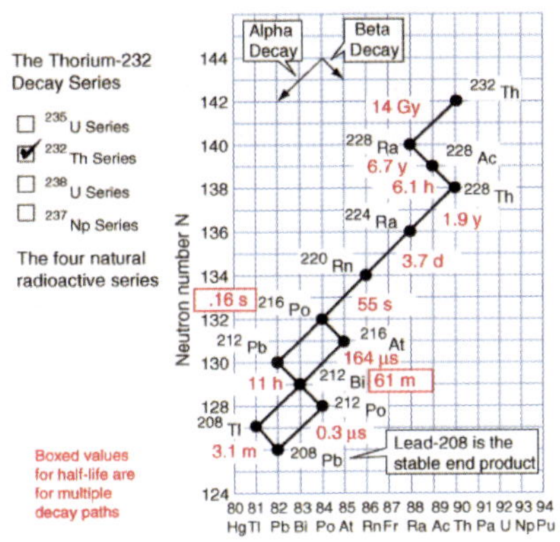

Uranium-238 Series (4n+2)

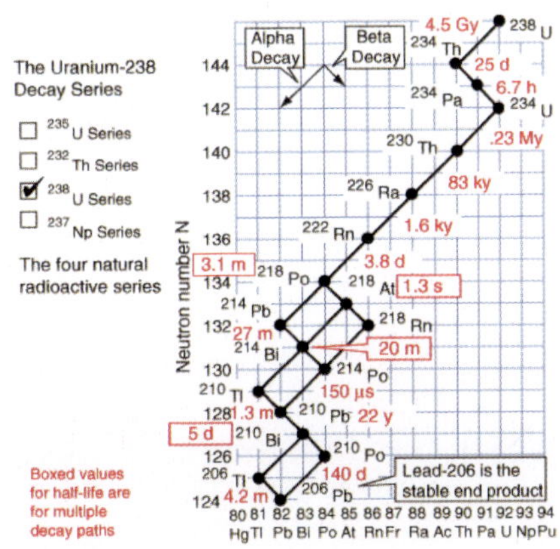

Neptunium-237 Series (4n+1)

The members of this series are not presently found in nature because the half-life of the longest lived isotope in the series is short compared to the age of the earth.

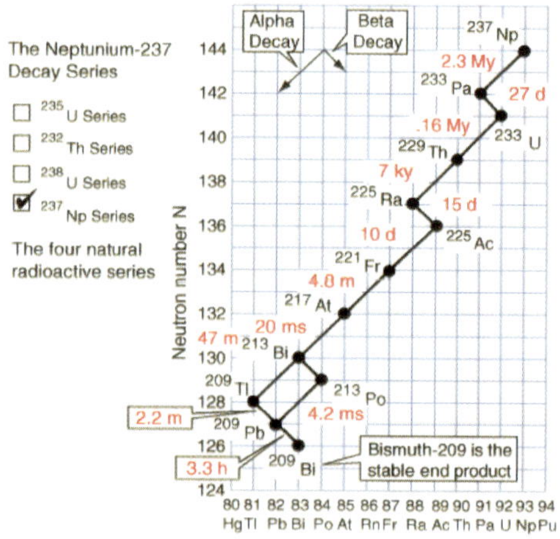

Radioactive Decay Rates

The *decay rate*, or *activity*, of a radioactive substance is characterized by:

1. Constant quantities:

- The *half-life*—$t_{1/2}$, is the time taken for the activity of a given amount of a radioactive substance to decay to half of its initial value.
- The *decay constant*— λ, "lambda" the reciprocal of the mean lifetime, sometimes referred to as simply *decay rate*.
- The *mean lifetime*— τ, "tau" the average lifetime (1/e life) of a radioactive particle before decay.

Although these are constants, they are associated with the statistical behavior of populations of atoms. In consequence, predictions using these constants are less accurate for minuscule samples of atoms.

In principle a half-life, a third-life, or even a (1/√2)-life, can be used in exactly the same way as half-life; but the mean life and half-life $t_{1/2}$ have been adopted as standard times associated with exponential decay.

2. Time-variable quantities:

- *Total activity*— *A*, is the number of decays per unit time of a radioactive sample.
- *Number of particles*—*N*, is the total number of particles in the sample.
- *Specific activity*—S_A, number of decays per unit time per amount of substance of the sample at time set to zero ($t = 0$). "Amount of substance" can be the mass, volume or moles of the initial sample.

These are related as follows:

$$t_{1/2} = \frac{ln(2)}{\lambda} = \tau\, ln(2)$$

$$A = -\frac{dN}{dt} = \lambda N$$

$$S_A a_0 = -\frac{dN}{dt}\Big|_{t=0} = \lambda N_0$$

where N_0 is the initial amount of active substance — substance that has the same percentage of unstable particles as when the substance was formed.

Mathematics of Radioactive Decay

Universal Law of Radioactive Decay

Radioactivity is one very frequently given example of exponential decay. The law describes the statistical behaviour of a large number of nuclides, rather than individual atoms. In the following formalism, the number of nuclides or the nuclide population *N*, is of course a discrete variable (a natural number)—but for any physical sample *N* is so large that it can be treated as a continuous variable. Differential calculus is used to model the behaviour of nuclear decay.

The mathematics of radioactive decay depend on a key assumption that a nucleus of a radionuclide has no "memory" or way of translating its history into its present behavior. A nucleus does not "age" with the passage of time. Thus, the probability of its breaking down does not increase with time, but stays constant no matter how long the nucleus has existed. This constant probability may vary greatly between different types of nuclei, leading to the many different observed decay rates. However, whatever the probability is, it does not change. This is in marked contrast to complex objects which do show aging, such as automobiles and humans. These systems do have a chance of breakdown per unit of time, that increases from the moment they begin their existence.

One-decay Process

Consider the case of a nuclide *A* that decays into another *B* by some process $A \rightarrow B$ (emission of other particles, like electron neutrinos $\bar{\text{ï}}_e$ and electrons e^- as in beta decay, are irrelevant in what follows). The decay of an unstable nucleus is entirely random in time so it is impossible to predict when a particular atom will decay. However, it is equally likely to decay at any instant in time. Therefore, given a sample of a particular radioisotope, the number of decay events $-dN$ expected to occur in a small interval of time dt is proportional to the number of atoms present *N*, that is

$$-\frac{dN}{dt} \propto N.$$

Particular radionuclides decay at different rates, so each has its own decay constant λ. The expected decay $-dN/N$ is proportional to an increment of time, dt:

$$-\frac{dN}{N} = \lambda dt$$

The negative sign indicates that N decreases as time increases, as the decay events follow one after another. The solution to this first-order differential equation is the function:

$$N(t) = N_0 e^{-\lambda t} = N_0 e^{-t/\tau},$$

Where N_0 is the value of N at time $t = 0$, with the decay constant expressed as λ or $1/\tau$

We have for all time t:

$$N_A + N_B = N_{\text{total}} = N_{Ao},$$

where N_{total} is the constant number of particles throughout the decay process, which is equal to the initial number of A nuclides since this is the initial substance.

If the number of non-decayed A nuclei is:

$$N_A = N_{Ao} e^{-\lambda t}$$

then the number of nuclei of B, i.e. the number of decayed A nuclei, is

$$N_B = N_{Ao} - N_A = N_{Ao} - N_{Ao} e^{-\lambda t} = N_{Ao}\left(1 - e^{-\lambda t}\right).$$

The number of decays observed over a given interval obeys Poisson statistics. If the average number of decays is $\langle N \rangle$, the probability of a given number of decays N is

$$P(N) = \frac{\langle N \rangle^N \exp(-\langle N \rangle)}{N!}.$$

Chain-decay Processes

Chain of Two Decays

Now consider the case of a chain of two decays: one nuclide A decaying into another B by one process, then B decaying into another C by a second process, i.e. $A \to B \to C$. The previous equation cannot be applied to the decay chain, but can be generalized as follows. Since A decays into B, *then* B decays into C, the activity of A adds to the total number of B nuclides in the present sample, *before* those B nuclides decay and reduce the number of nuclides leading to the later sample. In other words, the number of second generation nuclei B increases as a result of the first generation nuclei decay of A, and decreases as a result of its own decay into the third generation nuclei C. The sum of these two terms gives the law for a decay chain for two nuclides:

$$\frac{dN_B}{dt} = -\lambda_B N_B + \lambda_A N_A.$$

The rate of change of N_B, that is dN_B/dt, is related to the changes in the amounts of A and B, N_B can increase as B is produced from A and decrease as B produces C.

Re-writing using the previous results:

$$\frac{dN_B}{dt} = -\lambda_B N_B + \lambda_A N_{Ao} e^{-\lambda_A t}$$

The subscripts simply refer to the respective nuclides, i.e. N_A is the number of nuclides of type A, N_{Ao} is the initial number of nuclides of type A, λ_A is the decay constant for A - and similarly for nuclide B. Solving this equation for N_B gives:

$$N_B = \frac{N_{Ao}\lambda_A}{\lambda_B - \lambda_A}\left(e^{-\lambda_A t} - e^{-\lambda_B t}\right).$$

In the case where B is a stable nuclide ($\lambda_B = 0$), this equation reduces to the previous solution:

$$\lim_{\lambda_B \to 0}\left[\frac{N_{Ao}\lambda_A}{\lambda_B - \lambda_A}\left(e^{-\lambda_A t} - e^{-\lambda_B t}\right)\right] = \frac{N_{Ao}\lambda_A}{0 - \lambda_A}\left(e^{-\lambda_A t} - 1\right) = N_{Ao}\left(1 - e^{-\lambda_A t}\right),$$

as shown above for one decay. The solution can be found by the integration factor method, where the integrating factor is $e^{\lambda_B t}$. This case is perhaps the most useful, since it can derive both the one-decay equation (above) and the equation for multi-decay chains (below) more directly.

Chain of any Number of Decays

For the general case of any number of consecutive decays in a decay chain, i.e. $A_1 \rightarrow A_2 \cdots \rightarrow A_i \cdots \rightarrow A_D$, where D is the number of decays and i is a dummy index ($i = {}_{1, 2, 3, \ldots D}$), each nuclide population can be found in terms of the previous population. In this case $N_2 = 0$, $N_3 = 0, \ldots, N_D = 0$. Using the above result in a recursive form:

$$\frac{dN_j}{dt} = -\lambda_j N_j + \lambda_{j-1} N_{(j-1)o} e^{-\lambda_{j-1} t}.$$

The general solution to the recursive problem is given by *Bateman's equations*:

Bateman's equations

$$N_D = \frac{N_1(0)}{\lambda_D}\sum_{i=1}^{D}\lambda_i c_i e^{-\lambda_i t}$$

$$c_i = \prod_{j=1, i\neq j}^{D}\frac{\lambda_j}{\lambda_j - \lambda_i}$$

Alternative Decay Modes

In all of the above examples, the initial nuclide decays into just one product. Consider the case of one initial nuclide that can decay into either of two products, that is $A \rightarrow B$ and $A \rightarrow C$ in parallel. For example, in a sample of potassium-40, 89.3% of the nuclei decay to calcium-40 and 10.7% to argon-40. We have for all time t:

$$N = N_A + N_B + N_C$$

which is constant, since the total number of nuclides remains constant. Differentiating with respect to time:

$$\frac{dN_A}{dt} = -\left(\frac{dN_B}{dt} + \frac{dN_C}{dt}\right)$$

$$-\lambda N_A = -N_A\left(\lambda_B + \lambda_C\right)$$

defining the *total decay constant* λ in terms of the sum of *partial decay constants* λ_B and λ_C:

$$\lambda = \lambda_B + \lambda_C.$$

Notice that

$$\frac{dN_A}{dt} < 0, \frac{dN_B}{dt} > 0, \frac{dN_C}{dt} > 0.$$

Solving this equation for N_A:

$$N_A = N_{A0} e^{-\lambda t}.$$

where N_{A0} is the initial number of nuclide A. When measuring the production of one nuclide, one can only observe the total decay constant λ. The decay constants λ_B and λ_C determine the probability for the decay to result in products B or C as follows:

$$N_B = \frac{\lambda_B}{\lambda} N_{A0} \left(1 - e^{-\lambda t}\right),$$

$$N_C = \frac{\lambda_C}{\lambda} N_{A0} \left(1 - e^{-\lambda t}\right).$$

because the fraction λ_B/λ of nuclei decay into B while the fraction λ_C/λ of nuclei decay into C.

Corollaries of the Decay Laws

The above equations can also be written using quantities related to the number of nuclide particles N in a sample;

- The activity: $A = \lambda N$.
- The amount of substance: $n = N/L$.
- The mass: $M = A_r n = A_r N/L$.

where $L = 6.022 \times 10^{23}$ is Avogadro's constant, A_r is the relative atomic mass number, and the amount of the substance is in moles.

Decay Timing: Definitions and Relations

Time Constant and Mean-life

For the one-decay solution $A \rightarrow B$:

$$N = N_0 e^{-\lambda t} = N_0 e^{-t/\tau},$$

the equation indicates that the decay constant λ has units of t^{-1}, and can thus also be represented as $1/\tau$, where τ is a characteristic time of the process called the *time constant*.

In a radioactive decay process, this time constant is also the mean lifetime for decaying atoms.

Each atom "lives" for a finite amount of time before it decays, and it may be shown that this mean lifetime is the arithmetic mean of all the atoms' lifetimes, and that it is τ, which again is related to the decay constant as follows:

$$\tau = \frac{1}{\lambda}.$$

This form is also true for two-decay processes simultaneously $A \rightarrow B + C$, inserting the equivalent values of decay constants (as given above)

$$\lambda = \lambda_B + \lambda_C$$

into the decay solution leads to:

$$\frac{1}{\tau} = \lambda = \lambda_B + \lambda_C = \frac{1}{\tau_B} + \frac{1}{\tau_C}$$

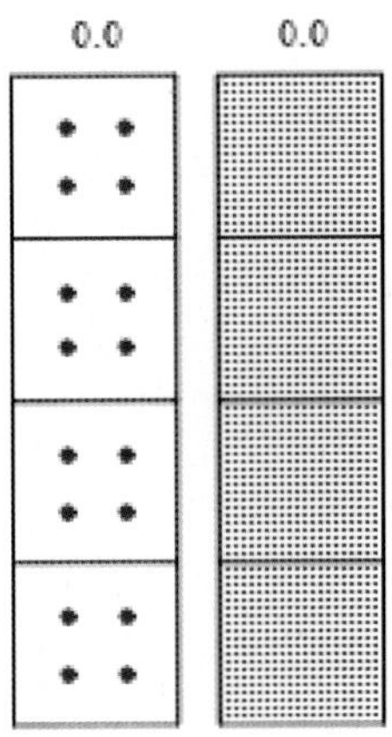

Simulation of many identical atoms undergoing radioactive decay, starting with either 4 atoms (left) or 400 (right). The number at the top indicates how many half-liveshave elapsed.

Half-life

A more commonly used parameter is the half-life. Given a sample of a particular radionuclide, the half-life is the time taken for half the radionuclide's atoms to decay. For the case of one-decay nuclear reactions:

$$N = N_0 e^{-\lambda t} = N_0 e^{-t/\tau},$$

the half-life is related to the decay constant as follows: set $N = N_0/2$ and $t = T_{1/2}$ to obtain

$$t_{1/2} = \frac{\ln 2}{\lambda} = \tau \ln 2.$$

This relationship between the half-life and the decay constant shows that highly radioactive substances are quickly spent, while those that radiate weakly endure longer. Half-lives of known radionuclides vary widely, from more than 10^{19} years, such as for the very nearly stable nuclide ^{209}Bi, to 10^{-23} seconds for highly unstable ones.

The factor of ln(2) in the above relations results from the fact that the concept of "half-life" is merely a way of selecting a different base other than the natural base *e* for the lifetime expression. The time constant τ is the e -1 -life, the time until only $1/e$ remains, about 36.8%, rather than the 50% in the half-life of a radionuclide. Thus, τ is longer than $t_{1/2}$. The following equation can be shown to be valid:

$$N(t) = N_0 e^{-t/\tau} = N_0 2^{-t/t_{1/2}}.$$

Since radioactive decay is exponential with a constant probability, each process could as easily be described with a different constant time period that (for example) gave its "(1/3)-life" (how long until only 1/3 is left) or "(1/10)-life" (a time period until only 10% is left), and so on. Thus, the choice of τ and $t_{1/2}$ for marker-times, are only for convenience, and from convention. They reflect a fundamental principle only in so much as they show that the *same proportion* of a given radioactive substance will decay, during any time-period that one chooses.

Mathematically, the nth life for the above situation would be found in the same way as above—by setting $N = N_0/n$, $t = T_{1/n}$ and substituting into the decay solution to obtain,

$$t_{1/n} = \frac{\ln n}{\lambda} = \tau \ln n.$$

Example

A sample of ^{14}C has a half-life of 5,730 years and a decay rate of 14 disintegration per minute (dpm) per gram of natural carbon.

If an artifact is found to have radioactivity of 4 dpm per gram of its present C, we can find the approximate age of the object using the above equation:

$$N = N_0 e^{-t/\tau},$$

$$\text{where: } \frac{N}{N_0} = 4/14 \approx 0.286,$$

$$\tau = \frac{T_{1/2}}{\ln 2} \approx 8267 \text{ years},$$

$$t = -\tau \ln \frac{N}{N_0} \approx 10356 \text{ years}.$$

Conservation Laws in Nuclear Decay

In analyzing nuclear reactions, we apply the many conservation laws. Nuclear reactions are subject to classical conservation laws for charge, momentum, angular momentum, and energy(including rest energies). Additional conservation laws, not anticipated by classical physics, are:

- Law of Conservation of Lepton Number
- Law of Conservation of Baryon Number
- Law of Conservation of Electric Charge

Certain of these laws are obeyed under all circumstances, others are not. We have accepted conservation of energy and momentum. In all the examples given we assume that the number of protons and the number of neutrons is separately conserved. We shall find circumstances and conditions in which this rule is not true. Where we are considering non-relativistic nuclear reactions, it is essentially true. However, where we are considering relativistic nuclear energies or those involving the weak interactions, we shall find that these principles must be extended.

Some conservation principles have arisen from theoretical considerations, others are just empirical relationships. Notwithstanding, any reaction not expressly forbidden by the conservation laws will generally occur, if perhaps at a slow rate. This expectation is based on quantum mechanics. Unless the barrier between the initial and final states is infinitely high, there is always a non-zero probability that a system will make the transition between them.

For purposes of analyzing non-relativistic reactions, it is sufficient to note four of the fundamental laws governing these reactions:

1. Conservation of nucleons. The total number of nucleons before and after a reaction are the same.
2. Conservation of charge. The sum of the charges on all the particles before and after a reaction are the same.
3. Conservation of momentum. The total momentum of the interacting particles before and after a reaction are the same.
4. Conservation of energy. Energy, including rest mass energy, is conserved in nuclear reactions.

Radioactive Decay Law

Calculations of the decay of radioactive nuclei are relatively straightforward, owing to the fact that there is only one fundamental law governing all decay process. This law states that the probability per unit time that a nucleus will decay is a constant, independent of time. This constant is called the decay constant and is denoted by λ, “lambda”.

The radioactive decay of certain number of atoms (mass) is exponential in time.

Radioactive decay law: $N = N.e^{-\lambda t}$

The rate of nuclear decay is also measured in terms of half-lives. The half-life is the amount of time it takes for a given isotope to lose half of its radioactivity. If a radioisotope has a half-life of 14 days, half of its atoms will have decayed within 14 days. In 14 more days, half of that remaining half will decay, and so on. Half lives range from millionths of a second for highly radioactive fission products to billions of years for long-lived materials(such as naturally occurring uranium). Notice that short half lives go with large decay constants. Radioactive material with a short half life is much more radioactive (at the time of production) but will obviously lose its radioactivity rapidly. No matter how long or short the half life is, after seven half lives have passed, there is less than 1 percent of the initial activity remaining.

The radioactive decay law can be derived also for activity calculations or mass of radioactive material calculations:

(Number of nuclei) $N = N.e^{-\lambda t}$ (Activity) $A = A.e^{-\lambda t}$ (Mass) $m = m.e^{-\lambda t}$,

where N (number of particles) is the total number of particles in the sample, A (total activity) is the number of decays per unit time of a radioactive sample, m is the mass of remaining radioactive material.

Isotope	Half life	Decay constant (s^{-1})
Uranium 238	4.5×10^9 years	5.0×10^{-18}
Plutonium 239	2.4×10^4 years	9.2×10^{-13}
Carbon 14	5570 years	3.9×10^{-12}
Radium 226	1622 years	1.35×10^{-11}
Free neutron 239	15 minutes	1.1×10^{-3}
Radon 220	52 seconds	1.33×10^{-2}
Lithium 8	0.84 seconds	0.825
Bismuth 214	1.6×10^{-4} seconds	4.33×10^3
Lithium 8	6×10^{-20} seconds	1.2×10^{19}

Table of examples of half lives and decay constants. Notice that short half lives go with large decay constants. Radioactive material with a short half life is much more radioactive but will obviously lose its radioactivity rapidly.

Decay Heat in Reactor

Time after shutdown	Decay heat as a fraction of nominal power	Decay heat for typical 3000MW reactor
0	0.065	195
1 sec	0.055	165
10 sec	0.05	150
1 min	0.035	105
10 min	0.02	60
30 min	0.015	45
1 hour	0.012	36
12 hours	0.006	18
1 day	0.005	15
10 days	0.0025	7.5
30 days	0.002	6
60 days	0.0015	4.5
1 year	0.0009	2.7

When a reactor is shut down, fission essentially ceases, but decay energy is still being produced.

The energy produced after shutdown is referred to as decay heat. The amount of decay heat production after shutdown is directly influenced by the power history (fission products accumulation) of the reactor prior to shutdown and by the level of fuel burnup (actinidies accumulation – especially in case of spent fuel handling). A reactor operated at full power for 10 days prior to shutdown has much higher decay heat generation than a reactor operated at low power for the same period. On the other hand, when the reactor changes its power from 50% to 100% of full power, the ratio of decay heat to neutron power drops to roughly half its previous level, and then builds up slowly as the fission product inventory adjusts to the new power.

The decay heat produced after a reactor shutdown from full power is initially equivalent to about 6 to 7% of the rated thermal power. Since radioactive decay is a random process at the level of single atoms, it is governed by the radioactive decay law. Note that, irradiated nuclear fuel contains a large number of different isotopes that contribute to decay heat, which are all subject to the radioactive decay law. Therefore a model describing decay heat must consider decay heat to be a sum of exponential functions with different decay constants and initial contribution to the heat rate. Fission fragments with a short half-life are much more radioactive (at the time of production) and contribute significantly to decay heat, but will obviously lose its share rapidly. On the other hand, fission fragments and transuranic elements with a long half-life are less radioactive (at the time of production) and produces less decay heat, but will obviously lose its share more slowly. This decay heat generation rate diminishes to about 1% approximately one hour after shutdown.

Alpha Decay

Alpha decay is a nuclear decay process where an unstable nucleus changes to another element by shooting out a particle composed of two protons and two neutrons. This ejected particle is known as an alpha particle and is simply a heliumnucleus. Alpha particles have a relatively large mass and a positive charge. This large mass means alpha particles can't go very far through the air, or get very deep into a solid. Alpha particles only affect surfaces, so alpha decay is rarely used in external medical radiation therapy.

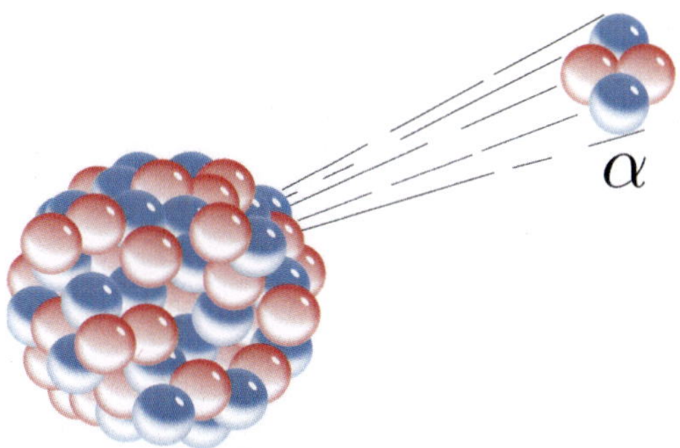

A model of alpha decay, showing the ejection of an alpha particle from a nucleus.

Alpha decay was originally distinguished from other forms of radiation by Ernest Rutherford by observing the deflection of the radiation through a magnetic field. Alpha decay deflects in the way you would expect a positive particle to since alpha particles have a +2e+2e charge.

The general equation representing alpha decay is:

$$ {}^{A}_{Z}X \rightarrow {}^{A-4}_{Z-2}Y + {}^{4}_{2}He $$

where:

- ${}^{A}_{Z}X$ is the parent nucleus, the starting nucleus.
- A Is the total number of nucleons (the number of neutrons plus the number of protons).
- Z Is the total number of protons.
- ${}^{A-4}_{Z-2}Y$ is the daughter nucleus, the ending nucleus
- ${}^{4}_{2}He$ is the released alpha particle

Alpha Decay Examples

Below are given some alpha decay examples:

The first example is the unstable seaborgium. The seaborgium atom u nderwent a radioactive decay and in the process converted to the rutherfordium by emitting an alpha particle as seen in the equation below:

$$ {}^{263}Sg_{106} \text{ --> } {}^{259}Rf_{104} + \alpha $$

We compare the difference in the atomic number and mass number of the seaborgium (Sg) and rutherfordium (Rf) is two and four respectively. The difference is due to the fact that the radioactive decay emits an alpha particle which has atomic mass and mass number of 2 and 4 respectively. So in the process the total number of sub atomic particles is conserved.

The second example is the unstable astatine. The astatine atom underwent a radioactive decay and in the process converted to the bismuth by emitting an alpha particle as seen in the equation below.

$$ {}^{211}At_{83} \text{ --> } {}^{207}Bi_{81} + \alpha $$

The bismuth and astatine are two places apart in the periodic table with the difference of 4 and 2 in their respective mass number and atomic mass.

The Alpha Particle Decay of Radon 198

Below is the isotope radon-198 will alpha decay to polonium-194. Alpha decay equation for radon-198 is:

$$ {}^{198}Rn_{86} \text{ ----> } {}^{194}Po_{84} + {}^{4}He_{2} $$

The radon has been shown on the left and the alpha particle, which is a helium nucleus is shown of the right with the polonium.

Alpha Particle Decay of Uranium 238

Alpha particles played an vital role in nuclear physics before the discovery of charged particle accelerators and were widely used in research.

The procedure of alpha decay is a nuclear reaction that can be written as:

$$^{A}_{Z}(Z)_{N} \text{--------->} \; ^{A-4}_{Z-2}(X)^{2-}_{N-2} + {}^{4}_{2}He^{2+} + Q\alpha$$

Where we have chosen to write out all of the superscripts and subscripts. Thus the alpha (αα)-decay of ^{238}U can be written as:

$$^{238}U \text{-------->} \; ^{234}Th^{2-} + {}^{4}He^{2+} + Q.$$

The Q-value is positive (exothermic) for impulsive alpha decay. The helium nucleus appears with a substantial velocity and is entirely ionized and the atomic electrons on the daughter are upset by the sudden change but the whole process preserves electrical charge.

Alpha Decay Problems

Below are given some problems on alpha decay:

Solved Examples

Question : Write the alpha decay equations for these following nuclides given below:

$^{256}Lr_{103}$

$^{231}Pa_{91}$

$^{225}Ac_{89}$

$^{211}Fr_{87}$

$^{185}Au_{79}$

Solution:

$^{256}Lr_{103}$ ----> $^{252}Md_{101}$ + $^{4}He_{2}$

$^{231}Pa_{91}$ -----> $^{227}Ac_{89}$ + $^{4}He_{2}$

$^{225}Ac_{89}$ ------> $^{221}Fr_{87}$ + $^{4}He_{2}$

$^{211}Fr_{87}$ ------> $^{207}At_{85}$ + $^{4}He_{2}$

$^{185}Au_{79}$ ------> $^{181}Ir_{77}$ + $^{4}He_{2}$

Question : Write the alpha decay equations for the particles shown below:

$^{232}U_{92}$

$^{149}Gd_{64}$

$^{232}Th_{90}$

$^{175}Pt_{78}$

$^{237}Np_{93}$

Solution:

$^{233}U_{92}$ ----> $^{229}Th_{90}$ + $^{4}He_{2}$

$^{149}Gd_{64}$ -----> $^{145}Sm_{62}$ + $^{4}He_{2}$

$^{232}Th_{90}$ -----> $^{228}Ra_{88}$ + $^{4}He_{2}$

$^{175}Pt_{78}$ ------> $^{171}Os_{76}$ + $^{4}He_{2}$

$^{237}Np_{93}$ ------> $^{233}Pa_{91}$ + $^{4}He_{2}$

Mechanism

The nuclear force holding an atomic nucleus together is very strong, in general much stronger than the repulsive electromagnetic forces between the protons. However, the nuclear force is also short range, dropping quickly in strength beyond about 1 femtometre, while the electromagnetic force has unlimited range. The strength of the attractive nuclear force keeping a nucleus together is thus proportional to the number of nucleons, but the total disruptive electromagnetic force trying to break the nucleus apart is roughly proportional to the square of its atomic number. A nucleus with 210 or more nucleons is so large that the strong nuclear force holding it together can just barely counterbalance the electromagnetic repulsion between the protons it contains. Alpha decay occurs in such nuclei as a means of increasing stability by reducing size.

One curiosity is why alpha particles, helium nuclei, should be preferentially emitted as opposed to other particles like a single proton or neutron or other atomic nuclei.[note 1] Part of the answer comes from conservation of wave function symmetry, which prevents a particle from spontaneously changing from exhibiting Bose–Einstein statistics (if it had an even number of nucleons) to Fermi–Dirac statistics (if it had an odd number of nucleons) or vice versa. Single proton emission, or the emission of any particle with an odd number of nucleons would violate this conservation law. The rest of the answer comes from the very high binding energy of the alpha particle. Computing the total disintegration energy given by the equation:

$$E = (m_{i} - m_{f} - m_{p})c^{2}$$

Where is the initial mass of the nucleus, m_{f} is the mass of the nucleus after particle emission, and m_{p} is the mass of the emitted particle, shows that alpha particle emission will usually be possible just with energy from the nucleus itself, while other decay modes will require additional energy. For example, performing the calculation for uranium-232 shows that alpha particle emission would need only 5.4 MeV, while a single proton emission would require 6.1 MeV. Most of this disintegration energy becomes the kinetic energy of the alpha particle itself, although to preserve conservation of momentum part of this energy becomes the recoil of the nucleus itself. However, since the mass numbers of most alpha emitting radioisotopes exceed 210, far greater than the mass number of the alpha particle the part of the energy going to the recoil of the nucleus is generally quite small.

These disintegration energies however are substantially smaller than the potential barrier provided by the nuclear force, which prevents the alpha particle from escaping. The energy needed is generally in the range of about 25 MeV, the amount of work that must be done against electromagnetic repulsion to bring an alpha particle from infinity to a point near the nucleus just outside the range of the nuclear force's influence. An alpha particle can be thought of as being inside a potential barrier whose walls are 25 MeV. However, decay alpha particles only have kinetic energies of 4 MeV to about 9 MeV, far less than the energy needed to escape.

Quantum mechanics, however, provides a ready explanation, via the mechanism of quantum tunnelling. The quantum tunnelling theory of alpha decay, independently developed by George Gamow and Ronald Wilfred Gurney and Edward Condon in 1928, was hailed as a very striking confirmation of quantum theory. Essentially, the alpha particle escapes from the nucleus by quantum tunnelling its way out. Gurney and Condon made the following observation in their paper on it:

> It has hitherto been necessary to postulate some special arbitrary 'instability' of the nucleus; but in the following note it is pointed out that disintegration is a natural consequence of the laws of quantum mechanics without any special hypothesis. Much has been written of the explosive violence with which the α-particle is hurled from its place in the nucleus. But from the process pictured above, one would rather say that the α-particle almost slips away unnoticed.

The theory supposes that the alpha particle can be considered an independent particle within a nucleus that is in constant motion, but held within the nucleus by nuclear forces. At each collision with the potential barrier of the nuclear force, there is a small non-zero probability that it will tunnel its way out. An alpha particle with a speed of 1.5×10^7 m/s within a nuclear diameter of approximately 10^{-14} m will collide with the barrier more than 10^{21} times per second. However if the probability of escape at each collision is very small, the half-life of the radioisotope will be very long, since it is the time required for the total probability of escape to reach 50%. As an extreme example, the half-life of the isotope bismuth-209 is 1.9×10^{19} years.

Working out the details of the theory leads to an equation relating the half-life of a radioisotope to the decay energy of its alpha particles, a theoretical derivation of the empirical Geiger–Nuttall law.

Safety

Although not very penetrating, the ingestion of a substance that undergoes alpha decay is harmful as the ejected alpha particles can damage internal tissues very easily despite its short range. This damage is a result of contact with membranes and living cells. Figure shows a diagram that illustrates the different levels of penetration for different types of radiation.

Overall, the health effects of alpha particles vary with how the exposure takes place. If the alpha emitter is inhaled, swallowed, or absorbed into the blood stream there can be lasting biological damage. This damage increases a persons risk of cancer. Alpha radiation is known to cause lung cancer in humans if the alpha emitter is inhaled. The inhalation of radon, an alpha emitter, is one of the biggest sources of alpha decay related illness in humans.

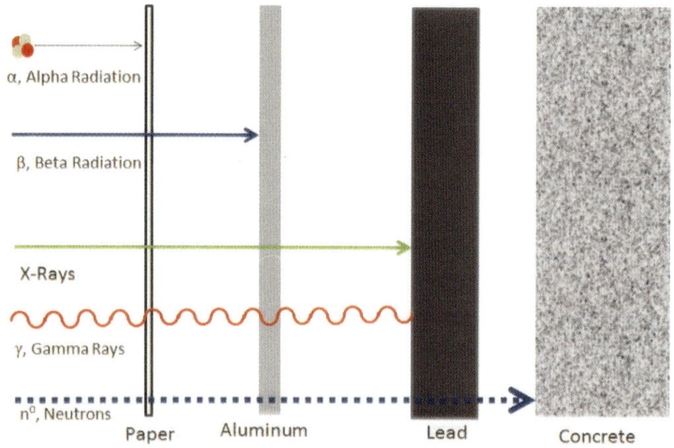

Different penetration levels of different products of decay, with gamma being one of the most highly penetrating and alpha being one of the least penetrating.

Applications and Importance

Radioactive elements that undergo alpha decay are used in smoke detectors. Americium is one frequently used element as it is a major alpha particle source. Inside the smoke detertor alpha particles are released. This in turn ionizes the air inside the detector. Smoke in the detector absorbs this alpha radiation, so if smoke is present the ionization is altered and the alarm is triggered.

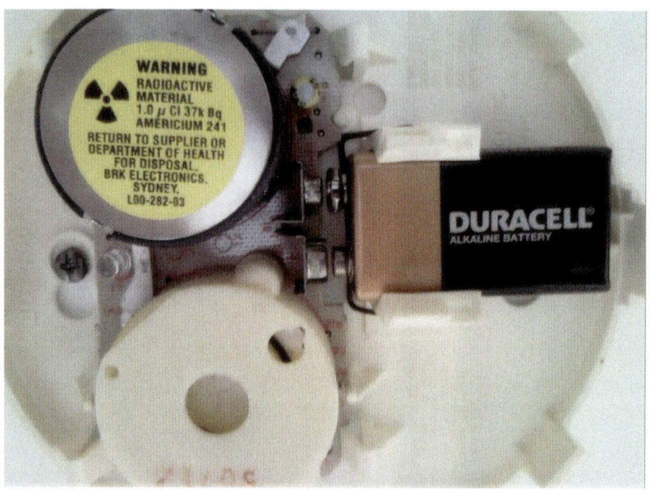

The inner workings of a smoke detector.

As well, alpha particles are used in a process known as Alpha Particle X-Ray Spectroscopy (APXS). This process is used to determine the elemental composition of rocks and soil. NASA used this method in its missions to Mars, including the Pathfinder missions, to determine what elements existed Martian rocks.

Alpha particles do also have some use in the medical field. A new cancer treatment known as targeted alpha therapy or TAT uses alpha decay to kill cancer cells. Lead-212 is ingested and travels to the site of the tumor, then giving off alpha radiation which kills all the cells in the area.

Beta Decay

Beta decay is a nuclear decay process where an unstable nucleus transmutes and ejects particles to become more stable. There are two different types of beta decay - beta minus and beta plus. In both of these decays, a nucleon in the nucleus is transformed into a different type of nucleon, releasing particles in the process. Both beta minus and beta plus decay are moderately penetrating (ie the radiation can go deep inside a solid object). There's a closely related process called electron capture, where an electron is captured in the nucleus which acts just like beta plus.

All atoms heavier than ordinary hydrogen have a nucleus consisting of neutrons and protons (neutral and positively charged particles, respectively), surrounded by negative electrons; these orbital electrons are not involved in the electron emission associated with beta decay. In electron emission, also called negative beta decay (symbolized β^--decay), an unstable nucleus emits an energetic electron (of relatively small mass) and an antineutrino (with little or possibly no rest mass), and a neutron in the nucleus becomes a proton that remains in the product nucleus. Thus, negative beta decay results in a daughter nucleus, the proton number (atomic number) of which is one more than its parent but the mass number (total number of neutrons and protons) of which is the same. For example, hydrogen-3 (atomic number 1, mass number 3) decays to helium-3 (atomic number 2, mass number 3). The energy lost by the nucleus is shared by the electron and the antineutrino, so that beta particles (the electrons) have energy ranging from zero to a distinct maximum that is characteristic of the unstable parent.

In positron emission, also called positive beta decay (β^+-decay), a proton in the parent nucleus decays into a neutron that remains in the daughter nucleus, and the nucleus emits a neutrino and a positron, which is a positive particle like an ordinary electron in mass but of opposite charge. Thus, positive beta decay produces a daughter nucleus, the atomic number of which is one less than its parent and the mass number of which is the same. Positron emission was first observed by Irène and Frédéric Joliot-Curie in 1934.

In electron capture, an electron orbiting around the nucleus combines with a nuclear proton to produce a neutron, which remains in the nucleus, and a neutrino, which is emitted. Most commonly the electron is captured from the innermost, or K, shell of electrons around the atom; for this reason, the process often is called K-capture. As in positron emission, the nuclear positive charge and hence the atomic number decreases by one unit, and the mass number remains the same.

Each chemical element consists of a set of isotopes the nuclei of which have the same number of protons but differ in the number of neutrons. Within each set the isotopes of intermediate mass are stable or at least more stable than the rest. For each element, the lighter isotopes, those deficient in neutrons, generally tend toward stability by positron emission or electron capture, whereas the heavier isotopes, those rich in neutrons, usually approach stability by electron emission.

In comparison with other forms of radioactivity, such as gamma or alpha decay, beta decay is a relatively slow process. Half-lives for beta decay are never shorter than a few milliseconds.

The general equation representing beta minus decay is:

$${}_{Z}^{A}X_{N} \rightarrow {}_{Z+1}^{A}Y_{N-1} + e^{-} + \bar{\nu}$$

where:

- ${}^{A}_{Z}X_{N}$ is the parent nucleus
- ${}^{A}_{Z+1}Y_{N-1}$ is the daughter nucleus
- e^{-} is the released beta particle, an electron
- $\bar{\nu}$ is the released anti-neutrino

Beta plus decay comes from a nucleus with too many protons. In this type of decay, a proton from the nucleus is transformed into a neutron and a positron (which is simply a "positive version" of the electron). To ensure rules of particle physics hold, a tiny particle known as a neutrino is also released. The general equation representing beta positive decay is:

$$ {}^{A}_{Z}X_{N} \rightarrow {}^{A}_{Z-1}Y_{N+1} + e^{+} + \nu $$

where:

- ${}^{A}_{Z}X_{N}$ is the parent nucleus
- ${}^{A}_{Z-1}Y_{N+1}$ is the daughter nucleus
- e^{+} is the released beta particle, a positron
- ν is the released neutrino

In both beta minus and beta plus decay it is the weak nuclear force that results in the changing of a nucleon into a different nucleon.

In all cases where β^{+}

decay (positron emission) of a nucleus is allowed energetically, so too is electron capture allowed. This is a process during which a nucleus captures one of its atomic electrons, resulting in the emission of a neutrino:

$$ {}^{A}_{Z}X + e^{-} \rightarrow {}^{A}_{Z-1}X' + \nu_{e} $$

An example of electron capture is one of the decay modes of krypton-81 into bromine-81:

$$ {}^{81}_{36}Kr + e^{-} \rightarrow {}^{81}_{35}Br + \nu_{e} $$

All emitted neutrinos are of the same energy. In proton-rich nuclei where the energy difference between the initial and final states is less than $2m_{e}c^{2}$,

β^{+}decay is not energetically possible, and electron capture is the sole decay mode.

If the captured electron comes from the innermost shell of the atom, the K-shell, which has the highest probability to interact with the nucleus, the process is called K-capture. If it comes from the L-shell, the process is called L-capture, etc.

Electron capture is a competing (simultaneous) decay process for all nuclei that can undergo β^+ decay. The converse, however, is not true: electron capture is the *only* type of decay that is allowed in proton-rich nuclides that do not have sufficient energy to emit a positron and neutrino.

Nuclear Transmutation

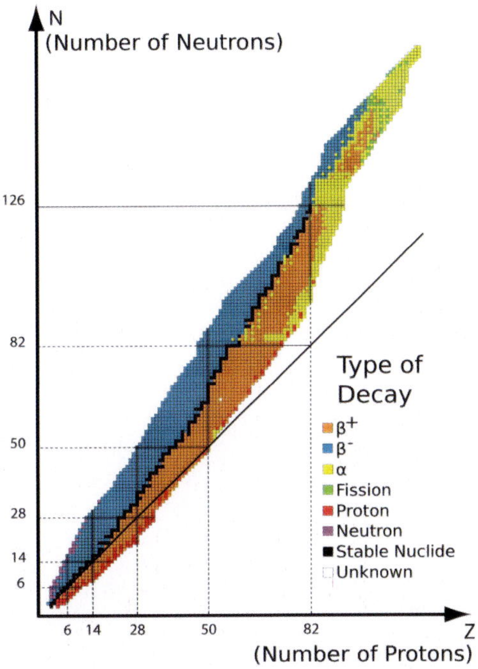

If the proton and neutron are part of an atomic nucleus, the above described decay processes transmute one chemical element into another. For example:

$$ {}^{137}_{55}\text{Cs} \rightarrow {}^{137}_{56}\text{Ba} + e^- + \bar{\nu}_e \quad \text{(beta minus decay)} $$

$$ {}^{22}_{11}\text{Na} \rightarrow {}^{22}_{10}\text{Ne} + e^+ + \nu_e \quad \text{(beta plus decay)} $$

$$ {}^{22}_{11}\text{Na} + e^- \rightarrow {}^{22}_{10}\text{Ne} + \nu_e \quad \text{(electron capture)} $$

Beta decay does not change the number (A) of nucleons in the nucleus, but changes only its charge Z. Thus the set of all nuclides with the same A can be introduced; these *isobaric* nuclides may turn into each other via beta decay. For a given A there is one that is most stable. It is said to be beta stable, because it presents a local minima of the mass excess: if such a nucleus has (A, Z) numbers, the neighbour nu clei $(A, Z-1)$ and $(A, Z+1)$ have higher mass excess and can beta decay into (A, Z), but not vice versa. For all odd mass numbers A, there is only one known beta-stable isobar. For even A, there are up to three different beta-stable isobars experimentally known; for example, ${}^{96}_{40}\text{Zr}$, ${}^{96}_{42}\text{Mo}$, and ${}^{96}_{44}\text{Ru}$ are all beta-stable. There are about 355 known beta-decay stable nuclides.

Competition of Beta Decay Types

Usually, unstable nuclides are clearly either "neutron rich" or "proton rich", with the former undergoing beta decay and the latter undergoing electron capture (or more rarely, due to the higher energy requirements, positron decay). However, in a few cases of odd-proton, odd-neutron radionuclides, it may be energetically favorable for the radionuclide to decay to an even-proton, even-neutron isobar either by undergoing beta-positive or beta-negative decay. An often-cited example is the single isotope ${}^{64}_{29}\text{Cu}$ (29 protons, 35 neutrons), which illustrates three types of beta decay in competition. Copper-64 has a half-life of about 12.7 hours. This isotope has one unpaired proton and one unpaired neutron, so either the proton or the neutron can decay. This particular nuclide (though not all nuclides in this situation) is almost equally likely to decay through proton decay by positron emission (18%) or electron capture (43%) to ${}^{64}_{28}\text{Ni}$, as it is through neutron decay by electron emission (39%) to ${}^{64}_{30}\text{Zn}$.

Stability of Naturally Occurring Nuclides

Most naturally occurring nuclides on earth are beta stable. Those that are not have half-lives ranging from under a second to periods of time significantly greater than the age of the universe. One common example of a long-lived isotope is the odd-proton odd-neutron nuclide ${}^{40}_{19}\text{K}$, which undergoes all three types of beta decay (β^-, β^+ and electron capture) with a half-life of 1.277×10^9 years.

Conservation Rules for Beta Decay

Baryon Number is Conserved

$$B = \frac{n_q - n_{\bar{q}}}{3}$$

where

n_q is the number of constituent quarks, and

$n_{\bar{q}}$ is the number of constituent antiquarks.

Beta decay just changes neutron to proton or, in the case of positive beta decay (electron capture) proton to neutron so the number of individual quarks doesn't change. It is only the baryon flavor that changes, here labelled as the isospin.

Up and down quarks have total isospin $I = \frac{1}{2}$ and isospin projections

$$I_z = \begin{cases} \frac{1}{2} & \text{up quark} \\ -\frac{1}{2} & \text{down quark} \end{cases}$$

All other quarks have $I = 0$.

In general

$$I_z = \frac{1}{2}(n_u - n_d)$$

Lepton Number is Conserved

$$L \equiv n_\ell - n_{\bar{\ell}}$$

so all leptons have assigned a value of +1, antileptons −1, and non-leptonic particles 0.

$$\begin{array}{ccccccccc} & \text{n} & \rightarrow & \text{p} & + & \text{e}^- & + & \bar{\nu}_e \\ L: & 0 & = & 0 & + & 1 & - & 1 \end{array}$$

Angular Momentum

For allowed decays, the net orbital angular momentum is zero, hence only spin quantum numbers are considered.

The electron and antineutrino are fermions, spin-1/2 objects, therefore they may couple to total $S = 1$ (parallel) or $S = 0$ (anti-parallel).

For forbidden decays, orbital angular momentum must also be taken into consideration.

Energy Release

The Q value is defined as the total energy released in a given nuclear decay. In beta decay, Q is therefore also the sum of the kinetic energies of the emitted beta particle, neutrino, and recoiling nucleus. (Because of the large mass of the nucleus compared to that of the beta particle and neutrino, the kinetic energy of the recoiling nucleus can generally be neglected.) Beta particles can therefore be emitted with any kinetic energy ranging from 0 to Q. A typical Q is around 1 MeV, but can range from a few keV to a few tens of MeV.

Since the rest mass of the electron is 511 keV, the most energetic beta particles are ultrarelativistic, with speeds very close to the speed of light.

β– Decay

Consider the generic equation for beta decay

$$ {}^{A}_{Z}\text{X} \rightarrow {}^{A}_{Z+1}\text{X}' + \text{e}^- + \bar{\nu}_e .$$

The Q value for this decay is

$$Q = \left[m_N \left({}^{A}_{Z}X \right) - m_N \left({}^{A}_{Z+1}X' \right) - m_e - m_{\bar{\nu}_e} \right] c^2 ,$$

where $m_N\left({}^{A}_{Z}X\right)$ is the mass of the nucleus of the ${}^{A}_{Z}X$ atom, m_e is the mass of the electron, and $m_{\bar{\nu}_e}$ is the mass of the electron antineutrino. In other words, the total energy released is the mass energy of the initial nucleus, minus the mass energy of the final nucleus, electron, and antineutrino. The mass of the nucleus m_N is related to the standard atomic mass m by

$$m\left({}^{A}_{Z}X\right)c^2 = m_N\left({}^{A}_{Z}X\right)c^2 + Zm_e c^2 - \sum_{i=1}^{Z} B_i.$$

That is, the total atomic mass is the mass of the nucleus, plus the mass of the electrons, minus the sum of all *electron* binding energies B_i for the atom. This equation is rearranged to find $m_N\left({}^{A}_{Z}X\right)$, and $m_N\,{}^{A}_{Z+1}X'$ is found similarly. Substituting these nuclear masses into the Q-value equation, while neglecting the nearly-zero antineutrino mass and the difference in electron binding energies, which is very small for high-Z atoms, we have

$$Q = \left[m\left({}^{A}_{Z}X\right) - m\left({}^{A}_{Z+1}X'\right)\right]c^2$$

This energy is carried away as kinetic energy by the electron and neutrino.

Because the reaction will proceed only when the Q-value is positive, β^- decay can occur when the mass of atom ${}^{A}_{Z}X$ is greater than the mass of atom ${}^{A}_{Z+1}X'$.

β+ decay

The equations for β^+ decay are similar, with the generic equation

$${}^{A}_{Z}X \rightarrow {}^{A}_{Z+1}X'\ e^- + \text{v}_e$$

giving

$$Q = \left[m_N\left({}^{A}_{Z}X\right) - m_N\left({}^{A}_{Z-1}X'\right) - m_e - m_{\upsilon_e}\right]c^2.$$

However, in this equation, the electron masses do not cancel, and we are left with

$$Q = m\left[\left({}^{A}_{Z}X\right) - m\left({}^{A}_{Z-1}X'\right) - 2m_e\right]c^2$$

Because the reaction will proceed only when the Q-value is positive, β^+ decay can occur when the mass of atom ${}^{A}_{Z}X$ exceeds that of ${}^{A}_{Z-1}X'$ by at least twice the mass of the electron.

Electron Capture

The analogous calculation for electron capture must take into account the binding energy of the electrons. This is because the atom will be left in an excited state after capturing the electron, and the binding energy of the captured innermost electron is significant. Using the generic equation for electron capture

$$ {}^{A}_{Z}X + e^{-} \rightarrow {}^{A}_{Z-1}X' + \nu_{e} $$

we have

$$ Q = \left[m_N\left({}^{A}_{Z}X \right) + m_e - m_N\left({}^{A}_{Z-1}X' \right) - m_{\nu_e} \right] c^2, $$

which simplifies to

$$ Q = \left[m\left({}^{A}_{Z}X \right) - m\left({}^{A}_{Z-1}X' \right) \right] c^2 - B_n, $$

where B_n is the binding energy of the captured electron.

Because the binding energy of the electron is much less than the mass of the electron, nuclei that can undergo β^+ decay can always also undergo electron capture, but the reverse is not true.

Beta Emission Spectrum

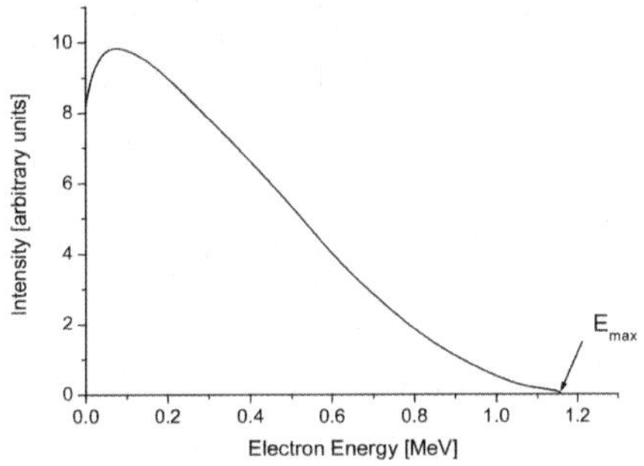

Beta spectrum of ^{210}Bi. E_{max}=Q=1.16 MeV is the maximum energy

Beta decay can be considered as a perturbation as described in quantum mechanics, and thus Fermi's Golden Rule can be applied. This leads to an expression for the kinetic energy spectrum *N(T)* of emitted betas as follows:

$$ N(T) = C_L(T)F(Z,T)pE(Q-T)^2 $$

where *T* is the kinetic energy, C_L is a shape function that depends on the forbiddenness of the decay (it is constant for allowed decays), *F(Z, T)* is the Fermi Function with *Z* the charge of the final-state nucleus, $E=T+mc^2$ is the total energy, $p=\sqrt{(E/c)^2-(mc)^2}$ is the momentum, and *Q* is the Q value of the decay. The kinetic energy of the emitted neutrino is given approximately by Q minus the kinetic energy of the beta.

Fermi Function

The Fermi function that appears in the beta spectrum formula accounts for the Coulomb attraction / repulsion between the emitted beta and the final state nucleus. Approximating the associated wavefunctions to be spherically symmetric, the Fermi function can be analytically calculated to be:

$$F(Z,T)=\frac{2(1+S)}{\Gamma(1+2S)^2}(2p\rho)^{2S-2}e^{\pi\eta}\,|\Gamma(S+i\eta)|^2,$$

where $S=\sqrt{1-\alpha^2 Z^2}$ (α is the fine-structure constant), $\eta=\pm\,\alpha ZE/pc$ (+ for electrons, – for positrons), $\rho=r_N/\hbar$ (r_N is the radius of the final state nucleus), and Γ is the Gamma function.

For non-relativistic betas ($Q \ll m_e c^2$), this expression can be approximated by:

$$F(Z,T)\approx\frac{2\pi\eta}{1-e^{-2\pi\eta}}.$$

Other approximations can be found in the literature.

Kurie Plot

A Kurie plot (also known as a Fermi–Kurie plot) is a graph used in studying beta decay developed by Franz N. D. Kurie, in which the square root of the number of beta particles whose momenta (or energy) lie within a certain narrow range, divided by the Fermi function, is plotted against beta-particle energy. It is a straight line for allowed transitions and some forbidden transitions, in accord with the Fermi beta-decay theory. The energy-axis (x-axis) intercept of a Kurie plot corresponds to the maximum energy imparted to the electron/positron (the decay's Q-value). With a Kurie plot one can find the limit on the effective mass of a neutrino.

Helicity (Polarization) of Neutrinos, Electrons and Positrons Emitted in Beta Decay

After the discovery of parity non-conservation, it was found that, in beta decay, electrons are emitted mostly with negative helicity, i.e., they move, naively speaking, like left-handed screws driven into a material (they have negative longitudinal polarization). Conversely, positrons have mostly positive helicity, i.e., they move like right-handed screws. Neutrinos (emitted in positron decay) have positive helicity, while antineutrinos (emitted in electron decay) have negative helicity.

The higher the energy of the particles, the higher their polarization.

Types of Beta Decay Transitions

Beta decays can be classified according to the angular momentum (L-value) and total spin (S-value) of the emitted radiation. Since total angular momentum must be conserved, including orbital and spin angular momentum, beta decay occurs by a variety of quantum state transitions to various nuclear angular momentum or spin states, known as "Fermi" or "Gamow-Teller" transitions. When beta decay particles carry no angular momentum (L=0), the decay is referred to as "allowed", otherwise it is "forbidden".

Other decay modes, which are rare, are known as bound state decay and double beta decay.

Fermi Transitions

A Fermi transition is a beta decay in which the spins of the emitted electron (positron) and anti-neutrino (neutrino) couple to total spin $S=0$, leading to an angular momentum change Ä$J=0$ between the initial and final states of the nucleus (assuming an allowed transition). In the non-relativistic limit, the nuclear part of the operator for a Fermi transition is given by

$$\mathcal{O}_F = G_V \sum_a \hat{\tau}_{a\pm}$$

with G_V the weak vector coupling constant, $\tau_\pm$ the isospin raising and lowering operators, and a running over all protons and neutrons in the nucleus.

Gamow-Teller Transitions

A Gamow-Teller transition is a beta decay in which the spins of the emitted electron (positron) and anti-neutrino (neutrino) couple to total spin $S=1$, leading to an angular momentum change Ä$J=0,\pm1$ between the initial and final states of the nucleus (assuming an allowed transition). In this case, the nuclear part of the operator is given by

$$\mathcal{O}_{GT} = G_A \sum_a \hat{\sigma}_a \hat{\tau}_{a\pm}$$

with G_A the weak axial-vector coupling constant, and σ the spin Pauli matrices, which can produce a spin-flip in the decaying nucleon.

Forbidden Transitions

When $L > 0$, the decay is referred to as "forbidden". Nuclear selection rules require high L-values to be accompanied by changes in nuclear spin (J) and parity (π). The selection rules for the Lth forbidden transitions are:

$$\Delta J = L-1, L, L+1; \Delta\pi = (-1)^L,$$

where $\Delta\pi$=1 or −1 corresponds to no parity change or parity change, respectively. The special case of a transition between isobaric analogue states, where the structure of the final state is very similar to the structure of the initial state, is referred to as "superallowed" for beta decay, and proceeds very quickly. The following table lists the ΔJ and $\Delta\pi$ values for the first few values of L:

Forbiddenness	ΔJ	$\Delta\pi$
Superallowed	0	no
Allowed	0, 1	no

First forbidden	0, 1, 2	yes
Second forbidden	1, 2, 3	no
Third forbidden	2, 3, 4	yes

Rare Decay Modes

Bound-state β– Decay

A very small minority of free neutron decays (about four per million) are so-called "two-body decays", in which the proton, electron and antineutrino are produced, but the electron fails to gain the 13.6 eV energy necessary to escape the proton, and therefore simply remains bound to it, as a neutral hydrogen atom. In this type of beta decay, in essence all of the neutron decay energy is carried off by the antineutrino.

For fully ionized atoms (bare nuclei), it is possible in likewise manner for electrons to fail to escape the atom, and to be emitted from the nucleus into low-lying atomic bound states (orbitals). This cannot occur for neutral atoms with low-lying bound states which are already filled by electrons.

Bound-state β decays were predicted by Daudel, Jean, and Lecoin in 1947, and the phenomenon in fully ionized atoms was first observed for $^{163}Dy^{66+}$ in 1992 by Jung et al. of the Darmstadt Heavy-Ion Research group. Although neutral ^{163}Dy is a stable isotope, the fully ionized $^{163}Dy^{66+}$ undergoes β decay into the K and L shells with a half-life of 47 days.

Another possibility is that a fully ionized atom undergoes greatly accelerated β decay, as observed for ^{187}Re by Bosch et al., also at Darmstadt. Neutral ^{187}Re does undergo β decay with a half-life of 42×10^9 years, but for fully ionized $^{187}Re^{75+}$ this is shortened by a factor of 10^9 to only 32.9 years. For comparison the variation of decay rates of other nuclear processes due to chemical environment is less than 1%.

Double Beta Decay

Some nuclei can undergo double beta decay (ββ decay) where the charge of the nucleus changes by two units. Double beta decay is difficult to study, as the process has an extremely long half-life. In nuclei for which both β decay and ββ decay are possible, the rarer ββ decay process is effectively impossible to observe. However, in nuclei where β decay is forbidden but ββ decay is allowed, the process can be seen and a half-life measured. Thus, ββ decay is usually studied only for beta stable nuclei. Like single beta decay, double beta decay does not change *A*; thus, at least one of the nuclides with some given *A* has to be stable with regard to both single and double beta decay.

"Ordinary" double beta decay results in the emission of two electrons and two antineutrinos. If neutrinos are Majorana particles (i.e., they are their own antiparticles), then a decay known as neutrinoless double beta decay will occur. Most neutrino physicists believe that neutrinoless double beta decay has never been observed.

Safety

Beta radiation is slightly more penetrating than alpha radiation, but still not nearly as penetrating as gamma radiation. Generally speaking, because beta radiation isn't extremely penetrating it is mainly an issue when ingested. If a beta source enters the body, it causes tissue damage and can increase the risk of cancer. Figure shows the relative levels of penetration of a variety of different radiation types.

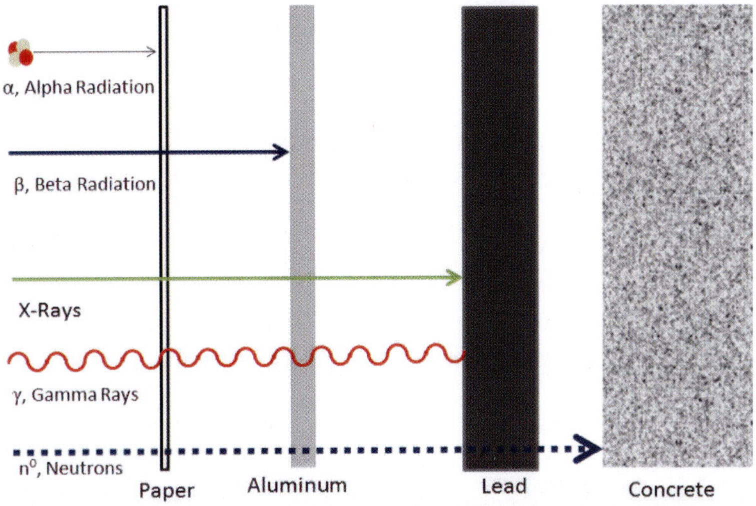

Different penetration levels of different products of decay, with gamma being one of the most highly penetrating and alpha being one of the least penetrating.

Exposure to beta radiation can cause a wide variety of health effects. Generally speaking, exposure to beta decay sources are chronic in nature. Chronic effects are the result of low-level exposures to beta radiation over an extended period of time, and can take anywhere between 5 and 30 years to develop. The most prominent side effect of exposure is cancer. Some beta emitters are distributed throughout the body - such as carbon-14 (which occurs naturally at levels that cause no harm to the human body)- while others accumulate in specific organs. An example of this would be iodine-131, which concentrates in the thyroid gland and increases the risk of thyroid cancer.

Applications and Importance

Elements that have beta decay can have useful medical applications. Radionuclide therapy (RNT) or radiotherapy is a cancer treatment that uses beta decay. In this process, lutetium-177 or yttrium-90 is attached to a molecule and ingested. Once inside the body, this molecule travels to the cancer cells. The radioactive atoms then undergo a decay process, releasing beta particles and killing nearby cancer cells.

Additionally, carbon dating relies on the properties of beta decay. To determine the approximate age of artifacts, wood, and animal remains the ratio of carbon-14 to carbon-12 in an object must be determined. Carbon-14 is generated from sunlight in the atmosphere from nitrogen-14, which plants breathe in with photosynthesis, and thus there is a certain amount of carbon-14 in organic remains. Plants are eaten by animals and get carbon-14 as well. When an organic body begins to decay, some of this carbon-14 becomes nitrogen-14 (through a beta decay process), and over the years the amount of carbon-14 in the sample is depleted. By looking at the ratio of carbon-14 to carbon-12, the approximate age of the artifact can be determined.

Gamma Decay

Gamma decay is one type of radioactive decay that a nucleus can undergo. What separates this type of decay process from alpha or beta decay is that no particles are ejected from the nucleus when it undergoes this type of decay. Instead, a high energy form of electromagnetic radiation - a gamma ray photon - is released. Gamma rays are simply photons that have extremely high energies which are highly ionizing. As well, gamma radiation is unique in the sense that undergoing gamma decay does not change the structure or composition of the atom. Instead, it only changes the energy of the atom since the gamma ray carries no charge nor does it have an associated mass.

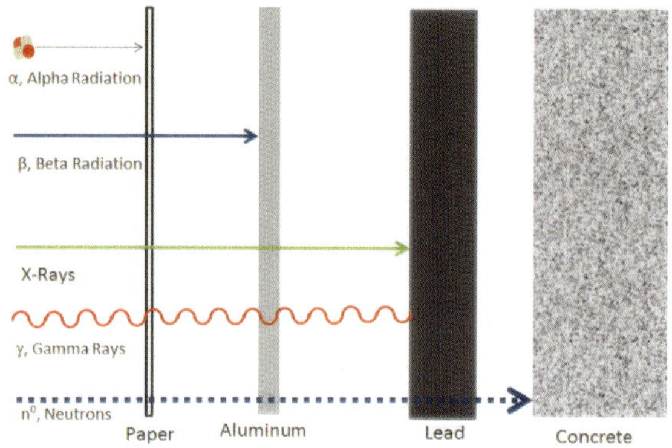

Different penetration levels of different products of decay, with gamma being one of the most highly penetrating.

In order for a nucleus to undergo gamma decay, it must be in some sort of excited energetic state. Experiments have shown that protons and neutrons are located in discrete energy states within the nucleus, not too different from the excited states that electrons can occupy in atoms. Thus if a proton or a neutron inside of the nucleus jumps up to an excited state - generally following an alpha or beta decay - the new daughter nucleus must somehow release energy to allow the proton or neutron to relax back down to ground state. When the nucleon makes this transition from a high to a low energy state, a gamma photon is emitted. The general equation that represents this process is:

$$A^* \rightarrow A + \gamma$$

where:

- A^* is the excited atom
- A is a relaxed state of the initial excited atom
- γ is the released gamma ray photon

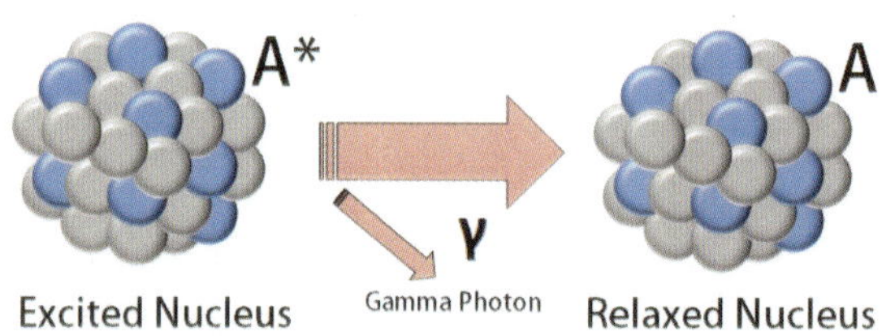

Figure Diagram showing gamma decay of a nucleus.

Knowing that an atom undergoes gamma radiation is important, but it is also possible to determine the frequency of the released gamma radiation if the initial and final states of the nucleon inside the nucleus are known. The equation representing the frequency of the gamma radiation is:

$$E_i - E_f = hf$$

where:

- E_i is the initial, higher energy state of the nucleon
- E_f is the final, lower energy state of the nucleon
- h is Planck's constant
- f is the frequency of the emitted radiation

In addition to radioactive nuclei, there are many objects in space that emit high levels of gamma radiation.

Applications and Importance

Gamma rays can at times be harmful due to the fact that they are generally very high energy and therefore penetrate matter very easily. Since it penetrates so easily, it is some of the most useful radiation for medical purposes.

Some of the most widely used gamma emitters are cobalt-60, cesium-137, and technetium-99m. Cesium is used widely in radiotherapy - the treatment of cancer using gamma rays - as well as being used to measure soil density at construction sites and to investigate the subterranean layers of the Earth in oil wells. Cobalt is used to sterilize medical equipment and irradiate food, killing bacteria and pasteurizing the food. Technetium-99m (which has a shorter half-life than technetium-99) is the most widely used for diagnostic medical tests to investigate the brain, bone, and internal organs. As well, exposure to gamma radiation can improve the durability of wood and plastics, and is thus used to toughen flooring in high-traffic areas.

In addition, uranium-238 and uranium-235 - used in fuel for nuclear power plants - undergo both alpha and gamma decays when used. Immediately following the fission process, gamma rays are released, resulting in high levels of radiation present around the reactor. However, safety precautions are in line to ensure that workers do not get close enough to this radioactive area to be harmed.

Spontaneous Fission

Spontaneous fission is type of Nuclear decay in which certain unstable nuclei of heavier elements split into two nearly equal fragments (nuclei of lighter elements) and liberate a large amount of energy. Spontaneous fission, discovered (1941) by the Russian physicists G.N. Flerov and K.A. Petrzhak in uranium-238, is observable in many nuclear species of mass number 230 or more. Among these nuclides, those with lower mass numbers generally have longer half-lives. Uranium-238 has a half-lifeof about 10^{16} years when it decays by spontaneous fission, whereas fermium-256 decays with a half-life of about three hours.

Nuclides that undergo spontaneous fission also are subject to alpha decay (emission from the nucleus of a helium nucleus). In uranium-238, alpha decay is about 2 million times more probable than is spontaneous fission, whereas in fermium-256, 3 percent of the nuclei undergo alpha decay and 97 percent undergo spontaneous fission.

The fission that occurs in nuclear reactors and explosive devices is induced by the neutron bombardment of certain types of nuclei.

Neutron Emission

The neutron emission is one of the Nuclear decays, by which unstable nuclei may reach the stability. In general, this type of radioactive decay may occur, when nuclei contain significant excess of neutrons or excitation energy. In this type of decay a neutron is simply ejected from the nucleus.

Although the neutron emission is usually associated with nuclear decay, it must be also mentioned in connection with neutron nuclear reactions. Some neutrons interacts with a target nucleus via a compound nucleus. Among these compound nucleus reactions are also reactions, in which a neutron is ejected from nucleus and they may be referred to as neutron emission reactions. The point is that compound nuclei lose its excitation energy in a way, which is identical to the radioactive decay. Very important feature is the fact the mode of decay of compound nucleus does not depend on the way the compound nucleus was formed.

The compound nucleus reactions, in which neutron emission occurs, are:

- Elastic Scattering Reaction. In some cases, if the kinetic energy of an incident neutron just right to form a resonance, the neutron may be absorbed and a compound nucleus may be formed. This interaction is more unusual (in comparison with potential scattering) and is also known as resonance elastic scattering. Due to formation of the compound nucleus, initial and final neutron are not the same and this reactions may be also referred to as one type of neutron emission reaction.
- Inelastic Scattering Reaction. In this case, the connection with neutron emission is more obvious. In an inelastic scattering reaction between a neutron and a target nucleus some energy of the incident neutron is absorbed to the recoiling nucleus and the nucleus remains in the excited state. The neutron is emitted then with a lower kinetic energy. If the kinetic energy of an incident neutron is sufficient the double, triple, or more, neutron emission may take place. These events are referred to as (n, 2n), (n, 3n) or (n, ...n) reactions. The probability of such reactions increases with increasing incident neutron energies.
- Nuclear Fission. The fission reaction is very specific reaction and is of importance in many fields of nuclear engineering. It is known the fission reaction produces fission neutrons that are of importance in any chain-reacting system. But not all neutrons are released at the

same time following fission. Even the nature of creation of these neutrons is different. From this point of view we usually divide the fission neutrons into two following groups:

- Prompt Neutrons. Prompt neutrons are emitted directly from fission and they are emitted within very short time of about 10^{-14} second.
- Delayed Neutrons. Delayed neutrons are emitted by neutron rich fission fragments that are called the delayed neutron precursors. These precursors usually undergo beta decay but a small fraction of them are excited enough to undergo neutron emission. The fact the neutron is produced via this type of decay and this happens orders of magnitude later compared to the emission of the prompt neutrons, plays an extremely important role in the control of the reactor.

Spontaneous Neutron Emission

As a consequence of the Pauli exclusion principle nuclei with an excess of protons or neutrons have a higher average energy per nucleon. Nuclei with a sufficient excess of neutrons have a greater energy than the combination of a free neutron and a nucleus with one less neutron, and therefore can decay by neutron emission. Nuclei which can decay by this mode are described as lying beyond the neutron drip line.

Two examples of isotopes that emit neutrons are beryllium-13 (decaying to beryllium-12 with a mean life 2.7×10^{-21} s) and helium-5 (helium-4, 7×10^{-22} s).

In tables of nuclear decay modes neutron emission is commonly denoted by the abbreviation *n*.

Neutron emitters to the left of lower dashed line

Z→	0	1	2												
n↓	n	H	He	3	4	5									
0		^{1}H		Li	Be	B	6								
1	^{1}n	^{2}H	^{3}He	^{4}Li	^{5}Be	^{6}B	C	7							
2		^{3}H	^{4}He	^{5}Li	^{6}Be	^{7}B	^{8}C	N	8						
3		^{4}H	^{5}He	^{6}Li	^{7}Be	^{8}B	^{9}C	^{10}N	O	9					
4		^{5}H	^{6}He	^{7}Li	^{8}Be	^{9}B	^{10}C	^{11}N	^{12}O	F	10			13	
5		^{6}H	^{7}He	^{8}Li	^{9}Be	^{10}B	^{11}C	^{12}N	^{13}O	^{14}F	Ne	11	12	Al	
6		^{7}H	^{8}He	^{9}Li	^{10}Be	^{11}B	^{12}C	^{13}N	^{14}O	^{15}F	^{16}Ne	Na	Mg	^{19}Al	14
		7	^{9}He	^{10}Li	^{11}Be	^{12}B	^{13}C	^{14}N	^{15}O	^{16}F	^{17}Ne	^{18}Na	^{19}Mg	^{20}Al	Si
		8	^{10}He	^{11}Li	^{12}Be	^{13}B	^{14}C	^{15}N	^{16}O	^{17}F	^{18}Ne	^{19}Na	^{20}Mg	^{21}Al	^{22}Si
			9	^{12}Li	^{13}Be	^{14}B	^{15}C	^{16}N	^{17}O	^{18}F	^{19}Ne	^{20}Na	^{21}Mg	^{22}Al	^{23}Si
				10	^{14}Be	^{15}B	^{16}C	^{17}N	^{18}O	^{19}F	^{20}Ne	^{21}Na	^{22}Mg	^{23}Al	^{24}Si
				11	^{15}Be	^{16}B	^{17}C	^{18}N	^{19}O	^{20}F	^{21}Ne	^{22}Na	^{23}Mg	^{24}Al	^{25}Si
				12	^{16}Be	^{17}B	^{18}C	^{19}N	^{20}O	^{21}F	^{22}Ne	^{23}Na	^{24}Mg	^{25}Al	^{26}Si
						13	^{19}C	^{20}N	^{21}O	^{22}F	^{23}Ne	^{24}Na	^{25}Mg	^{26}Al	^{27}Si
						14	^{20}C	^{21}N	^{22}O	^{23}F	^{24}Ne	^{25}Na	^{26}Mg	^{27}Al	^{28}Si

Photoneutron Emission

Some nucleides, e.g. Be-9, can be induced to eject a neutron by gamma radiation.

Beta-delayed Neutron Emission

Neutron emission usually happens from nuclei that are in an excited state, such as the excited $^{17}O^*$ produced from the beta decay of ^{17}N. The neutron emission process itself is controlled by the nuclear force and therefore is extremely fast, sometimes referred to as "nearly instantaneous". This process allows unstable atoms to become more stable. The ejection of the neutron may be as a product of the movement of many nucleons, but it is ultimately mediated by the repulsive action of the nuclear force that exists at extremely short-range distances between nucleons. The life time of an ejected neutron inside the nucleus before it is emitted is usually comparable to the flight time of a typical neutron before it leaves the small nuclear "potential well", or about 10^{-23} seconds.

Delayed Neutrons in Reactor Control

Most neutron emission outside prompt neutron production associated with fission (either induced or spontaneous), is from neutron-heavy isotopes produced as fission products. These neutrons are sometimes emitted with a delay, giving them the term delayed neutrons, but the actual delay in their production is a delay waiting for the beta decay of fission products to produce the excited-state nuclear precursors that immediately undergo prompt neutron emission. Thus, the delay in neutron emission is not from the neutron-production process, but rather its precursor beta decay, which is controlled by the weak force, and thus requires a far longer time. The beta decay half lives for the precursors to delayed neutron-emitter radioisotopes, are typically fractions of a second to tens of seconds.

Nevertheless, the delayed neutrons emitted by neutron-rich fission products aid control of nuclear reactors by making reactivity change far more slowly than it would if it were controlled by prompt neutrons alone. About 0.65% of neutrons are released in a nuclear chain reaction in a delayed way due to the mechanism of neutron emission, and it is this fraction of neutrons that allows a nuclear reactor to be controlled on human reaction time-scales, without proceeding to a prompt critical state, and runaway melt down.

Neutron Emission in Fission

Induced Fission

A synonym for such neutron emission is "prompt neutron" production, of the type that is best known to occur simultaneously with induced nuclear fission. Induced fission happens only when a nucleus is bombarded with neutrons, gamma rays, or other carriers of energy. Many heavy isotopes, most notably californium-252, also emit prompt neutrons among the products of a similar spontaneous radioactive decay process, spontaneous fission.

Spontaneous Fission

Spontaneous fission happens when a nucleus splits into two smaller nuclei and generally one or more neutrons.

Proton Emission

Proton emission is a form of radioactive decay characterized by the ejection of a proton from a nucleus. It is a rare decay process, not generally observed in naturally occurring isotopes. Proton emission can occur after beta decay from a high-lying excited state in a nucleus or from the ground state of a proton-rich nucleus. The decay is possible when the proton is unbound, or for negative proton separation energy.

The first proton emission to be observed was from an isomer of cobalt-53. Today, 25 isotopes have been identified that exhibit proton emission activity. Research in proton emission has guided the understanding of nuclear masses, structure and nuclear deformation as well as serves as a study of quantum tunneling.

Positron Emission

Positron emission is a radioactive decay, which is characterized by the conversion of a proton into a neutron inside a radionuclide nucleus, with the ejection of a positron and an electron neutrino. This process is also known as beta decay or β^+ decay. The decay is mediated by the weak force. It occurs in proton-rich radionuclides and results in nuclear transmutation. Positron emission was first witnessed in 1934 when aluminium was bombarded with alpha particles. Some of the isotopes that exhibit beta decay and emit positrons are potassium-40, carbon-11, oxygen-15, nitrogen-13, among others.

Internal Conversion

The excess energy of radioactive nuclei in excited states is usually relieved through gamma emission. However, if the wave function of an orbital electron is such that it can exist close to or in the nucleus, the excess energy can be transferred directly to the orbital electron. Hence, as an alternative to gamma emission, the excited nucleus may return to a lower state or to the ground state by ejecting an orbital electron. This is known as internal conversion and results in the emission of a mono-energetic electron. Following the internal conversion, outer orbital electrons fill the deeper atomic energy levels, leading to characteristic X-ray emission or, alternatively, to the emission of Auger electrons.

The internal conversion process can be described by:

$$^{A}_{Z}P^{*} \rightarrow [^{A}_{Z}\mathrm{P}]^{*} + e^{-}$$

The ratio of internal conversion to gamma emission photons is known as the internal conversion coefficient α where

$$\alpha = \frac{N_e}{N_\gamma}$$

where N_e is the number of de-excitations via electron emission and N_γ the number of de-excitations via gamma emission. The kinetic energy E_e of the ejected electron is almost equal to the excitation energy E^* less the binding energy of the orbital electron E_B, i.e.

$$E_e = E^* - E_B$$

The internal conversion coefficent alpha depends on the atomic number and the excitation energy, i.e.

$$\alpha \propto \frac{Z^3}{E^*}$$

Internal conversion is more important in heavy nuclei in particular in the decay of low-lying excited states (small E^*). Gamma emission predominated in light nuclei.

Example:

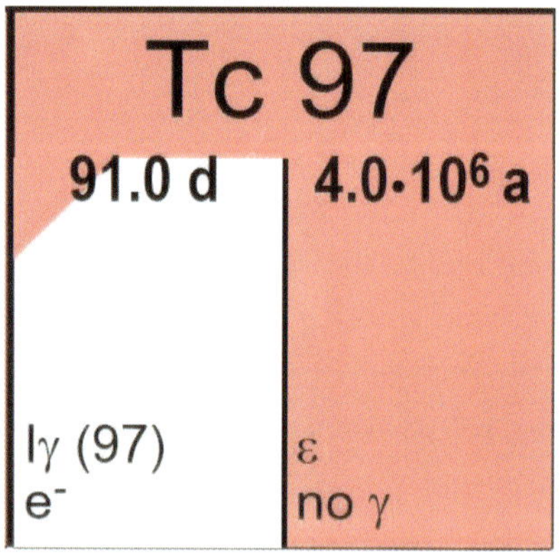

Tc97m

Consider the decay of the isomeric state Tc97m (half-life 91 days). The box is mainly white indicating isomeric transition as the main decay mode and the small red triangle indicates electron capture to the daughter Mo97. The nuclide data shows Iγ (97) and e^-. The symbol e^- indicates that the converison coefficient is > 1 i.e. the state de-excites mainly by electron conversion rather than by gamma emission.

In addition, the fact that the energy of the isomeric transition is in backets i.e. (97) implies that gamma emission occurs in less that 1% of the de-excitations. Hence the electron conversion occurs is over 99% over the de-excitations. The actual conversion coefficient is not given in the chart, but can be obtained from the literature e.g nuclear Data sheets or from Nucleonica.

The conversion coefficient is α = 311 i.e 311 times more conversion electrons than gamma photons. These conversion electrons are seen in the beta spectrum. This information can be found in the references in Nucleonica. Together with the original research papers, information is given on branching ratios etc.

References

- Best, Lara; Rodrigues, George; Velker, Vikram (2013). "1.3". Radiation Oncology Primer and Review. Demos Medical Publishing. ISBN 978-1-62070-004-4.

- Radioactive-decay, Radioactive-series: newworldencyclopedia.org, Retrieved 28 April 2018
- Sansare, K.; Khanna, V.; Karjodkar, F. (2011). "Early victims of X-rays: a tribute and current perception". Dentomaxillofacial Radiology. 40 (2): 123–125. doi:10.1259/dmfr/73488299. ISSN 0250-832X. PMC 3520298 . PMID 21239576
- Bolt, B. A.; Packard, R. E.; Price, P. B. (2007). "John H. Reynolds, Physics: Berkeley". The University of California, Berkeley. Retrieved 2007-10-01
- Gamma-decay: energyeducation.ca, Retrieved 18 July 2018
- Bijker, R.; Santopinto, E. (2015). "Valence and sea quarks in the nucleon". Journal of Physics: Conference Series. 578: 012015. arXiv:1412.5559 . doi:10.1088/1742-6596/578/1/012015
- Mould, Richard F. (1995). A century of X-rays and radioactivity in medicine : with emphasis on photographic records of the early years (Reprint. with minor corr ed.). Bristol: Inst. of Physics Publ. p. 12. ISBN 978-0-7503-0224-1
- Beta-decay, science: britannica.com, Retrieved 08 May 2018
- Clarke, R.H.; J. Valentin (2009). "The History of ICRP and the Evolution of its Policies"(PDF). Annals of the ICRP. ICRP Publication 109. 39 (1): 75–110. doi:10.1016/j.icrp.2009.07.009. Retrieved 12 May 2012
- Beta-decay, science: britannica.com, Retrieved 08 May 2018
- Wang, B.; Yan, S.; Limata, B.; et al. (2006). "Change of the 7Be electron capture half-life in metallic environments". The European Physical Journal A. 28 (3): 375–377. Bibcode:2006EPJA...28..375W. doi:10.1140/epja/i2006-10068-x. ISSN 1434-6001

Permissions

All chapters in this book are published with permission under the Creative Commons Attribution Share Alike License or equivalent. Every chapter published in this book has been scrutinized by our experts. Their significance has been extensively debated. The topics covered herein carry significant information for a comprehensive understanding. They may even be implemented as practical applications or may be referred to as a beginning point for further studies.

We would like to thank the editorial team for lending their expertise to make the book truly unique. They have played a crucial role in the development of this book. Without their invaluable contributions this book wouldn't have been possible. They have made vital efforts to compile up to date information on the varied aspects of this subject to make this book a valuable addition to the collection of many professionals and students.

This book was conceptualized with the vision of imparting up-to-date and integrated information in this field. To ensure the same, a matchless editorial board was set up. Every individual on the board went through rigorous rounds of assessment to prove their worth. After which they invested a large part of their time researching and compiling the most relevant data for our readers.

The editorial board has been involved in producing this book since its inception. They have spent rigorous hours researching and exploring the diverse topics which have resulted in the successful publishing of this book. They have passed on their knowledge of decades through this book. To expedite this challenging task, the publisher supported the team at every step. A small team of assistant editors was also appointed to further simplify the editing procedure and attain best results for the readers.

Apart from the editorial board, the designing team has also invested a significant amount of their time in understanding the subject and creating the most relevant covers. They scrutinized every image to scout for the most suitable representation of the subject and create an appropriate cover for the book.

The publishing team has been an ardent support to the editorial, designing and production team. Their endless efforts to recruit the best for this project, has resulted in the accomplishment of this book. They are a veteran in the field of academics and their pool of knowledge is as vast as their experience in printing. Their expertise and guidance has proved useful at every step. Their uncompromising quality standards have made this book an exceptional effort. Their encouragement from time to time has been an inspiration for everyone.

The publisher and the editorial board hope that this book will prove to be a valuable piece of knowledge for students, practitioners and scholars across the globe.

Index